Genders and Sexualities in the Social Sciences
Series Editors: **Victoria Robinson**, University of Sheffield, UK and **Diane Richardson**, University of Newcastle, UK

Editorial Board: **Raewyn Connell**, University of Sydney, Australia, **Kathy Davis**, Utrecht University, The Netherlands, **Stevi Jackson**, University of York, UK, **Michael Kimmel**, State University of New York, Stony Brook, USA, **Kimiko Kimoto**, Hitotsubashi University, Japan, **Jasbir Puar**, Rutgers University, USA, **Steven Seidman**, State University of New York, Albany, USA, **Carol Smart**, University of Manchester, UK, **Liz Stanley**, University of Edinburgh, UK, **Gill Valentine**, University of Leeds, UK, **Jeffrey Weeks**, South Bank University, UK, **Kath Woodward**, The Open University, UK

Titles include:

Jyothsna Belliappa
GENDER, CLASS AND REFLEXIVE MODERNITY IN INDIA

Edmund Coleman-Fountain
UNDERSTANDING NARRATIVE IDENTITY THROUGH LESBIAN AND GAY YOUTH

Niall Hanlon
MASCULINITIES, CARE AND EQUALITY
Identity and Nurture in Men's Lives

Brian Heaphy, Carol Smart and Anna Einarsdottir (*editors*)
SAME SEX MARRIAGES
New Generations, New Relationships

Sally Hines and Yvette Taylor (*editors*)
SEXUALITIES
Past Reflections, Future Directions

Meredith Nash
MAKING 'POSTMODERN' MOTHERS
Pregnant Embodiment, Baby Bumps and Body Image

Meredith Nash
REFRAMING REPRODUCTION
Conceiving Gendered Experiences

Barbara Pini and Bob Pease (*editors*)
MEN, MASCULINITIES AND METHODOLOGIES

Victoria Robinson and Jenny Hockey
MASCULINITIES IN TRANSITION

Yvette Taylor, Sally Hines and Mark E. Casey (*editors*)
THEORIZING INTERSECTIONALITY AND SEXUALITY

Yvette Taylor and Michelle Addison (*editors*)
QUEER PRESENCES AND ABSENCES

Kath Woodward
SEX POWER AND THE GAMES

Genders and Sexualities in the Social Sciences
Series Standing Order ISBN 978–0–230–27254–5 hardback
978–0–230–27255–2 paperback
(*outside North America only*)

You can receive future titles in this series as they are published by placing a standing order. Please contact your bookseller or, in case of difficulty, write to us at the address below with your name and address, the title of the series and the ISBN quoted above.

Customer Services Department, Macmillan Distribution Ltd, Houndmills, Basingstoke, Hampshire RG21 6XS, England

Reframing Reproduction
Conceiving Gendered Experiences

Edited by

Meredith Nash
University of Tasmania, Australia

Selection, introduction, conclusion, and editorial matter © Meredith Nash 2014
Individual chapters © Respective authors 2014

All rights reserved. No reproduction, copy or transmission of this publication may be made without written permission.

No portion of this publication may be reproduced, copied or transmitted save with written permission or in accordance with the provisions of the Copyright, Designs and Patents Act 1988, or under the terms of any licence permitting limited copying issued by the Copyright Licensing Agency, Saffron House, 6–10 Kirby Street, London EC1N 8TS.

Any person who does any unauthorized act in relation to this publication may be liable to criminal prosecution and civil claims for damages.

The authors have asserted their rights to be identified as the authors of this work in accordance with the Copyright, Designs and Patents Act 1988.

First published 2014 by
PALGRAVE MACMILLAN

Palgrave Macmillan in the UK is an imprint of Macmillan Publishers Limited, registered in England, company number 785998, of Houndmills, Basingstoke, Hampshire RG21 6XS.

Palgrave Macmillan in the US is a division of St Martin's Press LLC, 175 Fifth Avenue, New York, NY 10010.

Palgrave Macmillan is the global academic imprint of the above companies and has companies and representatives throughout the world.

Palgrave® and Macmillan® are registered trademarks in the United States, the United Kingdom, Europe and other countries.

ISBN 978–1–137–26712–2

This book is printed on paper suitable for recycling and made from fully managed and sustained forest sources. Logging, pulping and manufacturing processes are expected to conform to the environmental regulations of the country of origin.

A catalogue record for this book is available from the British Library.

A catalog record for this book is available from the Library of Congress.

Typeset by MPS Limited, Chennai, India.

Transferred to Digital Printing in 2014

Contents

List of Tables and Figures	vii
Acknowledgements	viii
Notes on Contributors	x
Introduction: Conceiving of Postmodern Reproduction *Meredith Nash*	1

Part I   Contested 'Choices' and Challenges

1 Towards a More Inclusive Framework for Understanding Fertility Barriers   23
   Katherine M. Johnson, Julia McQuillan, Arthur L. Greil, and Karina M. Shreffler

2 Constructions of the 'Best Interests of the Child' in New South Wales Parliamentary Debates on Surrogacy   39
   Catherine Ruth Collins, Damien W. Riggs, and Clemence Due

3 'Diseases', 'Defects', 'Abnormalities', and 'Conditions': Discursive Tensions in Prenatal Screening   54
   Meredith Vanstone, Elizabeth Anne Kinsella, and Jeff Nisker

4 The Limits of 'Choice': Abortion and Entrepreneurialism   69
   Kate Gleeson

5 Gaps in Post-Birth Care in Neoliberal Times: Evidence from Canada   84
   Cecilia Benoit, Camille Stengel, Rachel Phillips, Maria Zadoroznyj, and Sarah Berry

Part II   Reproductive Bodies and Identities

6 Unborn Assemblages: Shifting Configurations of Embryonic and Foetal Embodiment   101
   Deborah Lupton

7 Picturing Postpartum Body Image: A Photovoice Study   115
   Meredith Nash

8 'My Doctor Told Me I Can Still Have Children But ... ':
 Contradictions in Women's Reproductive Health
 Experiences after Spinal Cord Injury 135
 Heather Dillaway and Catherine Lysack

9 Taking a Long View of the 'Right Time' for Fatherhood 150
 Fiona Shirani

10 Anticipating and 'Experiencing' Birth: Men, Essentialisms,
 and Reproductive Realms 165
 Tina Miller

Part III The (Global) Reproductive Marketplace

11 Putting 'Daddy' in the Cart: Ordering Sperm Online 185
 Lisa Jean Moore and Marianna Grady

12 Reciprocity in the Donation of Reproductive Oöcytes 203
 Margaret Boulos, Ian Kerridge, and Catherine Waldby

13 Expressed Breast Milk as Commodity: Disembodied
 Motherhood and Involved Fatherhood 221
 Victoria Team and Kath Ryan

14 What Does Not Kill You Makes You Stronger: Young
 Women's Online Conversations about Quitting the Pill 236
 Elizabeth Arveda Kissling

Conclusion: Where Do We Go From Here? 251
Meredith Nash

Bibliography 255

Index 293

List of Tables and Figures

Tables

3.1	Participant demographics	59
5.1	Selected characteristics of study population compared to the Victoria Census Metropolitan Area (CMA), 2006 (Income in Canadian dollars (CAD))	89
11.1	Variations in web presence by each of the five sperm banks studied	194

Figures

1.1	Fertility barriers by race among a sample of US women (N = 4680)	32
1.2	Fertility barriers by education among a sample of US women (N = 4680)	33
1.3	Fertility barriers by self-reported abortion among a sample of US women (N = 4680)	35
1.4	Fertility barriers by parenthood status among a sample of US women (N = 4680)	36
7.1	Zoe	125
7.2	Judy's closet	127
7.3	Zoe's track pants	128
7.4	Lena's jeans	130
7.5	Lena wearing her size 16 jeans	131

Acknowledgements

The idea for this book emerged out of an undergraduate class that I taught for the first time in 2012 at the University of Tasmania. When I started to develop the materials for *Sociology of Reproduction*, I was unable to find a text to use in teaching that offered a framework for thinking about reproduction in postmodernity as well as cutting edge case studies and examples. The only solution seemed to be to edit a collection on the topic myself – what turned out to be an ambitious undertaking. I have many people to thank for getting me across the finish line.

Firstly, I thank each of the authors for their unique contributions to this book. The editorial process was made much easier thanks to their enthusiasm, commitment, and intellectual generosity for the duration of this project.

I am similarly indebted to the scholars who acted as peer reviewers for the chapters in this volume. Their constructive feedback strengthened this book in innumerable ways. Many thanks to Ann Victoria Bell (University of Michigan), Deb Dempsey (Swinburne University), Jan Draper (The Open University), Sarah Earle (The Open University), Fiona Giles (University of Sydney), Jessica Gunson (University of Adelaide), Catherine Kevin (Flinders University), Rebecca Kukla (Georgetown University), Gayle Letherby (Plymouth University), Susan Markens (CUNY-Lehman College), Jenni Millbank (University of Technology Sydney), Jennifer Morgan (University of Melbourne), Debbie Payne (Auckland University of Technology), Rhonda Shaw (Victoria University of Wellington), Niamh Stephenson (University of New South Wales), Nicola Swain (University of Otago), Jane Ussher (University of Western Sydney), and Clare Williams (Brunel University).

This book would not have been possible without the support of the School of Social Sciences at the University of Tasmania where I am currently employed as a Lecturer in Sociology. The School generously provided me with time, resources, and assistance to prepare this book. In particular, the research assistance of Susan Banks was invaluable. I am also grateful to my colleagues Michelle Phillipov and Eliza Burke for reading earlier versions of this book.

Special thanks also go to my students for constantly invigorating my sociological imagination.

I would also like to thank the editors, staff, and reviewers from Palgrave Macmillan. I am most grateful to my publisher, Philippa Grand, for believing in the value of this project.

Special thanks to my mum who taught me about feminist consciousness-raising from an early age and who supports me unfailingly in everything that I do.

I would like to express my deepest gratitude to my husband Chris for laughter and love.

Finally, I am grateful to the publishers of the following articles, who granted permission to reproduce parts of them for this book:

Reprinted by permission of BioMed Central: Benoit, C., Stengel, C., Philips, R., Zadoroznyj, M., & Berry, S. (2012) 'Privitisation and marketisation of post-birth care: The hidden costs for new mothers', *International Journal for Equity in Health*, 11(61), http://www.biomedcentral.com/content/pdf/1475-9276-11-61.pdf.

Reprinted by permission of Taylor & Francis, Ltd.: Nash, M. (2013) 'Shapes of motherhood: Exploring postnatal body image through photographs', *Journal of Gender Studies*, DOI: 10.1080/09589236.2013.797340. *Journal of Gender Studies* is available online at: www.tandfonline.com. The published article is available at the following address: http://www.tandfonline.com/doi/abs/10.1080/09589236.2013.797340#.Ugm8q1K0J8E.

Reprinted by permission of Taylor & Francis, Ltd.: Ryan, K., Team, V., & Alexander, J. (2013) 'Expressionists of the 21st century: The commodification and commercialization of expressed breast milk', *Medical Anthropology: Cross-Cultural Studies in Health and Illness*, DOI: 10.1080/01459740.2013.768620. *Medical Anthropology: Cross-Cultural Studies in Health and Illness* is available online at: www.tandfonline.com. The published article is available at the following address: http://www.tandfonline.com/doi/abs/10.1080/01459740.2013.768620#.Ugm3sFK0J8E\.

Notes on Contributors

Elizabeth Arveda Kissling is Professor of Women's Studies and of Communication at Eastern Washington University in the US, with interests in women's health, sexuality, and feminism. She is especially interested in how these issues are represented in mass media and the relationship between media and subjectivity. She is the author of *Capitalizing on the Curse: The Business of Menstruation* (2006) and numerous scholarly articles.

Cecilia Benoit is Professor in the Department of Sociology and Scientist at the Centre for Addictions Research of British Columbia. She is involved in a variety of projects that employ mixed methodologies to investigate the health of different vulnerable groups, including Aboriginal women in Vancouver's Downtown Eastside, young people confronting health stigmas linked to obesity and asthma, street-involved youth in transition to adulthood, workers in lower prestige service occupations, adults in the sex industry, and pregnant and early parenting women dealing with addiction and other challenges.

Sarah Berry is a PhD candidate and Lecturer in Sociology at McGill University. Using a range of mixed-method approaches, her research has focused on media and mental health, historical and contemporary understandings of stigma and madness, and maternal health and postnatal care. For her dissertation, she worked with the Mental Health Commission of Canada's Anti-Stigma Initiative, Opening Minds, on the largest content analysis of news media representations of mental illness ever conducted.

Margaret Boulos is a PhD candidate in the Department of Sociology and Social Policy at the University of Sydney. Her dissertation examines the social significance of providing reproductive tissues to stem cell research in Australia, particularly in light of calls to pay women to provide their oocytes for somatic cell nuclear transfer research. Her research interests include social constructions of gift-giving practices and the politics of scientific research in contemporary societies.

Catherine Ruth Collins recently completed a Master of Psychology (Clinical) degree at the University of Adelaide and currently works as a psychologist in the area of child protection. Her current research focuses on the psychological impact of pregnancy loss on a woman's adult relationships and on her relationship with children born before or after the loss.

Heather Dillaway is Associate Professor in the Department of Sociology at Wayne State University in Detroit, Michigan. She received her PhD in Sociology from Michigan State University in 2002. Her broad research interests lie within the study of women's reproductive health and structural inequalities including how women's experiences of menopause and midlife are shaped by their social locations and contemporary social contexts. She is also engaged in a research project on the reproductive health experiences of women with spinal cord injuries.

Clemence Due is Research Fellow in the School of Psychology at the University of Adelaide, and has published extensively in the areas of refugee studies, family studies, and media studies.

Kate Gleeson is Australian Research Council Fellow in Politics at Macquarie University, where she also lectures in the Masters of Politics and Public Policy, mostly in areas of gender, sexuality, and policy. She has published widely on the history of sexual regulation in Australia and the UK, including the regulation of abortion, homosexual sex, prostitution, and rape.

Marianna Grady is a recent graduate of Purchase College, State University of New York where she majored in Art History and Gender Studies. Her senior thesis, 'The Cyborg Motif: Photographic Depictions of Female Hysteria at the Salpêtrière', used cyborg theory and visual culture studies to explore the photographic techniques utilised by Dr Jean-Martin Charcot to document female hysteria at the Salpêtrière during the mid-to-late nineteenth century. She is currently doing a year of volunteer service in Seattle through AmeriCorps and the QuEST Fellowship programme.

Arthur L. Greil is Professor of Sociology at Alfred University. He is a co-investigator on the National Study of Fertility Barriers, funded by the National Institute of Child Health and Human Development. He has written extensively on psychosocial aspects of infertility. He has also written on fertility intentions, pregnancy loss, and the importance of motherhood. In addition, he does research in the sociology of religion.

Katherine M. Johnson is Visiting Assistant Professor of Sociology at Tulane University. Her research addresses human reproduction, gender, and family issues, especially reproductive technologies, reproductive rights, and postmodern family creation. She is also interested in bioethical issues in the context of health and reproduction and exploring possibilities for dialogue between social sciences and bioethics.

Ian Kerridge is Director and Associate Professor in Bioethics at the Centre for Values, Ethics and the Law in Medicine at the University of Sydney and staff haematologist/bone marrow transplant physician at Royal North Shore Hospital, Sydney. He is the author of over 150 papers in peer-reviewed journals and five textbooks of ethics, most recently *Ethics and Law for the Health Professions* (2013). His current research interests in ethics include the philosophy of medicine, stem cells, end-of-life care, synthetic genomics, public health, drug policy and organ donation.

Elizabeth Anne Kinsella is Associate Professor in the faculties of Health Sciences, Education, and Women's Studies and Feminist Research at the University of Western Ontario, London, Ontario, Canada. Her research is in the area of health professional education and practice, reflective practice, ethics education, embodiment and qualitative research. Recent projects include an edited book entitled *Phronesis as Professional Knowledge: Practical Wisdom in the Professions*, a project examining the use of the arts in ethics education, and a project investigating ethical tensions and their negotiation in the practices of health care practitioners.

Deborah Lupton is Professor in the Department of Sociology and Social Policy at the University of Sydney. She is the author/co-author of 13 books and many journal articles and book chapters on topics ranging from the sociology of medicine and public health; risk; the emotions; embodiment; HIV/AIDS; reproduction and parenting cultures; food; and obesity politics. Her latest books are *Medicine as Culture* (Third edition, 2012), *Fat* (2012), *Risk* (Second edition, 2013), and *The Social Worlds of the Unborn* (2013).

Catherine Lysack is Deputy Director of the Institute of Gerontology and Professor of Occupational Therapy and Gerontology at Wayne State University in Detroit, Michigan. Her major research interest is in the social and environmental influences on health and understanding how older adults and people with disability redevelop active and meaningful lives in the community after illness and injury.

Julia McQuillan is Chair and Professor of Sociology at the University of Nebraska-Lincoln. She studies social inequality in health and in workplace contexts. She was co-investigator on the National Study of

Fertility Barriers funded by the National Institute of Child Health and Human Development, and of the National Science Foundation institutional ADVANCE grant at the University of Nebraska. She has several publications applying social psychological and structural theories of inequality conditions to health conditions.

Tina Miller is Professor of Sociology at Oxford Brookes University in the UK. Her research and teaching interests include motherhood and fatherhood transitions, gender and identities, reproductive health, narratives, qualitative research methods, and ethics and she has published in all these areas. Tina has lived and worked in the Solomon Islands and Bangladesh as well as Oxford.

Lisa Jean Moore is Professor of Sociology and Gender Studies at Purchase College, State University of New York. She is the author of *Sperm Counts: Overcome by Man's Most Precious Fluid* (2007), and the co-author of *Missing Bodies: The Politics of Visibility* (2009) and *Gendered Bodies: Feminist Perspectives* (2007). Her latest co-authored book is *Buzz: Urban Beekeeping and the Power of the Bee* (2013).

Meredith Nash is Lecturer in Sociology at the University of Tasmania in Australia. Her research focuses on the gendered body as a way of understanding the relationships between people, place, politics, and culture. She is the author of *Making 'Postmodern' Mothers: Pregnant Embodiment, Baby Bumps and Body Image* (Palgrave Macmillan, 2012).

Jeff Nisker is Professor of Obstetrics and Gynecology and Co-ordinator of Health Ethics and Humanities at the Schulich School of Medicine and Dentistry, University of Western Ontario, London, Ontario, Canada. He holds/held CIHR and Genome Canada grants exploring ethics and social issues, including social determinants of health, in reproductive medicine, genetics, and exposure of pregnant women to environmental toxins; and a CIHR/Health Canada grant exploring public engagement in genetic science through theatre. He is currently researching ways to promote equality in access to emerging assistive technologies for disabled persons using his theatre for health policy development strategies.

Rachel Phillips is Postdoctoral Fellow at the Centre for Addictions Research British Columbia where she coordinates a research project examining the development of a health service programme for pregnant and parenting women affected by substance use. Her research also focuses on stigma and other determinants of health for vulnerable groups. She is currently involved in a Canadian Institute of Health

Research-funded project examining the role of supervisors in contexts of vulnerability and resiliency to violence among Canadian sex workers.

Damien W. Riggs is Senior Lecturer in Social Work at Flinders University, and the author of over 100 articles in the fields of gender and sexuality studies, critical race and whiteness studies, and family studies, including *What about the Children! Masculinities, Sexualities and Hegemony* (2010) and *Priscilla, (white) Queen of the Desert: Queer Rights/Race Privilege* (2006).

Kath Ryan is Head of the Centre for Postgraduate and Higher Degree Studies in the School of Nursing and Midwifery, La Trobe University, Australia. Her research interests include women's health, breastfeeding, consumer participation, and personal experiences of health and illness.

Fiona Shirani is Research Associate in the School of Social Sciences at Cardiff University. She has worked on a number of qualitative research projects and has been part of the UK-wide qualitative longitudinal Timescapes network. In 2011, she completed her PhD titled 'The "Right Time for Fatherhood?" A Temporal Study of Men's Transition to Parenthood'. Her research interests and areas of publication include: time (particularly life transitions and imagined futures), families and relationships, and qualitative longitudinal methods.

Karina M. Shreffler is Associate Professor of Human Development and Family Science at Oklahoma State University. She studies work and family, fertility, and reproductive health issues, including the mental health consequences of infertility, sterilisation regret, and pregnancy loss.

Camille Stengel is European Commission Erasmus Mundus Fellow in the dual-degree programme Doctorate in Cultural and Global Criminology. She is based at the University of Kent in the UK and at Eötvös Loránd University in Hungary. Previous research experience includes a position as a research associate at the Centre for Addictions Research British Columbia, a research assistant at the International Health Development Research Centre, an intern at the International Drug Policy Consortium, and a conference and media officer at Harm Reduction International.

Victoria Team is Research Fellow at Mother and Child Health Research Centre in the School of Nursing and Midwifery, La Trobe University and Editorial Assistant for *Medical Anthropology: Cross-Cultural Studies in Health and Illness*. Her research interests are in the area of women's health. Her publications focus on skin cancer prevention, immigration, and caregiving.

Meredith Vanstone is Assistant Professor in the Department of Clinical Epidemiology and Biostatistics, and a member of the Centre for Health Economics and Policy Analysis at McMaster University in Hamilton, Ontario, Canada. Her research interests include social and ethical issues in health technology assessment and ethical issues surrounding the policy and implementation of new reproductive technologies.

Catherine Waldby is Professorial Future Fellow in the Department of Sociology and Social Policy, Sydney University, and Visiting Professor at the Centre for Biomedicine and Society, Brunel University, London. She researches and publishes in social studies of biomedicine and the life sciences. She is the author of several books including *AIDS and the Body Politic: Biomedicine and Sexual Difference* (1996) and *Clinical Labor: Tissue donors and Research Subjects in the Global Bioeconomy* (with Melinda Cooper, 2014).

Maria Zadoroznyj is Associate Professor of Sociology at the University of Queensland in Australia. Maria has a long-standing interest in the provision of hospital and community-based maternity services and their implications for the health and well-being of new mothers and their families. Her research focuses on how health and social policies shape the provision of care to families following the birth of a child. Her current research projects include a national evaluation of Australia's first paid parental leave scheme, and a state-wide evaluation of community-based postnatal contact services in Queensland, Australia.

Introduction: Conceiving of Postmodern Reproduction

Meredith Nash

When Ann Oakley wrote about embodied experiences of gender and reproduction in the mid-1970s in Britain, men were not routinely present for the birth of their children, foetuses existed in only two dimensions on the ultrasound screen, women could not easily 'elect' to have a caesarean section, and buying sperm online was inconceivable. Today, this is all possible. In fact, as this book will make apparent, quite a lot has changed in the last 40 years or so in the West[1] – reproduction has evolved alongside neoliberalisation, commodification, advances in reproductive technologies, shifting notions of 'gender 'and 'sex', and the rise of postfeminism.

As the role and meaning of reproduction in the everyday lives of women and men in the industrialised world change over time, the nature of our intellectual engagement with reproduction within sociology has also developed. The title of this book captures my desire to 'reframe' or redefine the study of reproduction within sociology – an endeavour that involves positioning reproduction more clearly as a branch of knowledge in its own right, critically engaging with interdisciplinary feminist perspectives and post-structuralist gender theory, and acknowledging the necessity of emergent methodological approaches in understanding the complex 'choices', anxieties, and challenges that come alongside postmodern reproduction for both women and men. Beyond its obvious association with reproduction, I use the term 'conceiving' in the title to underline the *process* of 'reframing' reproduction. This term points to the importance of the generative dynamics of feminism(s) in shaping the reproductive realm along with the interdisciplinary perspectives necessary in describing those dynamics throughout the book.

Why am I posing such a challenge? Historically, mainstream sociology has largely neglected to acknowledge human reproduction as a distinct subject for analysis despite a precedent stemming from Engels ((1972)[1884]) or the legacy of interdisciplinary feminist contributions to the development of reproduction as an area of sociological enquiry (Stacey and Thorne, 1985). Although feminist theory has challenged androcentric biases embedded in how sociologists have traditionally understood bodies and power (e.g. de Beauvoir, 1949), sociology has an uneasy relationship with feminism when it comes to generating discourse on reproduction. Feminist perspectives are often *implied* as opposed to guiding discussions of reproduction (Annandale and Clark, 1996, p. 17).

Over the last 20 years, there has been a limited range of scholarship directly developing this observation that feminist critiques of power relations are not foregrounded in sociological theorising about reproduction (Anderson, 2005; Annandale, 2009; Malson and Swann, 2003; Ray, 2006; Thorne, 2006). The implication of this critique is twofold. From one perspective, developments in post-structuralist feminist theory have led to the deconstruction of essentialist notions of gendered/sexed identities. This has had significant impact across a range of disciplines including sociology. However, this has also arguably resulted in a de-politicisation of 'reproduction' and 'reproductive bodies' in sociology. Against a backdrop of postmodernity and 'postfeminism', the denaturalisation of sex/gender and a strong focus on individualisation may make it seem as though reproduction has a less prominent place in contemporary western women's lives and that this is 'unproblematically liberating' (Malson and Swann, 2003, p. 195). In contrast, as the contributors in the book will show, reproduction is firmly present in women's (and men's lives) and continues to function as a regulatory ideal for women, especially (p. 197). In operating within and across multiple conceptual frameworks and keeping sight of the critical feminist edge inherent in reproductive studies, it may be possible to trigger a long overdue paradigm shift in researching reproduction in sociology (Thorne, 2006).

This book aims to explore new possibilities for conceptualising the relationship between gender and reproduction in postmodernity – a time of widespread social change and shifts in gender relations in the West. Given the multiply-contested nature of postmodern bodies in feminist thought (Alcoff, 1988; Bordo, 1990; Fraser and Nicholson, 1990), the aim of this book is to reflect on the ambivalent, negotiated, and fragmented embodied experiences of reproduction. In the twenty-first century, the contingencies and complexities associated with gendered experiences of reproduction highlight the necessity of theorising across disciplinary frameworks in order to challenge the dichotomous boundaries (e.g. public/private, mind/body, health/illness, nature/culture, and self/other) that have often characterised sociological research on reproduction. Binaries have never been able to adequately convey the range of human experiences of reproduction or the scope of changes wrought by significant developments in the technologies of gender and the 'choices' now faced by individuals in the reproductive arena.

In order to provide a glimpse into what a 'reframing' of reproduction might look like, I have gathered scholars from a variety of disciplines – sociology, gender studies, medicine, nursing, midwifery, law, psychology, and public health – to explore the themes of contested 'choices' and challenges, reproductive bodies and identities, and the global reproductive marketplace as some of the key nodes connecting sociology and feminist scholarship in terms of both theory building and empirical research. Stretching disciplinary boundaries, as Skeggs (2008, p. 684) has noted, is essential for invigorating our ideas. In orienting the discussion of reproduction within and beyond sociology, it becomes possible to gain a more nuanced understanding of the wider social, political, and economic issues posed by the forces of globalisation and neoliberalisation that affect human experiences of reproduction in the West today.

Throughout this book, the contributors examine the social, cultural, and moral contexts which underpin postmodern reproduction drawing on timely examples such as surrogacy, infertility, and commodified and technologised experiences of reproduction from the perspectives of individuals in a range of cultural positions (across ethnicity, gender, sexuality, class, and ability) and geographical locations in the West. In particular, the chapters consider the social aspects of how women and men feel, think, and act in relation to their reproductive 'choices' and increased uncertainty in postmodernity. How do rapid social and technological changes shape reproductive realms today? What is at stake? What problems are raised? What solutions are offered?

Throughout this book, the authors (re)consider what we know about postmodern reproduction and what the future holds. It is my hope that this volume can provide a starting point for conceiving of a 'new' sociology of reproduction for the twenty-first century and may allow us to identify some of the 'blindspots' in sociological studies of reproduction to date. I see this volume as having an intended audience of social scientists and gender and feminist scholars, but also of psychologists, health care practitioners, and students with an interest in reproduction.

To begin this endeavour, I canvass the key intellectual contributions of interdisciplinary feminist scholarship and activism that have contributed to the development of reproduction as a branch of knowledge within sociology and that are the keystones for the future development of the field. In the remainder of the chapter, I outline some of the central features of postmodern reproduction and provide an overview of the chapters in the book.

Feminist readings of gender, bodies, and biomedical power

Given the prominent place of reproduction in women's lives, the interface between gender, bodies, and power has been a primary area of focus within feminist scholarship for more than 30 years. One area in which feminist scholarship has been particularly fruitful is in the re-reading/ re-theorising of medical history and patriarchal medical practices from a range of disciplinary perspectives. Feminist scholars have argued convincingly that women's bodies have traditionally been represented as 'Other' – weaker, uncivilised, and threatening in medical discourses (Birke, 1999; Jordanova, 1989; Lupton, 2003; Martin, 1992; Ussher, 1992) and that this has provided the basis for cultural constructions of subordinated femininities. A key focus of these discussions has been on the range of pathologising activities (e.g. dissection, the development of medical tools like the speculum) used by male doctors over centuries to contain and regulate women's bodies especially in relation to gynaecology and obstetrics (Kapsalis, 1997; Showalter, 1990; Ussher, 2006).

Feminist scholars have argued that medical discourses have constructed female reproductive bodies as intrinsically pathological and that this resulted in their 'hysterisation' (Malson and Swann, 2003, p. 193; see also Ehrenreich and English, 1973). Susan Bordo (1993a) has contended that the emergence of hysteria (named after the Greek word for uterus) in the late nineteenth century is a primary example of the pathologisation of middle-class women's reproductive bodies and the entrenchment of a cultural view that women are fragile and dependent on men. In this way, feminist scholars have been deeply concerned with medicalisation or 'the process in which non-medical problems become defined and treated as medical problems' (Conrad, 1992, p. 209), arguing that it directly affects women's health and well-being (Ehrenreich and English, 1973). Ussher (1989) has noted that women have been socialised to see key biological events as primary sources of identity. Contemporary feminist analyses have examined the implications of this claim in relation to menstruation (e.g. Koeske, 1983; Swann, 1997; Ussher, 1989, 1992, 2006), menopause (e.g. Gannon and Stevens, 1998; Hunter and O'Dea, 1997; Perz and Ussher, 2008), pregnancy (e.g. Martin, 1992; Rothman, 1993; Young, 1984), and childbirth (e.g. Cartwright, 1979; Oakley, 1980).

More recently, feminist scholars have contended that women are not merely passive 'victims' of medicalisation but are active participants in medical power relations (Morgan, 1998; Oinas, 1998). Despite critiques by Deveaux (1994) and Ramazanoglu (1993), the influence of Foucault (1977, 2003b) has seen feminist arguments mounted in favour of women both accommodating and resisting medical power, specifically reproductive health (Abel and Browner, 1998; Bordo, 1993b; Cooper, 1994; Lorentzen, 2008).

According to feminist scholars, one of the primary effects of medical power relations has been that physicians have attempted to use the scientific method as a means of advancing objective 'truths' about women's bodies, discounting women's embodied or experiential knowledge (Lorentzen, 2008). Thus interdisciplinary feminist research has been instrumental in challenging the gender biases and 'objectivity' of scientific/medical knowledge (Birke, 1999). Some of the most powerful feminist philosophical contributions to understanding reproduction have questioned what 'counts' as scientific knowledge as a means of establishing the validity of women's experiential knowledge (Alcoff and Potter, 1993; Fox Keller, 1985; Harding, 1991). As I describe in the next section, the power of experiential knowledge in relation to women's reproductive health came to bear in feminist activism of the 1970s as women attempted to regain control of their bodies from biomedicine.

Feminist and women's health movements

In the late 1960s and early 1970s, 'reproduction' transformed from a seemingly bounded biological 'thing' to a multidimensional problem and a field for feminist action (Murphy, 2012, p. 6). This was a time when the concepts 'woman', 'bodies',' sex', 'reproduction', 'choice', 'freedom', 'power', and 'oppression' were questioned and politicised. Inspired by texts including *The Dialectic of Sex* (Firestone, 1970), *The Female Eunuch* (Greer, 1970), and *Vaginal Politics* (Frankfort, 1972), feminist and women's health movements in the West started to gain momentum around the question of how feminists could most effectively challenge and reinvent medical constructions of women's bodies and reproductive health. Feminists like Adrienne Rich (1977) were advocating that women should be involved in the *production* of knowledge about their own bodies. Throughout the 1970s in the West, feminists argued that women's reproductive essentialisation served as a fundamental obstacle to their advancement. Under the famous banner of 'the personal is political', feminist activists claimed that their personal experiences of family life and relationships, reproduction, sexuality, and health were political issues with consequences (Kline, 2010, p. 13).

Consciousness-raising was a primary example of the kinds of social processes that were instigated by feminism that were aimed at developing experiential knowledge, giving women a 'voice' and unifying their experiences (although this is contested in terms of race/ethnicity, class, sexuality, etc.),[2] and empowering women in their relationships to their bodies. Feminist scholars have argued that the 1971 publication of *Our Bodies, Our Selves* (OBOS) by the Boston Women's Health Collective was a powerful demonstration of US women's frustration with the medical gaze and of their desire to learn about their own bodies through direct observation both individually and collectively (Murphy, 2012). OBOS is viewed as 'revolutionary' because it enabled women to 'seize' the 'means of reproduction' including 'the practices and technologies used in the management of reproduction' such as the speculum (p. 49) (see also Kapsalis, 1997). A famous poster of Wonder Woman taking a speculum from the hands of a male doctor and shouting 'With my speculum, I am strong, I can fight' illustrates this view.[3]

'Seizing' control of reproduction for women in the 1970s primarily applied to the freedom to 'choose' whether or not to have children using new reproductive technologies. The widespread use of the pill and feminist advocacy for abortion rights signalled the major shift to technologised (reproductive) bodies and ultimately foregrounded the ideology of 'choice' in framing debates about reproduction from that point forward (Hayden and O'Brien Hallstein, 2010). Technology not only introduced a wide range of reproductive 'choices', but also contested the idea of a 'natural' body and what this meant. For example, the introduction of new reproductive technologies and techniques (e.g. forceps, anaesthesia, episiotomy, caesarean sections, and induction) in the 1970s had major implications for the management of pregnancy and birth (Martin, 1992). As a result, there was a push among some feminists to reclaim birth as 'natural' process (Arms, 1975; Gaskin, 1975). Ina May Gaskin, a foundational figure in the natural childbirth movement, was seen by some as an embodiment of the self-knowledge promoted by OBOS – that the experience of giving birth in itself makes women experts.

The consciousness-raising activities described in this section exemplify the outcomes of developments in feminism around choice and the medicalisation of reproductive care. Importantly, feminist grassroots organising was central in reshaping existing reproductive practices and technologies including the establishment of community-based women's health centres, as well as changes to national health policies and medical education (Murphy, 2012, p. 30).

The study of reproduction in sociology

Feminist health activism, coupled with the rise of women's studies programmes at US universities, and the development of interdisciplinary gender scholarship provided the backdrop for reproduction to emerge as an area of academic inquiry in sociology and other complementary disciplines from the 1970s onward (Boxer, 1982; Curthoys, 2000). Pioneering feminist scholars pushed forward the idea that 'reproduction is an entry point to the study of social life' (Ginsburg and Rapp, 1995, p. 2). Several landmark texts have contributed to the development of reproduction as a branch of knowledge in sociology, particularly in the areas of menstruation and menopause (Laws, 1990; Ussher, 2006); pregnancy and childbirth (Arms, 1975; Cartwright, 1979; Davis-Floyd, 1992; Kitzinger, 1978; Macintyre, 1977; Martin, 1992; Rothman, 1982); medical attitudes toward women's bodies (Corea, 1977; Ehrenreich and English, 1973; Fisher, 1986; Roberts, 1985; Scully, 1980); reproductive technologies (Arditti et al. 1984; Corea, 1985; Duden, 1993; Rothman, 1993; Taylor, 2008); motherhood (Chodorow, 1999; Miller, 2005; Ragoné, 1994); reproductive consumption/consumerism (Taylor et al. 2004); foetal personhood (Morgan, 2009; Morgan and Michaels, 1999); and the politics of women's healthcare (Broom, 1991; Daniels, 1993; Roberts, 1997).

The sociological study of reproduction gained momentum in the 1970s as feminist sociologists argued that gender should be conceptualised in social, not biological terms (Annandale, 2009, p. 40; Oakley, 1972). As Anderson (2005, p. 441) has observed, feminist sociology was deeply connected to feminist and women's health movements and as such, the keystones of feminist movements such as the importance of experiential knowledge and the politicisation of women's embodied experiences led feminist critiques of the discipline. Much of feminist sociological work at this time was about 'insisting that women's lives mattered, for society and sociology (Anderson, 2005, p. 439). To illustrate, Diana Scully and Pauline Bart published their foundational article, 'A Funny Thing Happened on the Way to the Orifice' (1973), a sociological critique of the representation of women in gynaecology textbooks.

In attempting to bring reproduction closer to the centre of social theory, feminist sociologists like Ann Oakley (1974a, 1974b) argued against the functionalist approaches of Comte, Durkheim, and Parsons who largely justified gender inequality in their articulations of social roles, and this had transformative effects on the ways in which gender, and by extension, reproduction were later conceptualised (Annandale, 2009, pp. 38–39). Oakley's 1975 transition to motherhood project in which she interviewed women in London during pregnancy and postpartum proved to be particularly influential in the development of reproduction as an area of study in sociology. Out of this research came two germinal books: *Becoming a Mother* (1979) (women's accounts of their experiences) and *Women Confined* (1980) (an academic text focussing on birth within the context of other sociological work on life events and transitions).

While the development of reproduction as branch of knowledge has been a remarkable achievement within sociology and elsewhere, the overwhelming message of work in this area seems to be that women are primarily 'reproducers' and 'mothers' and that research on reproduction has been conducted mostly by women. Men and their reproductive careers, in comparison, have been studied mostly in relation to sex/sexuality and they received little attention from sociologists until the 1990s when the influence of feminist post-structuralist theories of bodies and selves was impacting studies of masculinities (Lupton and Barclay, 1997; Marsiglio et al., 2000, 2013). Inhorn et al. (2009) have argued that men are the 'second sex' when it comes to reproduction and that there has been a similar dearth of anthropological research focussing on men's reproductive health (see also Dudgeon and Inhorn, 2003). Although men/masculinities have appeared in sociological accounts of sperm (Moore, 2007) and contraception (Walker, 2011), in decision-making around pregnancy and birth (Reed, 2005), in discussions of infertility (Throsby and Gill, 2004), in ultrasound screenings (Draper, 2003a), at the birth itself (Miller, 2010; Reed, 2005), and in relation to reproductive loss (Earle et al., 2013), how men negotiate and navigate gendered identity within reproduction has rarely been a focus of discussion (Inhorn, 2012; Marsiglio and Roy, 2012). Fathering has primarily been conceptualised through the 'breadwinner' ideology in which heterosexual men are viewed as being individually responsible for the financial stability of the household but largely detached from parenting (Williams, 2008). Recent scholarship has challenged the centrality of heterosexuality and/or biological fatherhood to hegemonic masculinity (Lewin, 2009; Riggs et al., 2010). Moreover, a number of scholars have raised valuable questions surrounding the nature of men's procreative consciousness over time (Miller, 2010), gay men's reproductive stratification (Riggs, 2010), and how men construct their identities as fathers (Hobson, 2002; Reed, 2005). Despite the developments in understanding men's reproductive experiences, as Marsiglio et al. (2013, p. 19) have argued, far more attention needs to be paid to 'how men emerge and express themselves as procreative beings'.

Characterising postmodern reproduction

At the beginning of this chapter, I noted that the meaning and experience of reproduction for individuals in the West has undergone significant change. Although the definition of postmodernism is contested (Nicholson, 1990), what is apparent is that postmodernism is characterised by instability, fragmentation, and flux, and 'involves a changing relation between our bodies and our worlds' (Halberstam, 1991, p. 447). Viewing reproduction through the lens of postmodernism allows us to further critique essentialisms and binaries. Furthermore, as feminist scholars have contended, there are useful affinities between postmodernism and feminism(s) within sociology (Yeatman, 1990, 1991). In this section, I will canvass five key interconnected nodes that characterise contemporary postmodern reproduction including: neoliberalisation, the development of reproductive technologies, commodification, shifting notions of 'gender 'and 'sex', and postfeminism.

Neoliberalisation

Neoliberalism is a controversial concept that has been widely associated with free-market policies, privatisation, competition, efficiency, and growth (Peck et al., 2009). Although it had its ideological 'home' in the US and UK in the 1970s and 1980s, neoliberalism has come to envelop much of the global South and various other countries under different historical and cultural circumstances (Tickell and Peck, 2003, p. 164; see also Peet, 2002). Thus neoliberalism may be thought of as a *process* that is constantly evolving and culturally nuanced in terms of its appropriation and development (Tickell and Peck, 2003, p. 165).

The neoliberal pursuit of capital accumulation over social reproduction is important because it shapes the lives and reproductive 'choices' available to individuals in the West. Although the notion of 'choice' was central in securing women's reproductive rights during feminist health movements of the 1970s, today, the notion of reproductive 'choice' has become complicated by market-based neoliberal agendas in which people lay claim to their reproductive rights as 'consumers', not citizens. As Baker (2008, p. 57) has argued, 'choice' is routinely used as an 'overarching and uncontextualised principle' to demonstrate the ways in which individual effort is essential in determining one's opportunities. Today, women and men are invited to become self-sufficient 'consumers' of goods and services in a global reproductive industry that encompasses both reproductive healthcare and technologies like in vitro fertilisation (IVF). Thus the increasing influence of neoliberal ideology on everyday life has meant that some groups have greater access to reproductive 'choices' while the 'choices' of others are limited – a feature of stratified reproduction (Colen, 1986). To maintain these reproductive hierarchies, various institutions (social, economic, and political) regularly intervene in matters relating to contraception, fertility, abortion, and maternal and child health.

One of the consequences of the 'neoliberalisation of life' more generally has been population decline – throughout the West, birth rates are dropping and the average age for parenthood is rising (Waldby and Cooper, 2008, p. 57). In Australia, for example, the fertility rate has been below replacement level since 1976 and the average age of mothers has risen to 30.6 years and 33 years for fathers (Australian Bureau of Statistics (ABS), 2011) (whereas in 1985 it was 27.3 years and 30.1 years respectively) (ABS, 2005). Similar patterns are apparent in the UK (Office for National Statistics, 2010) and in the US (Centers for Disease Control and Prevention, 2010). These patterns have a significant impact on economic growth as ageing populations increase and the demand for welfare and health care services increase. As Waldby and Cooper have noted (2008, p. 58), the simplistic notion that women should just have more babies obviates the more pressing concerns facing women and men as they 'prioritise economic security and career development over the production of large families'.

Reproductive technologies

Coupled with neoliberalisation, the increasing use of technology in relation to reproduction is one of the most profound changes to have occurred since the 1970s. Technology has both changed and unsettled western views about what is natural when it comes to reproduction, exemplifying the fragmented and uncertain processes of postmodernity. The definition of 'biological parent' has been modified such that now individuals are mostly comfortable with the idea of test-tube babies, IVF, and surrogacy. Similarly, technological interventions have become regular solutions for affluent people experiencing fertility issues. Moreover, women have experienced the benefits and burdens associated with technologies including the contraceptive pill, abortion, and foetal monitoring during pregnancy.

Yet, in the twenty-first century, it is also apparent that the same concerns and strategies for resistance around technologised vs. 'natural' bodies that feminists raised in the 1970s are still relevant. The rate of caesarean births has almost doubled in the last 20 years in the West (Lauer et al., 2010). New 'lifestyle' contraceptive pills that purport to free women of the 'hassles' of menstruation continue to bring with them a risk of serious side effects including blood clots, depression, and death (McDonough, 2013). Pregnancy has come to be administered under the category of 'risk' (Lupton, 1999). Scientists are hard at work to create artificial wombs (McKie, 2002). Foetuses have emerged as 'public' individuals with 'rights' that are supposed to be protected at all costs (Mitchell, 2001). From another perspective, the increasing interest in 'natural' approaches to pregnancy, birth, and parenting over the last decade is evidence of resistance to medicalisation (Bobel, 2002). Similarly, the rising numbers of childless women and men (for whatever reason) demonstrate resistance to western parenthood mandates (Inhorn and van Balen, 2002).

Commodification

The expansion of neoliberal ideologies that prioritise commodification/consumerism coupled with developments in medical research and reproductive technologies (e.g. contraception, IVF, abortion, surrogacy, gamete donation, etc.) has driven the development of a global industry that has been built around reproductive labour and the sale and/or distribution of biomedical 'products'. Today people are availing themselves of a range of services and products to enable their reproductive 'choices'. Pregnancy, for example, is an experience that is now deeply embedded within market forces, involving a wide variety of consumer choices and opportunities for affluent women from conception to birth (Nash, 2012a; Taylor, 2008). Similarly, it is now possible for individuals to purchase donor sperm online (Moore and Grady, this volume), travel to another country for IVF treatment (Martin, 2009), hire an Indian surrogate (Pande, 2010), or adopt a child from another country (Fronek and Cuthbert, 2012). The burgeoning market for these reproductive services has meant that single and lesbian women as well as gay men are able to become parents more easily than ever before. This has raised important questions for the field in terms of rethinking commodification and its relationship to definitions of family and parenthood (Dempsey, 2013; Ertman, 2003).

However, commodified reproduction also presents a number of challenges. As Spar (2006) has argued, the reproductive marketplace still remains largely unacknowledged given the moral and ethical quandaries that it raises surrounding bodily ownership and integrity as well as the potentially coercive and unethical aspects of the consumption of reproductive services (Inhorn, 2003; Rothman, 1989). Furthermore, national differences in legal, cultural, or religious traditions means that it is more widely accepted that individuals will travel long distances (often overseas) to access services and technologies that are unavailable at home, engaging in 'reproductive tourism' (Martin, 2009). For example, British couples have been known to travel to the Czech Republic for IVF treatment given shorter wait times, reduced costs, and better success rates (Culley et al., 2011). Reproductive tourism provides evidence for stratified reproduction as most 'tourists' are economically privileged. The ability to pay top dollar can also transcend categories of identity which would normally limit one's access to a particular reproductive treatment or service (such as sexuality or age) (Kroløkke et al., 2012).

While reproductive tourism can result in increased reproductive autonomy, there are also problems associated with these activities (Blyth and Farrand, 2005). Feminist scholars have argued that the global nature of the marketplace makes it more likely that marginalised women become donor bodies and/or perform reproductive labour (Kroløkke et al., 2012). The lack of consistent regulations around the consumption and provision of reproductive services has meant that Indian surrogates, for example, are more often subject to 'high-risk multiple egg transfers, receive less compensation, and no parental rights' compared to British surrogates (Martin, 2009, p. 254). However, rather than concluding that surrogates are invariably 'exploited' by intended parents, feminists have argued that transnational surrogacy is a social justice issue that requires a more nuanced approach to understanding why marginalised women become surrogates in the first place (Bailey, 2011).

'Gender 'and 'sex'

Postmodern and post-structuralist theoretical frameworks have become central in destabilising gender binaries that have implied essential differences between men's and women's reproductive bodies. Whereas the sex/gender distinction was viewed to be a necessity in liberating women from the bounds of patriarchal medicine in the 1970s, in postmodernity, there is less certainty around 'gender' and 'sex'. The terms 'woman' and 'man' have been theorised as regulatory ideals – they are neither inevitable nor guaranteed (Butler, 1990, 1993). Thus new spaces are emerging for multi-layered and dynamic performances of gender in the context of reproduction (Nash, 2012a).

To illustrate, fluid conceptions of 'gender 'and 'sex' have led to changes in western family life. Postmodern families are characterised by their permeable boundaries that allow a free exchange in ideas, values, and importantly, gender roles (Johns and Gyimóthy, 2003). Postmodern families are less likely to be headed by a male breadwinner, one of the more enduring features of social reproduction. As the costs of living increase, most middle-class households require two incomes (a 'dual breadwinner' model) and, as such, gender roles within households are shifting (Broomhill and Sharp, 2007, p. 87). Greater numbers of women and men are choosing to delay childbearing or remain childless and it is becoming more acceptable for men to be primary caregivers. We are also seeing an increasing diversity of family types across various social groups as heterosexual couples are less likely to marry and single parent and queer families are becoming more visible. In this way, the nuclear family has arguably become a 'cultural myth' (Denzin, 1987, p. 36).

However, as Annandale (2009, p. 95) has observed, post-structuralist gender theories are problematic to a certain extent because health and illness are ultimately 'drawn towards opposition'. As she notes, biological sex, in many ways, continues to form the basis of gendered experiences of reproduction (p. 99). The unhinging of gender from sex, especially in the context of reproduction/reproductive health, is politically problematic for feminists because deconstructing gender/sex binaries does not necessarily lead to a deconstruction of gender hierarchies. Although postmodernity has brought with it more flexible arrangements and increased reproductive 'choices', when coupled with declining fertility rates, this situation has *not* necessarily led to greater gender equality (McRobbie, 2009). On one hand, women in western societies are told that they can 'choose' whether or not to reproduce and when, to choose paid employment in either a full-time or part-time capacity, or to be full-time mothers. But these 'choices' still often come with the burden of the *expectations* associated with the male breadwinner model. Moreover, the concept of a ticking 'biological clock' is still more readily applied to women than men (even though biological clocks tick for men too – see Shirani, in this volume). (Biological) motherhood, for women, is still seen as being central to the performance of femininity.

Postfeminism

Neoliberal practices and values coupled with the disavowal of the constraints of 'sex' and 'gender' can make it seem as though feminism is no longer relevant in the context of women's health or reproduction, a postfeminist sensibility. McRobbie (2004) has associated the emergence of a de-politicised belief that feminism and gender are no longer 'relevant' with the term widely known as 'postfeminism'. McRobbie (2009, p. 11) understands postfeminism as an 'undoing' of feminist gains of the 1970s and 1980s in the name of modernisation – a necessity of neoliberalism. Postfeminism is notable in its association with whiteness and affluence, and is 'anchored in consumption as a strategy (and leisure as a site) for the production of the self' (Tasker and Negra, 2007, p. 2). Adkins (2002) and McRobbie (2009) have argued that the sociology of Beck and Beck-Gernsheim (2002) and Giddens (1991) has been premised on the idea that freedoms for women have been 'won' without accounting for existing structural inequalities. In other words, actions and situations in relation to reproduction which would have otherwise been positioned as sexist or patriarchal are now problematically located within a framework of 'empowerment' and agency under neoliberalism.

Organisation of the volume

As this collection encompasses many viewpoints from an exciting array of established and emerging scholars from around the world, in order to help the reader navigate these perspectives, the book is organised thematically into three sections that reflect the key characteristics of postmodern reproduction that I described earlier: 1) Contested 'choices' and challenges; 2) Reproductive bodies and identities; and 3) The (global) reproductive marketplace. Although the chapters are grouped into sections for clarity, the essays themselves make it evident that the fluidities and ambiguities surrounding reproduction require us to think across and beyond these divisions.

Part I: contested 'choices' and challenges

The chapters in this section interrogate the proposition that the neoliberalisation of life means that individuals are now 'empowered' or 'liberated' by the increased range of reproductive 'choices' and opportunities available to them.

In Chapter 1, Katherine M. Johnson, Julia McQuillan, Arthur L. Greil, and Karina M. Shreffler address the multiple meanings of 'infertility'. The authors contend that medical definitions of infertility do not adequately capture the experiences of most US women. Rather, the concept of 'fertility barriers' allows for a more nuanced understanding of a range of complex issues that affect women's reproductive 'choices' including both biomedical and situational barriers to childbearing. Importantly, they examine the ways in which indicators such as social class and race/ethnicity provide further evidence for stratified reproduction.

Chapter 2 shifts our focus to Australia – Catherine Ruth Collins, Damien W. Riggs, and Clemence Due examine constructions of the 'best interests of the child' in relation to a recent parliamentary debate about surrogacy in New South Wales. The authors point to the ways in which the 'child' and 'family' were used rhetorically by politicians to reinforce the view that surrogacy disrupts the primacy of the heteronormative family structures. Although non-heterosexual and single individuals are positioned through neoliberal discourses as welcome consumers of assisted reproductive technologies, the extracts from the debate position single and gay Australians as merely reproductive consumers who see parenthood as a 'lifestyle choice'. This chapter provides a valuable discussion of how cultural discourses about reproduction impact upon legislative contexts, highlighting the ways in which the reproductive choices of certain social groups are seen to require regulation.

Meredith Vanstone, Elizabeth Anne Kinsella, and Jeff Nisker examine the construction of choice in the context of prenatal screening in Canada in Chapter 3. Although 'choice' is a key tenet of western biomedical ethics, feminism, and neoliberal discourses, the focus on women's individual choices in relation to deciding whether to participate in prenatal screening neglects to consider the range of structural and societal influences that can constrain their choices. As the authors argue, although prenatal screening is one space in which 'informed choice' is framed positively, women's desires to conform to medicalised ideals of 'normal' warrant much deeper investigation into the meaning and enactment of reproductive choices.

In Chapter 4, Kate Gleeson explores choice in the context of a 2010 Australian legal case in which a young Queensland couple was arrested for procuring and self-administering RU-486, an abortion drug. Gleeson argues that Tegan Leach's decision to induce her own abortion exposes the limits of neoliberal framings of reproductive choice in Australia. Namely, as a 19–year-old woman, Leach epitomised the entrepreneurial female self, a position in which young women view their reproductive choices as enabling their freedom and 'empowerment' as self-constituting subjects of neoliberalism. Gleeson problematises the alignment of entrepreneurialism with postfeminism.

Since the 1970s, several industrialised countries have adopted neoliberal reform ideologies in the interests of efficiency and effectiveness. Some scholars have argued that these reforms have resulted in inequities in health care, especially in the primary care sector. At the same time, feminist health activists and scholars have called for the demedicalisation of pregnancy and childbirth and, by extension, postpartum care. In Chapter 5, Cecilia Benoit, Camille Stengel, Rachel Phillips, Maria Zadoroznyj, and Sarah Berry examine how policy reforms in one region of Canada altered women's access to health services in the postpartum period. Here, the authors explore the relationship between continuity of care, satisfaction with care services, and other non-clinical factors that impacted upon maternal health and well-being.

Part II: reproductive bodies and identities

The chapters in this section point to the diverse terrains upon which identity, embodiment, and reproduction are (re)constituted in postmodernity, especially in the context of neoliberalisation (described in Part I). This section also addresses feminist critiques surrounding the erosion of gender binaries and how this plays out in context of reproduction and health.

In Chapter 6, Deborah Lupton draws on the notion of 'assemblages' to explore the social worlds of the unborn and the contingent nature of unborn embodiment. Specifically, Lupton examines how multiple layers of meaning (historical, cultural, technological, and biomedical) are intertwined in the embodiment of the unborn in postmodernity. As she argues, unborn humans occupy multiple and competing categories of personhood. Yet the visualisation of unborn humans using medical technology disrupts unborn assemblages in various ways and this has had wide-ranging implications in terms of how unborn and maternal bodies are perceived and what political effects flow from this.

Documenting how digital photographs can be used to think critically about postnatal embodiment is my own contribution to this volume. In Chapter 7, my key contention is that participant-produced photographs can reveal important information about how Australian mothers negotiate a changed embodiment over time. The examples discussed demonstrate that digital cameras were tools that allowed women to portray themselves and their experiences of post-pregnancy in ways that would otherwise be impossible. Their visual accounts of their bodies reflect the more ambivalent and varied 'realities' of maternal embodiment in postmodernity compared to the limited range of popular images of postnatal bodies.

In Chapter 8, Heather Dillaway and Catherine Lysack examine the lived experiences of reproduction among a group of US women with spinal cord injuries (SCIs). Although women with SCIs often live 'normal' reproductive lives, their experiences of reproductive health care are complex and they are routinely complicated by a range of social and structural barriers, in addition to their physical impairments. Lay and medical attitudes to disability and motherhood present women with multiple hurdles in trying to make reproductive choices. Dillaway and Lysack argue that women with SCIs live on the margins of 'normal' and 'abnormal' as they navigate multiple layers of meaning around health, reproduction, disability, femininity, and motherhood.

In the last two chapters of this section, Fiona Shirani and Tina Miller capture British men's participation in the procreative realm, specifically around childbearing, and with a critical eye toward masculinities and life courses. Given that discussions of fertility decision-making and notions of biological clocks have primarily focussed on women, in Chapter 9, Shirani shows how men define the 'right time' for fatherhood. Drawing on a qualitative longitudinal study of men across the transition to first-time fatherhood, her chapter provides a valuable exploration of how men view their fertility timing decisions when embarking on parenthood and when revisited eight years later. Shirani highlights the importance of time as a tool for understanding men's life course transitions.

Following on from this, in Chapter 10, Tina Miller focuses on men's emerging paternal identities as they anticipate and prepare for the birth of their first children. Miller discusses the ways in which the men in her study used a range of discourses to narrate their transitions to fatherhood, both accommodating and resisting normative hegemonic masculine ideals. As she argues, examining men's expectations and experiences of the antenatal period and childbirth provides important insights into their performances of masculinity as well as their reproductive choices.

Part III: the (global) reproductive marketplace

The chapters in this final section take consumption as a starting point for examining how gendered experiences of reproduction are mediated through a global reproductive marketplace (Rothman, 2006). Rather than arguing whether the commodification of reproduction is inherently 'good' or 'bad', the chapters in this section examine reproductive commodification and consumption in different contexts 'as a critical framework for revealing the ways in which consumers adapt reproduction to their own desires and needs' (Fletcher, 2006, p. 41).

In Chapter 11, Lisa Jean Moore and Marianna Grady critically analyse the effects of Web 2.0 on the sperm banking industry – the authors analyse the social media strategies that US sperm banks are employing in order to stay relevant in a competitive global marketplace. Moore and Grady contend that the clever use of social media such as Facebook has been extremely effective in creating online communities in which users of the services can connect with one another in the process of putting (an imagined) 'daddy' into their virtual shopping cart. However, as the authors observe, the male donors themselves are silent in this process. Donors are disembodied from their sperm and sperm banks sell a narrative of masculinity and fatherhood to consumers.

In Chapter 12, Margaret Boulos, Ian Kerridge, and Catherine Waldby focus on questions of altruism, anonymity, and reciprocity in their study of identified egg donation in Sydney. The authors propose that the 'gift' economy of egg donation creates intimate social relationships between Australian women but does not necessarily lead to expectations of reciprocity. Specifically, the authors note the importance of understanding both the nature of the relationships forged between women and the context of the practice of donation. The authors also describe how the gift exchange operates in the context of neoliberal consumerist agendas.

In Chapter 13, Victoria Team and Kath Ryan outline the ways in which British women commodify expressed breast milk (EBM). Team and Ryan argue that EBM as a product has a variety of social and economic values and that these influence women's decisions about milk expression and EBM feeding. For example, EBM has economic value because it allows women to return to the paid workforce. EBM also has social value because it allows men to feed babies and to become more involved in parenting responsibilities. However, the commercialisation of EBM aligns with neoliberal agendas which can result in women becoming increasingly detached from breastfeeding in contrast to public health messages that implore its importance for mothers and babies.

Finally, in Chapter 14, Elizabeth Arveda Kissling contends that neoliberal and postfeminist sensibilities are enmeshed in young women's online conversations about quitting hormonal birth control in the US. As Kissling observes, the consumption of the pill is more than just a means of preventing pregnancy for women – its consumption is marketed as a lifestyle choice that actively consolidates a neoliberal subjectivity for women as sexual subjects in control of their own bodies. Yet the more debilitating side effects of hormonal birth control are rarely publicly acknowledged. As Kissling argues, although quitting the pill is a form of protest in itself, it does not create social change for women as part of feminist political action. Rather, quitting the pill becomes a consumer boycott that does not disrupt the neoliberal economy.

Acknowledgements

Many thanks to Eliza Burke, Michelle Phillipov, and Philip Martin for their valuable feedback on earlier versions of this chapter.

Notes

1. Throughout this book, Western Europe, North America, Australia, and New Zealand are associated with 'the West'. This term is utilised as a construction or a theory more than a particularity of time or place.
2. As Murphy (2012, p. 37) has noted, feminist health movements were 'deeply informed by whiteness'. The terms 'women' and 'we' were deployed through 'visions of unraced, unclassed sex' (p. 38). The writings of the Combahee River Collective and black feminists of the 1970s offered a very different view of US 'women's health' in light of civil rights movements (Breines, 2006).
3. The poster hung on the wall of the Los Angeles Feminist Women's Health Centre.

Part I
Contested 'Choices' and Challenges

1
Towards a More Inclusive Framework for Understanding Fertility Barriers

Katherine M. Johnson, Julia McQuillan, Arthur L. Greil, and Karina M. Shreffler

Introduction

An important focus for North American feminists has been addressing and improving women's reproductive autonomy (Lorber, 1989; Rapp, 2001; Roberts, 1997). Since the beginning of the second wave of feminist activism, there has been an 'explosion' of feminist research on reproduction (Feree and Hess, 2000; Ginsburg and Rapp, 1995). As feminist scholars and others have worked to bring human reproduction onto centre stage in the sociological enterprise, they have emphasised that reproduction is not simply a biological process but a socially constructed reality involving power relations over who controls women's bodies (Rothman, 1989). Many feminist scholars have made use of intersectionality theory (e.g. Collins, 1990), which emphasises the ways in which gender, race/ethnicity, class, and sexual orientation – among other factors – interact to create social realities. Indeed, although (western) women in postmodern society are represented as having enhanced reproductive choices, feminist scholars have critiqued this as highly individualistic, ignoring the larger social context that such choices are embedded in (Bird and Rieker, 2008; Lublin, 1998; Nash, 2012a). In this vein, the concept 'stratified reproduction' (Colen, 1986) has helped to illuminate how reproduction is structured across social and cultural boundaries, empowering privileged women and disempowering less privileged women (Bell, 2010; Roberts, 1997).

Yet despite these important developments in feminist theorising about fertility and reproduction, barriers to conception have often been on the margins of the feminist sociology of reproduction (Rapp, 2001; Sandelowski and de Lacey, 2002). As van Balen and Inhorn (2002) have noted, there is a 'scholarly lacuna' of social scientific work on infertility and childlessness. Earle et al. (2008) have similarly observed social science's emphasis on studying reproductive 'success' over reproductive 'failures', such as miscarriage, stillbirth, infant and maternal mortality, and disruptions to 'normal' reproduction more broadly. Two major exceptions to this generalisation are the significant bodies of work on assisted reproductive technologies (ARTs) (for a review, see Thompson, 2002) and on stratified reproduction in surgical sterilisation (Price, 2010).

In this chapter, we assert the importance of understanding the experience of infertility itself – and not just simply the use of reproductive technologies – as a sociological phenomenon. We argue that feminist theorising will benefit from thinking of infertility as one instance of 'fertility barriers', a term we employ to refer to a range of both biological and social factors that prevent women and men from having wanted children (Greil and McQuillan, 2010; Jacob et. al., 2007). A minority of women are infertile at any point in time, and a very small minority make use of ART; however, most women face obstacles to having the children they desire. Thus fertility barriers represent an important limitation to reproductive 'choice' in postmodern society. Our goal is to contribute to a more holistic picture of the factors that shape and constrain women's and men's reproductive lives in the contemporary US.

We employ a women-centred approach here because women's bodies have historically been, and continue to be, the primary site for blame or responsibility and treatment, even in situations of male-only infertility (Becker, 2000; Greil, 1991; Lorber, 1989). In the sections below, we work toward articulating a more inclusive 'fertility barriers' framework. First, we briefly visit themes in feminist research addressing infertility. We then provide an overview and critique of both the medical definition of infertility and social science terminology of 'involuntary childlessness'. We present our more inclusive typology of fertility barriers and illustrate its utility using data from the National Survey of Fertility Barriers (NSFB). We conclude by clarifying the linkages between the fertility barriers perspective and feminist concerns.

Theoretical and empirical background

Themes in feminist scholarship on infertility

Reflecting on earlier feminist thinking about reproductive technologies, Thompson (2002) distinguished between Phase One (early 1970s to the mid-1980s) and Phase Two (late 1980s and 1990s). During Phase One, most – but certainly not all – feminist writing criticised reproductive technology as reinforcing patriarchal control of women's bodies and conflating motherhood with womanhood. Reproductive technology was also depicted as pitting privileged women, who had access to these technologies, against less privileged women, who did not. Little attention was paid to the suffering of infertile women; some of the women interviewed by Greil (1991) felt that they were being sent the message that if they were 'true' feminists they should not desire children so much. Although Phase Two was not a clear break from Phase One, Thompson (2002) suggests that feminists began to focus more on the experiences of infertile women and shifted from certainty to ambivalence about reproductive technology as inherently 'good' or 'bad' for women. This accompanied a move among many feminists to go beyond thinking that the desire for motherhood was necessarily in conflict with feminism. Phase Two saw further development of themes concerning stratified reproduction.

Recent feminist work on infertility and the new reproductive technologies has echoed broader themes in the scholarship on reproduction. These themes include: understanding the experience of infertility in terms of patriarchy and the motherhood mandate (Becker, 2000; Greil, 2002; Inhorn, 1996; Remennick, 2000; Ulrich and Weatherall, 2000); problematising ideologies of biological motherhood as 'true motherhood' (Letherby, 1999); documenting resistance to dominant images of infertile women (Letherby, 2002a, 2002b; Parry, 2005; Riessman, 2000; Todorova and Kotzeva, 2003); addressing the moral economy of infertility treatment in terms of who can (and should) have access to reproduction (Agigian, 2004; Becker, 2000; Bell, 2010; Steinberg, 1997); theorising women's agency to make reproductive decisions in the midst of social and cultural constraints, such as the technological imperative (Beckman and Harvey, 2005); and addressing societal tensions between women's childbearing, educational and career opportunities, and general life course timing (Earle and Letherby, 2007; Friese et al., 2006; Martin, 2010). There is also a large body of feminist work that analyses and critiques reproductive technologies (Franklin, 1997; Thompson, 2002), including surrogacy/gestational carriers (Markens, 2007; Rothman, 1989; Teman, 2010), in-vitro fertilisation (Gerrits, 2008; Throsby and Gill, 2004) and gamete donation (Almeling, 2007; Tong, 1996). Yet there has been a dearth of research that directly addresses, and potentially alters, the definition of infertility itself as a medicalised phenomenon. Indeed, much social science research on infertility appears to proceed with the assumption that it is an agreed-upon medical phenomenon (see Sandelowski and de Lacey, 2002 regarding the invention of infertility as a discursive category). Our goal is to move forward the sociological discussion of infertility itself, by focusing on the problem of its very definition.

Defining infertility and childlessness: problems and politics

The contemporary understanding of infertility is that it is a medical problem, subject to the definitional and treatment authority of physicians (Conrad, 2007). A common medical definition of infertility is no conception after a year of regular, unprotected sex (American Society for Reproductive Medicine, 2008). Feminist scholarship has critiqued this definition for reinforcing a heteronormative paradigm of sex and reproduction (Agigian, 2004; Johnson, 2012) for focusing on couples as a unit of analysis, thereby obscuring gender asymmetries in treatment and diagnosis (Sandelowski, 1993), and for ignoring the more subjective implications of lack of conception for women under different social and cultural conditions (Inhorn et al., 2009; Sundby, 2002). The definition further assumes that women are either trying to become pregnant (if they are not using contraception) or trying to avoid pregnancy (Greil and McQuillan, 2010).

The medical conceptualisation of infertility better captures the experiences of White affluent women who tend to see their pregnancies as events that can and should be planned (Bell, 2010; McQuillan et al., 2011) and who have relatively better access to medical services (Greil et al., 2011) than other women. Feminists therefore assert that the act of medically defining infertility implicitly defines the intended population of infertile patients as partnered, heterosexual, White women who are intending to conceive a pregnancy and who have socio-economic resources to seek medical treatments when problems arise (Greil and McQuillan, 2010). This definition has material implications for women's lives because it is often used in creating legislation about infertility and in setting the terms of insurance coverage for treatment (Agigian, 2004; Johnson, 2012; King and Meyer, 1997).

Letherby (1999) recommends using the term 'infertility' to refer to the medically defined physical condition and using the term 'involuntarily childlessness' to describe the socially constructed experience that may accompany the medical condition. While this (re)asserts the social nature of the phenomenon, it ignores the experience of the roughly half of infertile women (who are not childless when they experience infertility) at the same time that it reifies distinctions between complex and potentially overlapping categories, such as *involuntarily childless*, *childless by choice*, or *childfree* (Abma and Martinez, 2006; Koropeckyj-Cox et al., 2007). The use of such categories implies that most adults are *either* parents, childfree, or involuntarily childless. Yet the line between 'voluntary' and 'involuntary' can be murky, hiding the reality that some women who have children face barriers to having additional children and obscuring the fact that many women have faced multiple barriers to fertility. Furthermore, the use of terms like voluntary and involuntary childlessness implies that women are either *trying* or *not trying* to become pregnant. However, many US women are uncertain about their fertility intentions (Hagewen and Morgan, 2005). Thus, the 'voluntary'/'involuntary' dichotomy masks a continuum of intentionality and a complex history of experiences. What is needed is a way to talk about obstacles to having children without oversimplifying a complex reality more than is necessary and without arbitrarily excluding some women from consideration.

Towards a more inclusive framework: from infertility to fertility barriers

As a way forward, we offer and develop the notion of 'fertility barriers' to better understand the full diversity and complexity of issues involved when childbearing does not or cannot proceed 'normally.' Based on prior empirical work developed with the data described below (Greil and McQuillan, 2010; Jacob et al., 2007), we outline six potential types of barriers that women face in having children. This list is not necessarily exhaustive, but it hopefully begins the conversation about fertility barriers in a way that highlights the variety of ways that women may face obstacles to having desired children.

Intent status and infertility

As we argued above, there is much to critique about the medical definition of infertility. Nonetheless, we believe that it is still useful to incorporate the medical notion of infertility into our framework. First, it does capture some women's experiences; second, the definition is firmly entrenched in medical literature and practice. In other work, we have split infertility into two categories: *infertile with intent* and *infertile without intent* (Greil et al., 2009). The former refers to women who were explicitly *trying* to get pregnant while meeting the medical definition of infertility. The latter refers to women who met the medical criteria, but did not express intention to conceive. We recognise that women often have difficulty with the idea of intended/unintended pregnancies (Moos et al., 1997) and therefore have employed measures which allow women to state that they were 'ok either way' rather than forcing them into the dichotomous categories of 'intended' and 'unintended'.

Other biomedical barriers

Not all women who face biomedical barriers to having desired children are infertile by the medical definition. Some women might be advised by physicians that a health condition could make it dangerous for them to be pregnant (e.g. anaemia). Still others may be advised that their medication interferes with or increases health risks during pregnancy or affects the ability to have a healthy child. Some women decide that their own or their partner's depression or disability will make raising a child too hard. All of these examples refer to health-related barriers to having desired children even though they do not clearly fit the medical definition of infertility.

Sterilisation regret

Sterilisation is neither inherently liberating nor oppressive; the meaning of sterilisation depends on its context (Schoen, 2005). Women's surgical sterilisation has become increasingly popular since the mid-1960s and is now the second most frequently used method of contraception in the US, used by 45.3 per cent of contracepting women (Mosher and Jones, 2010). Sterilisation can offer women a sense of reproductive control and empowerment if freely chosen, but sterilisation can also be the result of overt or subtle coercion (Schoen, 2005). Given the history of eugenic sterilisation in the US, especially for women of colour and poor women (Roberts, 1997), it is crucial to recognise sterilisation in the context of both stratified reproduction and contemporary fertility barriers. Sterilisation may also be the consequence of treating a health condition that might make pregnancy or childbearing difficult or impossible. Because sterilisation is relatively permanent contraception, women who use this method may later wish that they could conceive and therefore see sterilisation as preventing childbearing. Other women may undergo sterilisation surgery for a medically necessary reason despite desiring more children. In such cases, women who do not fit the medical definition of infertility may still face barriers to having desired children.

Miscarriage and stillbirth

Approximately 14 per cent of all clinically recognised pregnancies in the US result in miscarriage, or a loss during the first 20 weeks of pregnancy, and another 0.5 per cent result in stillbirth, a loss after the 20th week (Saraiya et al., 1999). Saraiya et al. (1999) also observe that physicians consider recurrent miscarriages as a type of fertility problem, but even a single pregnancy loss can prevent women from realising their fertility goals and can be experienced as highly distressing (Shreffler et al., 2011).

Situational barriers

We consider life situations that prevent women from having (more) biological children as another form of fertility barrier (Jacob et al., 2007). In contrast to the previously described categories, this explicitly addresses social as opposed to medical factors. Women with situational barriers would like to have children at some point but perceive their current situation as not right. Several of these women may eventually become permanently childless due to continually postponing childbearing; others will have fewer children than they intended. Situational barriers can include: relationship status, sexual orientation, conflicting fertility ideals between partners, interference from education/career, financial constraints, and competing social obligations, such as caring for sick or elderly family members. Some women simply express that they are not yet emotionally and psychologically 'ready' for children. Thus many women who are biologically 'fertile' nonetheless face barriers that prevent them from having desired children.

Multiple barriers

As should be obvious from the descriptions, these fertility barrier categories are not mutually exclusive. Many women will confront multiple barriers throughout their lives. For instance, because of structural constraints such as the demands of education or jobs, women may delay childbearing and then find it more difficult to conceive. Or women can have children but then suffer medical complications that prevent future conception. The experience of any one fertility barrier is likely shaped by prior reproductive episodes or events.

To inform thinking about feminist approaches to reproduction, we provide a description of the distribution of fertility barriers among US women. Consistent with feminist attention to social inequality, stratified reproduction, and the inclusion of a variety of women's reproductive experiences, we examine how fertility barriers differ by race/ethnicity and education. Also, because it is common to think of infertile women and women who have unintended pregnancies as two distinct groups, we compare women who had abortions and women who have not had abortions with regard to fertility barriers. Finally, because it is common to think of childless women and those with children as two distinct groups, we further analyse fertility barriers by parental status.

Practical application

Is the broader conceptualisation of fertility barriers useful for capturing some of the complexity of fertility experiences and stratified reproduction among US women? To answer this question we used the NSFB, a telephone survey of a random sample of 4,786 women designed to assess social and health factors related to reproductive choices and fertility. The telephone interviews were conducted between September 2004 and December 2006. Our analytic sample included all women who reported their race as White, Black, Hispanic, or Asian.

Concepts and measures
Fertility barriers

To construct the fertility barrier categories, we first created indicator variables for each type of barrier described above. Next, we used these variables to determine the unique combination of barriers that each woman had experienced. Theoretically, women could, over their lifetime, experience all of these barriers, making 720 possible combinations. In the NSFB sample, we found 33 unique combinations of multiple fertility barriers among women, which we merged into the 'multiple barriers' category to facilitate visualisation of the patterns. Thus the final fertility barriers measure had eight categories (the six described above plus 'no barriers' and 'multiple barriers'). All were based on lifetime prevalence and referred to whether or not a woman had *ever* experienced a particular barrier.

'No barriers' included 1,676 women (35.8 per cent of the sample) who did not report facing any of the six fertility barriers. It should be noted that this figure indicates that the *majority* of US women (64.2 per cent) *have* experienced a fertility barrier at some point in their lives. 'Infertile with intent' included 173 women (3.7 per cent) who tried for more than 12 months to conceive any pregnancy or who had ever tried to get pregnant for more than 12 months without conception and who encountered no other barrier. 'Infertile without intent' included 694 women (14.8 per cent) who reported regular, unprotected intercourse for more than a year without conception, who did *not* report that they were trying to get pregnant, and who reported that this was their only fertility barrier. Women who we call 'infertile without intent' can also be conceptualised as the 'hidden infertile' because they rarely seek medical help (Greil et. al., 2009). The majority of infertile women faced additional barriers and were placed in the 'multiple barriers' category.

The 'other biomedical problems' group included 28 women (0.6 per cent) who did not fit medical definitions of infertility and who did not report any other fertility barriers but who wanted (more) children and reported barriers such as medical conditions that made it inadvisable to become pregnant. The 'sterilisation regret group' included 114 women (2.4 per cent) who did not report experiencing infertility or any other fertility barrier but who said that their own or their partner's surgical sterilisation kept them from having wanted children. The 484 women (10.3 per cent) in the 'miscarriage or stillbirth' group had at least one miscarriage or stillbirth and did not meet the criteria for any other fertility problems.

Women (N = 116; 2.5 per cent) were included in the 'situational barriers' group if they did not meet the criteria for any other barrier, if they intended to have a child in the future, if they had not yet had a child, and if they provided a situational reason for not yet having a child. These women were unique because not having a child is one of the criteria for being placed in this group. Only women who said that they have no children were asked why, and only those who provided a situational barrier were included in this category. Thus the number of women who have experienced situational barriers at some point in their lives is probably much larger than this. All women (N = 1,395; 29.8 per cent) who fit into more than one category were placed in the 'multiple barriers' group.

Additional variables

Race/ethnicity was measured using the two standard US Census categories (United States Census Bureau, 2011). There is no single best way to categorise individuals who report multiple races/ethnicities; as is done in many US studies, we gave first priority to identification as 'Hispanic' and second priority to identification as 'Black'. White and Asian women were also included in the analysis, but those indicating 'other' race/ethnicities were excluded because there were too few women in the category. We recognise that all racial/ethnic groups contain heterogeneous subgroups but use these larger categories as indicators of gross distinctions that reflect patterns of racial formation in the US. Education was dichotomised into those who did versus did not have a bachelor's degree.

For each pregnancy women had, they were asked whether the outcome was a live birth, stillbirth, miscarriage, or abortion. We used whether or not a women had an abortion as a proxy for whether or not she had an unwanted pregnancy, although we recognise that a very small percentage of women may seek an abortion for health or medical reasons (Finer et al., 2005). Women who had a live birth or who adopted children were coded as parents, and all other women were coded as non-parents.

Patterns

Because prior research has found stratified reproduction by race/ethnicity and social class (Bell, 2010; Colen, 1986), we examined patterns of fertility barriers across these categories (see Figure 1.1). Patterns of

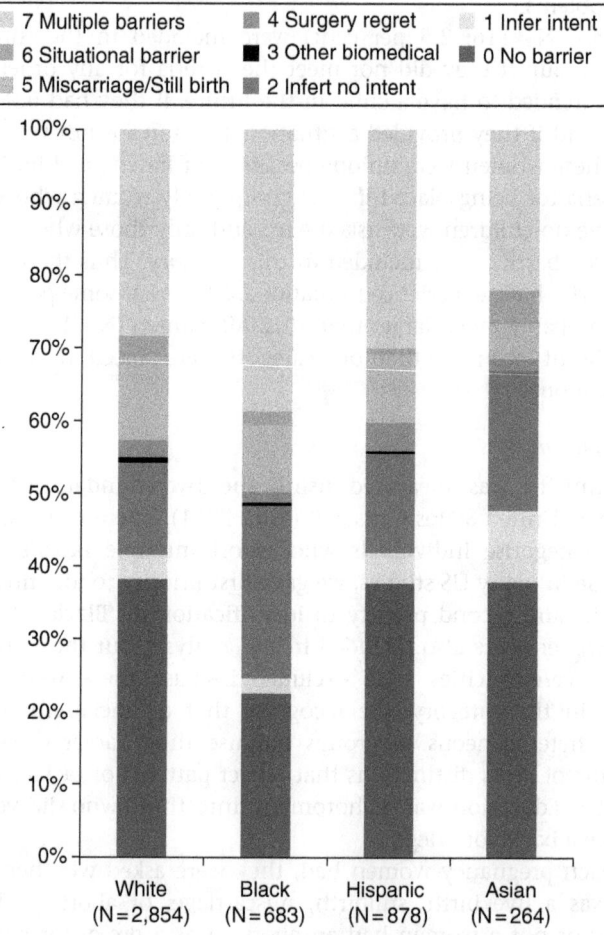

Figure 1.1 Fertility barriers by race among a sample of US women (N = 4680)

fertility barriers *do* differ significantly by race/ethnicity ($\chi^2 = 193.11$; df = 21; p = .000). A higher proportion of White, Hispanic, and Asian women reported no fertility barriers compared to Black women. Just over 20 per cent of Black women had no barriers, but nearly 40 per cent of White, Hispanic and Asian women had no barriers. Infertility with intent was less common, but infertility without intent was more common among Black women compared to women in the other categories. Hispanic women were more likely than women in other groups to regret sterilisation. Miscarriage was least common, but situational barriers were more common among Asian women compared to all other women. Black women were most likely and Asian women least likely to experience multiple barriers.

Patterns of fertility barriers among women also varied significantly by education ($\chi^2 = 195.19$; df = 7; p = .000) (see Figure 1.2). Contrary to

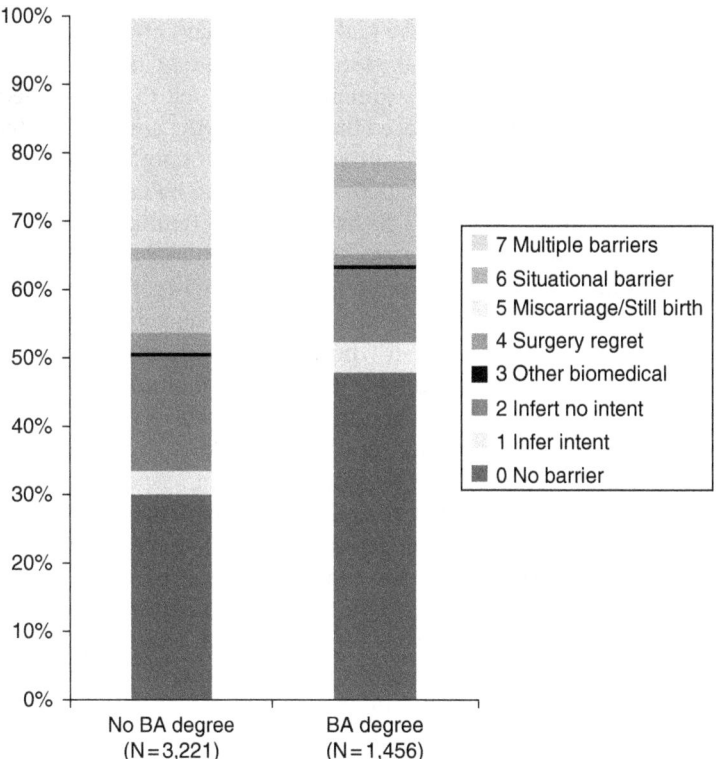

Figure 1.2 Fertility barriers by education among a sample of US women (N = 4680)

the popular image of infertility as primarily a problem for middle-class White women, this representative sample showed that barriers were more common among less educated women. A slightly higher percentage of women with higher education had infertility with intent, and fewer had infertility without intent than women with lower education. Surgery regret was also more common among women with lower levels of education. Women with less than a college degree were twice as likely to experience situational barriers as women with a college degree or greater. Women with less education were much more likely to experience multiple barriers than women with higher education. These patterns across racial/ethnic groups and social classes reaffirm the existence of stratified reproduction in fertility barriers: overall, women who had relatively lower social and economic resources to seek treatment are the same women who were more likely to experience fertility barriers.

Many women who have had abortions have had at least one unintended and/or unwanted pregnancy (Finer et al., 2005). Yet most of the women (62.4 per cent) who have had abortions have also faced fertility barriers (see Figure 1.3). This suggests that at some points during their reproductive careers, the same women who have felt that they needed to end pregnancies have also faced barriers to having children. Patterns of fertility barriers varied significantly by whether or not one has had an abortion ($\chi^2 = 79.50$; df = 7; p = .000). Women who had abortions (20.6 per cent) were much less likely to have faced no fertility barriers than women who did not have abortions (37.6 per cent). Women who had at least one abortion were much more likely to have experienced infertility, but the vast majority of infertile women who had an abortion were infertile without intent. Women who had an abortion (35.1 per cent) were somewhat more likely to have experienced multiple barriers than women who had not had an abortion. These patterns further affirm prior arguments that we should not view abortions and wanted pregnancies as discrete reproductive events applying to separate groups of women (Bessett, 2010; Finer et al., 2005). These patterns make it apparent that we cannot clearly separate out women who have had unwanted pregnancies from women who have faced fertility barriers.

Figure 1.4 displays differences in fertility barriers between parents and non-parents. Because infertility is often conflated with being involuntarily childless, the pattern of fertility barrier patterns by parenthood status may be surprising ($\chi^2 = 536.15$; df = 7; p = .000). Perhaps due to younger age and never having 'tested' their fertility, non-parents were more likely (47.0 per cent) to be in the 'no barrier' category than parents

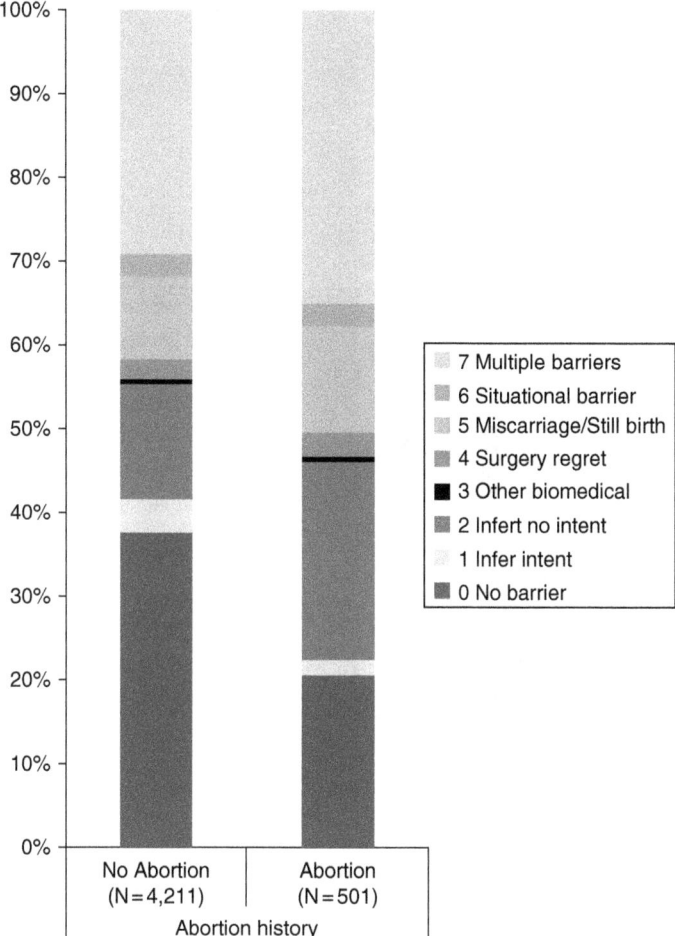

Figure 1.3 Fertility barriers by self-reported abortion among a sample of US women (N = 4680)

(33.9 per cent). In addition, in this survey only women without children were asked about situational barriers to having children. Although people typically associate fertility barriers with childlessness, the majority of women who were parents (76.1 per cent) had experienced at least one fertility barrier at some point in their lives. Most of the infertile women in our sample were parents, and parents were much more likely to report multiple barriers.

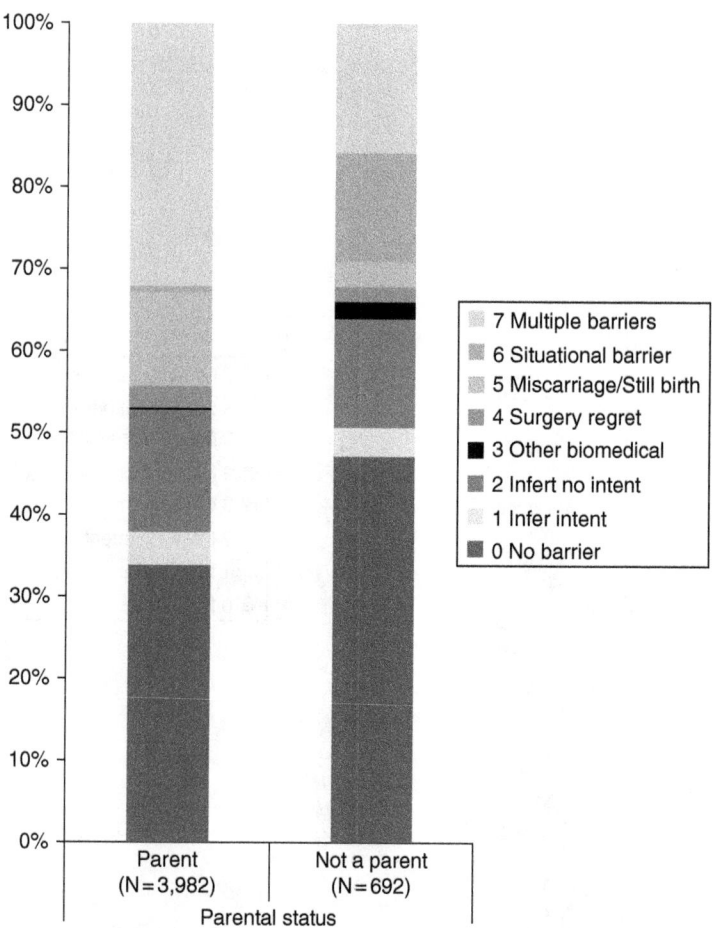

Figure 1.4 Fertility barriers by parenthood status among a sample of US women (N = 4680)

Discussion

One of the paradoxes in postmodern society is that the ideology of enhanced reproductive 'choice' coexists with material and ideological obstacles to exercising that 'choice' (Lublin 1998; Martin, 2010). This is as true for women who face obstacles to having desired children as it is for women who are trying to avoid unwanted births and for women who are trying to have the birth experience they desire. As we (and other scholars) have argued, the biomedical definition of infertility artificially creates the impression that some women face barriers to having desired children but that for most women having children is personally unproblematic. Furthermore, social discourses about reproductive technologies reinforce the notion that women and men can be classified into distinct categories: those who are infertile (i.e. privileged women), those who are 'too fertile' (i.e. unwed mothers, women of colour, and 'welfare mums'), and those who are 'dysfertile' (i.e. lesbians and gay men) (Ikemoto, 1996).

We have shown here that most US women face barriers to having wanted children and that many women confront multiple barriers. Investigation into patterns of barriers by race/ethnicity and social class reveals continuing evidence of stratified reproduction – the same women who might be viewed as 'too fertile' (Ikemoto, 1996) are actually more likely to face barriers compared to more privileged women. As such, the ability to act on one's 'choice' to have desired children is deeply constrained by the economic and social realities of postmodern society (Bird and Reiker, 2008).

Furthermore, we have shown that many women who have unwanted pregnancies *also* experienced fertility barriers and that most women with children have experienced fertility barriers. Although unwanted births and fertility barriers are often studied in isolation, for many women they are not experienced that way. And, even though individual barriers are often studied separately, women pass in and out of different fertility barrier categories during their reproductive careers.

Because the biomedical model of infertility and related reproductive issues have dominated the social science framing of fertility barriers (Greil et al., 2011; Macaluso et al., 2010), infertility has appeared disconnected from sterilisation regret, miscarriages, or situational barriers such as delaying conception to finish school, not finding a suitable partner for raising children, and violent relationships. We live in a time when there are biomedical solutions to social fertility barriers (e.g. ART for same sex couples), and social solutions to biomedical fertility barriers (e.g. adoption or choosing a childfree life). The broader fertility barriers framework we have presented here allows us to see these connections among fertility barriers as women move through their reproductive careers.

In this chapter, we have argued that feminist analyses of reproduction will benefit from a broader conceptualisation of fertility barriers which prevent women from having wanted children. However, we want to emphasise that the stance we take here is not pronatalist. We are not saying that all women want children or that they should want (biological) children. Such barriers are not always negative factors in women's lives: for some women, encountering fertility barriers leads to a meaningful life without children that was previously not a consideration in a pronatalist society (Gillespie, 2003; Parry, 2005). Finally, we suggest that our broader framework of fertility barriers specifically speaks to the problem of contextualising women's reproductive decision-making. We live in an era dominated by individualistic ideas of reproductive 'choice' and an array of new reproductive technologies to supposedly help women make those 'choices'.

Most feminist scholars urge reproductive rights to enhance reproductive autonomy for women. This includes avoiding undesired pregnancies, having birth options respected, and having desired children. A feminist perspective helps to challenge both the presumption that adult women should be or need to be mothers to be valued and the presumption that only privileged women should have help to overcome fertility barriers.

2
Constructions of the 'Best Interests of the Child' in New South Wales Parliamentary Debates on Surrogacy

Catherine Ruth Collins, Damien W. Riggs, and Clemence Due

Introduction

Across many sectors of Australian society, surrogacy continues to be viewed as a controversial mode of family formation. One possible reason for this is that women who carry and give birth to a child for another person disrupt normative understandings of what it means to be a mother, and what constitutes a family (see Riggs and Due, 2012 and Raymond, 1994 for a discussion of how motherhood is constructed in debates over surrogacy, and Jones, 1990 and Orloff, 1993 for examples of feminist writing concerning the vulnerabilities of all women in terms of their presumed reproductive capabilities and the subjugating effects this can have). One particular way in which controversy over surrogacy is voiced is via claims that surrogacy is not in 'the best interests of the child' (where children's best interests are normatively understood as served by having a mother and father who conceived and birthed the child). Yet as Jenkins (1998, p. 2) argues, such recourse to normative notions of 'the best interests of the child' serve 'as a "human shield" against criticism', due to the fact that what constitutes 'the best interests of the child' is rarely, if ever, clearly defined. Instead, claims about children's interests are deployed rhetorically to bolster the position of the speaker and their own conception of what the category 'child' means, rather than necessarily being about the actual needs of children (Baird, 2008).

With the above points about current controversies over surrogacy in mind, the present chapter focuses on one recent example of constructions of children's 'best interests', namely in politicians' debate on surrogacy legislation recently introduced and enacted in the Australian state of New South Wales (NSW). The NSW Surrogacy Bill was tabled with the stated aim of better managing altruistic surrogacy arrangements, which had previously been covered by three separate Acts:

1. The *Assisted Reproductive Technology Act 2007*, which prohibits onshore commercial surrogacy, makes commercial surrogacy agreements legally void and unenforceable, and requires genetic records to be kept in a central register;
2. The *Status of Children Act 1996*, which contains a presumption of parentage in relation to children born through a fertilisation procedure in favour of the birth mother; and
3. The *Adoption Act 2000*, which allows intending parents in a surrogacy arrangement to apply to adopt the child and thereby become the legal parents of the child.

The Bill followed from an inquiry conducted by the NSW Legislative Council's Standing Committee on Law and Justice. The Standing Committee held four days of public hearings before publishing its *Legislation on Altruistic Surrogacy in NSW* report in May 2009 (Robertson, 2009). The Bill that was subsequently passed by 53 votes to 27 votes did not make altruistic surrogacy in NSW any more or less legal than was the case under previous legislation. Rather, the aim of the new legislation was to enable legal parentage to be transferred to the intending parents, and in so doing improve outcomes for women who act as surrogates in an altruistic arrangement, the intending parents in such an arrangement, and the resultant child. The Bill also clarified prohibitions on the advertising of surrogacy arrangements, in addition to prohibiting residents of NSW from engaging in commercial arrangements either interstate or overseas.

Yet despite the relatively pragmatic focus of the Bill upon the regulation of surrogacy, much of the content of both the inquiry and subsequent debate in parliament was concerned with whether surrogacy should be allowed at all, and on numerous occasions speakers appealed to 'the best interests of the child' in order to support their argument. Our interest in this chapter, then, is how the category of 'the child' was deployed within parliamentary debate in relation to the Bill in ways that promoted very specific (and for the most part highly normative) understandings of children's best interests. Importantly, our interest is not to analyse individual politician's attitudes towards surrogacy *per se*, but rather our focus is on how, as culturally competent individuals who are well versed in public debates over surrogacy, politicians justified their stance on the topic by appealing to notions of 'the best interests of the child' as a taken-for-granted category. Before moving on to analyse extracts from the Hansard for one day of the debates, we first outline in more detail the legal context of surrogacy in Australia.

Background

Despite a process initiated in 2006 by then Attorney General Phillip Ruddock aimed at reviewing and unifying laws on surrogacy, wide variation remains across Australian states and territories (Page and Harland, 2011; see also Millbank, 2011 for detailed examinations of the current state of play of legislation relating to surrogacy across Australia). To summarise existing legislation briefly: altruistic surrogacy is legal across Australia (though there are some differences in who is deemed an eligible intended parent and what is deemed acceptable reimbursement of expenses to women who act as surrogates), whilst commercial surrogacy arrangements within Australia are considered a criminal offence. Further, some states and territories (including NSW, as discussed in relation to the aforementioned Bill), ban residents from international surrogacy arrangements (Page and Harland, 2011; Stuhmcke, 2011). Given this variation in legislation across Australia, some commentators have called for another review of legislation relating to surrogacy (see Stuhmcke, 2011), particularly with regard to ambiguity over what constitutes 'reasonable expenses' payable to women who act as surrogates within an arrangement considered to be altruistic.

In addition to debates concerning what types of surrogacy arrangements ought to be legal, there are also inconsistencies across Australia relating to who is legally allowed to be an intended parent, together with what rights intended parents receive in relation to the child's legal parentage. In regards to legislation concerned with intended parents, Millbank (2011, p. 4) argues that the 'increasingly complex web of eligibility rules' may not in fact function to safeguard interests of either children or their intended parents. Part of this is the product, Millbank suggests, of the fact that most of the Inquiries conducted in states and territories across Australia in relation to surrogacy legislation have been undertaken hastily, with the resulting legislation put together on the basis of abstract ideas, rather than the actual experiences of people involved in surrogacy arrangements. This is highlighted in differing legislation across Australia concerning who is able to be an intended parent (and therefore commission a surrogacy arrangement), where some states require that the intended parent(s) are heterosexual and/or in a relationship and/or are female, and with most states requiring that the intended parent(s) prove a 'need' for surrogacy or are infertile.

Such debates within Australia reflect international debates over who constitutes a 'proper' intended parent (an issue reflected again in the data we examine in this chapter). Following Sorin and Galloway (2006), we would suggest that the child/adult dyad is constituted within such debates as a standardised relational pair, one that always already evokes the image of children as innocent and in need of protection, the corollary being that adults are those best placed to determine children's best interests. Yet as Burr's (2000) examination of the British Report of the Committee of Inquiry into Human Fertilisation and Embryology Act (Warnock, 1984) indicates, not all adults are necessarily constructed as equally suitable to be responsible for looking after the best interests of children. Burr found that the concept of the best interests of the child in regards to surrogacy was used rhetorically to justify a position in which the only acceptable approach to conceiving children was within a two-parent, heterosexual relationship. Such instances of heteronormativity within parliamentary debates have also been seen in previous parliamentary debates within Australia, such as in the Inquiry preceding the *Sexuality Discrimination Bill* (1995) (Morgan, 1997), the *West Australian Acts Amendment (Lesbian and Gay Law Reform) Bill* (2001) (Summers, 2007), and proposed changes to the *Sex Discrimination Act* (1984) aimed at restricting reproductive technologies on the basis of marital status (Smith, 2003). The last two examples in particular highlight the use of ambiguous rhetoric concerning 'the best interests of the child' to justify particular arguments about rights for non-heterosexual people (as well as the rights of single men and women regardless of sexual orientation) in relation to children (including the right to adopt and to use reproductive technologies). In the analysis that follows, we provide another example of how debates over reproduction typically evoke a highly normative image of what an intended parent should look like, and how this reifies one family form (i.e., a heterosexual nuclear family where the children are born to the mother) over all others.

Method

Our data are the Hansard transcript of the debate that took place on 10 November 2010 in the Legislative Assembly of the Parliament of NSW, immediately prior to the vote in which the parliament passed the legislation. We chose the transcript of this day since it was the date on which the Bill was declared with amendments, and, as the final day of debating, built upon previous debating on the 28 October 2010 in the Legislative Assembly and three other days of debates in the Legislative Council, and thus, we would argue included summaries of the positions of each politician in terms of the Bill. Our analysis of the data is based on the approach to discourse analysis described by Wetherell and Potter (1992), which involves the identification of interpretive repertoires as a way of analysing the content of discourse. Wetherell and Potter define interpretative repertoires as 'broadly discernable clusters of terms, descriptions and figures of speech often assembled around metaphors or vivid images' (1992, p. 90). In addition to identifying the dominant interpretative repertoires in relation to constructions of children across the data set, we were interested to examine the ideological claims that supported each repertoire. As Billig (1991) suggests, when expressing their viewpoints, individuals draw upon forms of argumentation that they take to be common-sense, but which are historically and culturally specific and ideologically loaded. Thus, as the data demonstrate, ideologies are dilemmatic, containing competing arguments and characterised by contradiction. Political debates over topics considered contentious – such as surrogacy – are thus excellent sites in which to identify some of the interpretative repertoires and rhetorical devices through which particular argumentative positions are warranted.

Through our repeated readings of the data set, passages that made explicit or implicit reference to children, the rights of children, and the best interests of the child in the context of surrogacy were identified and extracted for analysis. Extracts from individuals both in support and in opposition to surrogacy were included, although it should be noted that politicians who supported surrogacy made reference to children and their rights less frequently than did those who opposed surrogacy. As a result, the extracts presented in the analysis are not representative of all of the debate over the Bill, which included discussion around a number of aspects of the Bill (such as the mechanism for transferring parental status from the woman acting as the surrogate to the commissioning parents).

Two interpretative repertoires relating to the best interests of children emerged from the data. These were: 1) The best interests of children are served by having a mother and a father; and 2) Surrogacy is a lifestyle choice that is not in the best interests of children. In the analysis that follows we examine a selection of representative extracts that highlight the deployment of these interpretative repertoires, each of which involves specific constructions of the category 'the child'.

Analysis

The best interests of children are served by having a mother and father

A predominant repertoire in the data emphasised the claim that all children need a mother and father, and that this is in children's best interests. This repertoire includes common-sense notions of conventional nuclear families, along with paired contrasts between heterosexual and homosexual couples, and between single-parent families and two-parent families. Overall, those who drew upon this repertoire emphasised heterosexual marriage as the only context in which children should be raised, as the first extract indicates:

> Extract 1
> I believe marriage should be seen as the setting which most fully acknowledges the dignity of the child, and which establishes the relationship of equality between a child and his or her parents, and respects the child's right to enjoy an immediate and enduring link with those natural parents. Surrogacy arrangements are particularly disturbing because they involve deliberately deciding to bring a child into existence with the intention of separating that child from his or her birth mother. I touch upon the concerns that relate, for example, to the birth mother and to the child. Surrogacy instrumentalises children by placing the process of their conception, birth and upbringing under a contract. A child becomes the object of an arrangement aimed at fulfilling the needs of the commissioning parents.
> New South Wales, Parliamentary Debates, Legislative Assembly, 10 November 2010, 27584 (John Aquilina, Parliamentary Secretary, Labor Party)

In this extract, the speaker presents marriage (presumably heterosexual given laws in Australia that prohibit non-heterosexual marriage) as being in the best interests of children, and contrasts this with surrogacy arrangements. Specifically, the speaker constructs marriage as enabling a 'relationship of equality' between children and parents, despite the fact that the parent/child relationship is typically unequal, regardless of family structure. By arguing that heterosexual marriage is in the best interests of children, the speaker implies, by contrast, that other kinds of relationships and family structures are not in the best interests of children. The heterosexual nuclear family is further depicted as being in the best interests of children by the use of positive vocabulary such as 'enjoy', 'relationship', 'enduring', 'natural', and 'immediate'. By contrast, families created using surrogacy are depicted negatively, specifically with the suggestion that surrogacy 'instrumentalises' children. Ironically, the speaker himself instrumentalises children in this extract by relying on the discursive power of the construct of 'the child' as a tool to support his argument. This is an example of the discursive resource of child fundamentalism (Baird, 2008, p. 293), in which a speaker's argument 'relies wholly or in part on an insistence on the child as an impermeable category that must be defended'.

The following extract also draws upon the interpretative repertoire of *children's best interests are served by having a mother and father*, albeit via a different formulation of the adult/child relationship:

Extract 2
The strongest argument for a Surrogacy Bill that resonates with me is the submission made in relation to child support. The argument was advanced that children could be excluded from the child support regime if the commissioning parents separated during the limbo phase of the adoption. It seems that this shortcoming in the current transferral mechanism cannot be overcome. It would be wrong to deny a right—a right that would otherwise be afforded to children who have been conceived naturally—to a child, simply because he or she was born in a surrogate circumstance through no fault of his or her own. It is in the best interests of the child to ensure that the law holds both his or her mother and father responsible for the child, for only they could make the decision to bring this child into this world.
New South Wales, Parliamentary Debates, Legislative Assembly, 10 November 2010, 27592 (Victor Dominello, Liberal party)

Contrary to the first extract in which the relationship between a child and its parent(s) is constructed as equal, this extract builds an account of the parent/child relationship in which the parent is in a position of authority. The construct of 'child' mobilised, therefore, is one of passive innocence ('she was born in a surrogate circumstance through no fault of her own'). This is in contrast to the construction of adults, who are depicted as possessing sole agency in the relationship, 'for only they could make the decision to bring this child into this world'.

The extract also orients to a normative family structure. That a family comprises a mother, a father, and a child is unquestioned and unchallenged, a finding which reflects feminist research concerning the continued prominence of the heterosexual, nuclear family as the 'best' context in which to raise children (Clarke, 2001). For example, the speaker states that 'the law holds both his or her mother and father responsible' and 'only they could make the decision to bring this child into this world'. Despite the fact that the debate is about surrogacy and many speakers acknowledged the potential for non-traditional family structures to emerge – such as families with two fathers – the understanding of family as constituted by a mother and father is so taken for granted that it is treated by this speaker as requiring no further elaboration. Another example of such a taken-for-granted assumption in this extract is the comparison made between children who are 'conceived naturally' and those who are 'born in a surrogate circumstance'. No attention is paid to the multiple ways in which conception amongst heterosexual couples occurs. Instead, the word 'natural' – and its implied counter 'unnatural' – is effectively deployed to implicitly construct a hierarchy of appropriate modes of reproduction.

The following extract again repeats the repertoire that *children's best interests are served by having a mother and father*:

Extract 3
I had trouble with the recent Adoption Amendment (Same Sex Couples) Bill 2010, which I did not agree to, not because I have any trouble with adults who adopt a gay lifestyle, that is their right, but because I believe that the issue is not one of gay rights but that the fundamental rights of a child need to be acknowledged. Associated with the paramount rights of the child is the right of a child to have a mother and a father. In an ideal world every child would be cared for by a mother and a father. The law should take every step it can to ensure that children do have that fundamental right.
New South Wales, Parliamentary Debates, Legislative Assembly, 10 November 2010, 27595 (Chris Hartcher, Liberal party)

This extract again links children's best interests to family structure. Specifically, the importance of a heterosexual partnership for bringing up children is emphasised. Whilst the opening statement 'not because I have any trouble with adults who adopt a gay lifestyle' acts as a disclaimer, this is followed by a clear statement *against* non-heterosexual parents by saying that it is 'the right of a child to have a mother and a father'. This argument is stated even more bluntly in the ideological claim that 'In an ideal world every child would be cared for by a mother and a father'. Repetition is a powerful rhetorical device in itself, but this sentence also works on another level, by appealing to the common-sense ideology of 'an ideal world' that is assumed by the speaker to be commonly held amongst their audience. Appealing to an ideal world works to undermine the legitimate counter-argument, namely that many children do not have two parents (due to, for example, singe parenthood by choice, divorce, death, abandonment, and other scenarios that do not occur in an 'ideal world'). There is no defence or argument presented as to why a mother and a father are necessary, and thus the statement is presented as rhetorically incontestable.

As a whole, the extracts in this interpretative repertoire rely on the notion that children need a mother and a father, an argument made to speak against surrogacy as a legitimate mode of family formation. Whilst this is illogical as it effectively excludes heterosexual parents who have children through surrogacy (parents who would otherwise, it appears, be welcomed by the speakers on the basis of their heterosexuality), the statements made by the speakers draw attention to the fact that they are not simply referring to the best interest of children as having a mother and a father, but more specifically they are referring to the best interests of children as being conceived through heterosex. This brings with it an implicit derogation of many family forms, an issue that is taken up more closely in the following repertoire.

Surrogacy as a lifestyle choice that is not in the best interests of children

This repertoire similarly encompassed talk about family structure and the relative rights of parents and children, but focuses specifically on the type of family that would be formed as a result of the surrogacy arrangement. Certain adults using surrogacy arrangements to form a family were described as making a 'lifestyle choice', and this was constructed pejoratively, as can be seen in the following extract:

Extract 4
I have a significant problem with clause 21 of the bill. I am not concerned if a mother cannot have a child because of a medical issue. I know some fantastic people who would make excellent parents but for medical reasons they cannot have children. I support their right to seek out medical intervention to have a child that I know would be raised in a safe and stable relationship. What has not been explained sufficiently is the phrase in the bill 'medical or social need'. I have a real problem with what 'social need' means. To me social need means that it becomes a lifestyle choice, and I have an issue with that at a very deep level. If people choose to have children,

adoption is already available to them. However, if a woman cannot physically have a baby surrogacy may be an option. In principle, I cannot support people going down the surrogacy path when the child will be created to meet a social need.

New South Wales, Parliamentary Debates, Legislative Assembly, 10 November 2010, 27594 (David Harris, Parliamentary Secretary, Labor Party)

In this extract it is argued that surrogacy is acceptable in some situations but not others. The speaker constructs a binary in which the 'need' for reproductive assistance is either 'medical' or 'social'. Medical need for surrogacy is constructed as clearly defined and morally trouble-free, and restricted to infertile heterosexual couples. For example, the speaker refers to a 'mother' who cannot have a child for medical reasons. The use of 'mother' rather than the possible alternatives of 'woman', 'person', 'parent' or 'couple' is noteworthy, and works in a number of ways to reinforce the construction of heterosexual infertility as a medical need for surrogacy. Firstly, the term is specifically female; secondly, it defines the person by her child-rearing role (even though she has not had a child, thus reflecting previous feminist work suggesting that women are frequently defined by their potential for reproduction, see Orloff 1993); and thirdly, as part of a standardised relational pair in the context of a heteronormative society, it is implicitly accompanied by the category 'father'.

The use of the word 'mother' also works to construct people in 'medical' need as people who are familiar and comfortable to the audience, and this is contrasted with the more distant, impersonal, term 'people' to describe those who are in 'social' need. The speaker suggests that such a woman might be someone 'I know', which also works to create an image of the close and known; the 'us' to be compared with the 'them' who are in 'social' need of surrogacy. The use of positive vocabulary to describe people who meet the 'medical need' criteria, such as 'fantastic people', 'excellent parents', and 'safe and stable relationships' is another discursive strategy that allows the speaker to make a strong and effective contrast with the other half of the dichotomy of need.

'Social need' is contrasted with 'medical need' in numerous ways. The speaker's own views on each exemplify this. He is 'not concerned' with infertile heterosexual couples using surrogacy for medical need, but he has a 'real problem with what social need means'. Medical need is an acceptable reason for 'women' to consider surrogacy, but social need is for 'people'. In a further contrast, adults who 'choose to have children' to address a social need can adopt, whereas for women addressing a 'physical' medical need 'surrogacy may be an option'. With these points in mind, it can be inferred that the implicit message of the statements made in this extract is that surrogacy is an option for infertile heterosexual couples, but not for gay individuals or couples (or indeed single fertile heterosexual people).

Importantly, the language of 'choice' is critical here. Social need is equated with 'lifestyle choice', which, as was discussed above, is a phrase that plays a powerful role in debates over surrogacy by raising the implication that a 'choice' brings with it no rights (to reproduce). Notably, neither in this extract, nor in the following, is the language of 'choice' with respect to reproduction applied to heterosexual couples. The phrases 'lifestyle choice' and 'people who choose to have children' work together to imply that people 'choose' to be gay and therefore they choose to be unable to have children. Similarly, this argument could also be extended to include (for example) women who are seen as being 'career oriented' and leave having children until they are 'older', and then may have to enter into a surrogacy arrangement. Again, the phrase 'lifestyle choice' implies that they have chosen to be unable to have children, and thus the line between medical and social needs could well be a blurry one. For the following speaker, this is what constitutes social need:

Extract 5
I also find the Surrogacy Bill 2010 difficult because, tested against the fundamental proposition that the rights of the child should be paramount, this legislation would allow adults, people over the age of 18 years, as the member for Wyong so well illustrated, the right to have a surrogate child simply to fulfil a social need. A social need is really an expression of a personal wish because there is no requirement financially, culturally or legally in our society that people have children. The Prime Minister herself is famous for her well-publicised decision not to have children and that is a right that is respected. Social need simply becomes a lifestyle choice and to subordinate the right of a child to a lifestyle choice is to me unacceptable.
   New South Wales, Parliamentary Debates, Legislative Assembly, 10 November 2010, 27595 (Chris Hartcher, Liberal party)

In this extract the extreme case formulation of children's rights being both 'fundamental' and 'paramount' maximises the role of children's rights in the justification of a position where surrogacy is depicted as a 'lifestyle choice'. The speaker equates social need with 'lifestyle choice', and in so doing builds an image of people choosing to have a child in the same way that they might choose consumer items. Having a child is constructed as a selfish act, and the speaker works to defend this position by stating that there is no 'requirement financially, culturally or legally' for people to have children. This is undoubtedly true, but it can equally be applied to the conception of all children, and thus it may be suggested that it is the nature of the family structure that will arise from a surrogacy arrangement that is the cause for concern (for example, whether it will be a homosexual or heterosexual parented family, or a single or dual parented family), rather than surrogacy as a mode of conception in itself.

What is interesting in this extract is how the construct of the rights of the child is used to build an emotive argument that surrogacy is an inappropriate mode of human reproduction. The speaker is not saying that surrogacy is unacceptable in his opinion *per se*, but rather he is saying that surrogacy is unacceptable *because* it 'subordinates the right of a child to a lifestyle choice'. This is rhetorically powerful because it implies that children conceived in this fashion will be damaged and disadvantaged, and this is a rhetorical position against which it is difficult to argue. It also positions adults who chose to use surrogacy as doing harm to children.

The key implication of the ideological position elaborated within extracts in this repertoire is that adults who make a 'lifestyle choice' and have children through surrogacy are 'merely' consumers – they have no real desire for children, but rather are just complimenting their 'lifestyle' by adding children, and that this is not in the best interests of children. This type of construction is not only offensive to parents who have their children through surrogacy, but it also has significant implications for the support that families formed through surrogacy receive, and the exclusion they may face as a result.

Conclusion

A key theme to emerge from the data was the normative understanding of family and reproduction that was evoked. Echoing previous findings from feminist research on understandings of family, many speakers adopted a taken-for-granted position such that family consists of two parents – one mother and one father – who together have conceived their children without assistance. This is the image of family that framed much of the debate prior to the enactment of the Bill, and consequently non-heterosexual parents and/or those who use assisted reproductive technology are depicted as unacceptable departures from this norm. Indeed, the notion that a family comprises two opposite-sex parents is so deeply rooted in Australian society that debate around the use of surrogacy by a single man or woman was missing from the data altogether.

One clear implication of this highly normative account of families as it appeared in politician's debates over the Bill was that it served to justify opposition to the use of surrogacy by non-heterosexual couples without the need to explicitly express disapproval in most instances. By drawing upon the interpretative repertoire of *children's best interests are served by having a mother and father*, speakers were able to imply that two mothers or two fathers would be against the best interests of the children, reflecting work by Clarke (2001). Thus, speakers could position themselves in a way that showed them to be concerned with the welfare of the child (which is a rhetorically impenetrable position) rather than as 'anti-gay'. In this regard, the debate in the NSW parliament echoed themes from the Warnock Report on surrogacy in the UK. In her analysis of the report, Burr (2000, p.109) suggests that the language of best interests is deployed in the Report to justify claims about the most suitable people to raise children, namely a 'heterosexual couple living together in a stable relationship'. Burr claims that the language of best interests is used to reinforce a heteronormative family structure and a narrow definition of appropriate procreation.

Interestingly, the speakers themselves often exploited the image of 'the child' to build an argument based on upholding children's rights. For example, children were typically constructed as innocent and demanding of protection, and as having the potential to be damaged due to their parent's selfish desires. A further point to note in relation to the analysis is that building an argument against surrogacy on the construction of children's best interests in relation to the symbolic child denies specific children an existence. At the same time, by focusing on the alleged perils of non-normative families for children's welfare, these same politicians are deflecting attention from the serious problems experienced by existing children in damaging (often two-parent, heterosexual) families.

A consistent feature arising from the analysis was the dilemmatic nature of the rhetoric surrounding children and their best interests. This was to be expected in political debate about a controversial social issue, but it was interesting to observe the range of binary positions that were adopted, each providing insight into the common-sense attitudes that circulate within Australian society. For example, children's rights were contrasted with adult's rights; parents were either 'good' and loving or 'bad' and neglectful; reproduction was either 'natural' or 'unnatural'; and the inability to have children was either medical or social. All these positions are made up of essentially individualistic stances – a good parent, a child's rights, and an infertile woman – and within such an ideological framework, discussion of family is problematic. In other words, surrogacy was constructed as fulfilling an adult's needs or creating a child, but absent from the ideology exhibited in this particular forum was the construction of surrogacy as family formation. The limited empirical research into surrogacy outcomes to date indicates that families formed through surrogacy are as functional and healthy as those formed in other ways, and that the quality of family life is what matters more than family structure (Golombok, 2000).

It is important to point out that the analysis above is but one interpretation of the debate, not least because the approach taken was to focus on one specific aspect of the discourse, that of the 'best interests of the child'. Similarly, a parliamentary debate demonstrates a very specific form of rhetoric conducted by a very specific group of people that may not necessarily be broadly representative of how surrogacy discourse is constructed in other settings. Through the conduit of the media, however, politicians are perhaps one of the most instrumental groups in shaping policy debate on controversial topics, particularly as has been the case with surrogacy, when changes to legislation are under consideration (Riggs and Due, 2012). Therefore an understanding of how key decision-makers build accounts either for or against a particular social issue has wider reach than the parliamentary chamber. The echoes of child welfare discourse in political settings that come from such sources as the surrogacy debates in the UK in the 1980s (Warnock, 1984), and contemporary discussions about child custody in Australia (Fogarty and Augoustinos, 2008), lend support to the assumption that consideration of the issues above has wider relevance.

What individuals really believe about surrogacy is not accessible from the approach adopted in this chapter, but what is critical is that the act of talking and constructing arguments around surrogacy very clearly influences society through the manufacture of laws, as well as through reinforcing modes of reproduction which reflect the norm of heterosexual 'nuclear' families. Hence, the way in which the topic is spoken about reflects the ways of talking that are socially available today in relation to modes of family formation, and also contributes to the production of legislative changes that further enshrine such modes as the norm. At stake in such discussions is whether or not new family forms made viable through reproductive technologies receive the same rights and recognitions as traditional family forms, an issue which has received much attention within feminist literature (Butler, 1990; Clarke, 2001; Kitzinger, 1996). In addition, whilst it was not the case in this instance, public debates over surrogacy hold the possibility of moving beyond representations of women's bodies as defined by their reproductive capabilities, and towards recognition of the diverse ways of parenting and raising children, together with alternative constructions of surrogacy and family that centre on the best interests of the family unit in whatever form it takes, rather than the best interests of either adults or children as separate from their families.

3
'Diseases', 'Defects', 'Abnormalities', and 'Conditions': Discursive Tensions in Prenatal Screening

Meredith Vanstone, Elizabeth Anne Kinsella, and Jeff Nisker

Introduction

Pregnant women in many industrialised countries are now offered the opportunity to receive a prenatal screening test for disability, and asked to make an informed decision about whether or not they wish to engage in this testing process, which may lead to future decisions about invasive diagnostic testing and pregnancy termination. In this chapter we examine the way in which choice is discursively constructed in the context of prenatal screening, contending that the terms used to describe prenatal screening and disability may have the effect of enabling certain courses of action and discouraging others, even when the test is offered under the guise of increasing individual choice.

We begin by introducing our approach to discourse analysis, and outlining the context in which informed choices for prenatal screening take place in Canada. Using data from three sources collected for a related study (Vanstone, 2012), we identify two different discourses related to informed choices about prenatal screening and explore the tensions between them. After comparing the ways that these discourses operate together and in conflict with each other, we consider the understandings that these particular discourses may infer about disability, pregnancy, and motherhood. Finally, we reflect on the potential implications of dominant discourses about informed choices about prenatal screening. In this endeavour, we hope to highlight the influence of social context on choices about prenatal screening, emphasising the importance of considering the impact of social context on reproductive choices when thinking sociologically about reproduction.

Shakespeare (1999) has written about 'gene rhetoric' and discourses of disability in scholarly literature which alternately portray prenatal diagnosis as a triumph of medicine over chaos to avert tragedy and suffering or as a fascist eugenic regime. This chapter explores the more subtle points of tension in written and spoken language about prenatal screening, tensions which may be related to, but do not directly drawn upon the stronger rhetoric identified by Shakespeare (1999).

Fairclough (1995) writes about the ways in which discourses are used selectively, depending on which discourses are available to the producers and interpreters of text (written or linguistic). When multiple discourses are combined, they may be in opposition. The producer of the text makes a series of language choices or acts, and through these speech acts may resist the established discourse by using alternative language or may work with opposing discourses by borrowing from both in a form of tenuous reconciliation (Fairclough, 1995). Drawing on these ideas, we contend that discursive contradictions in women's talk about prenatal screening show that some women struggle to articulate their thoughts and beliefs using the language afforded by dominant cultural discourses of prenatal screening. These dominant discourses may be seen to be implicitly informed by particular values and assumptions (Hodgson et al., 2005). If there is a conflict between dominant cultural values and an individual woman's personal values, it may be difficult for a woman to articulate her own perspectives and make choices concordant with her own values (Anderson, 1999).

Discursive tensions

Throughout this chapter we use the term 'discursive tensions' which we hope will evoke an understanding of the ways in which different discourses can both coexist and be in conflict with each other, such as when an individual uses a particular discourse that has inherent assumptions or values which are in conflict with the values of the speaker. We understand discourse in the Foucauldian sense, as a group of institutionalised statements (including any type of utterance) about a particular topic (or object) that functions socially by forming that topic (Foucault, 1972).

According to Blood (2005), discourse is the manifestation of thought into language and can both transmit and produce power, defining the ways we can talk or think about a topic, and therefore defining the truth of that topic. Discourse shapes and constrains our ways of understanding the world, by acting as a system that structures our perceptions of reality (Blood, 2005).

Prenatal screening and informed choices

Prenatal screening is a non-invasive, non-diagnostic test performed by ultrasound measurements and a series of blood tests in the first and second trimesters (Chitayat et al., 2011). Results from these different testing modalities are combined with the mother's age to produce a numerical probability of foetal anomaly (ibid.). Prenatal screening can detect a number of chromosomal anomalies (e.g. Down syndrome), incomplete neural tube defects (e.g. spina bifida), as well as an assortment of other conditions (ibid.). After receiving results from the first round of non-invasive, non-diagnostic screening tests, a woman may choose to participate in further testing which will yield a definitive diagnosis (e.g. amniocentesis). While therapeutic interventions are available for a few conditions identified by prenatal screening tests, in most instances test results provide information to consider when choosing between giving birth and terminating the pregnancy.

In Canada, as in many other countries, including the US (American College of Obstetricians and Gynecologists (ACOG), 2007), the Netherlands (Health Council of the Netherlands, 2006), the UK (United Kingdom National Screening Committee, 2010), Australia, and New Zealand (Royal Australian and New Zealand College of Obstetricians and Gynaecologists (RANZCOG), 2010) prenatal screening is provided through a process of informed decision-making, where each woman is encouraged to make her own choice about participation in the screening and diagnostic tests, and what to do with the results. While informed decision-making does not have a single, authoritative definition (Bekker et al., 1999), the most commonly used definition describes an informed decision as one which is based on relevant knowledge, consistent with the decision-maker's values, and behaviourally implemented (Marteau et al., 2001).

When making the decision of whether or not to participate in prenatal screening, women are faced with the task of identifying their own preferences and values in order to make a decision (Vanstone et al., 2012). These decisions may be constrained by the 'implicit expectations, subtle influences and restricted choices' that shape the decision-making process (Shakespeare, 2006, p. 88). Such influences may include the particular word choices used by the counselling health care provider (Hodgson et al., 2005) or patient education pamphlets (Dahl et al., 2011; Loeben et al., 1998; Vanstone and Kinsella, 2010); including the ways in which the test and the idea of making a choice are introduced (Pilnick, 2008; Pilnick, 2004). The potential influence of such aspects, and the broader social and cultural contexts they represent, may be reflected in discursive constructions. When one considers the social contexts in which women are asked to make a choice about prenatal screening, it is easy to recognise that it may be challenging for women to identify and articulate values and preferences that run counter to dominant discourses of 'healthy' pregnancy, 'normal' bodies, and 'good' mothers. This may be especially true when prenatal screening is presented without acknowledgement of the values embedded in the test (Weil, 2003). Medical discourses of objectivity may disguise the presence of embedded assumptions and values (Asch, 2000), giving them a subtle form of power, and making it more difficult for women to recognise, think beyond, or resist such assumptions and values (Anderson, 1999). For these reasons, choices may be unintentionally constrained or directed by prevalent discourses, and the imperative of informed decision-making may not be met.

The emphasis on informed choice in prenatal screening is built upon the bioethical principle of autonomy (Beauchamp and Childress, 2009), but the use of this principle in prenatal screening has been problematised by feminist scholars (Ho, 2008; McLeod, 2002; Seavilleklein, 2009; Sherwin, 1998). On one hand, autonomy affords protection to people who are vulnerable to coercion and other influences. Without strong respect for the principle of autonomy, patients and other vulnerable populations may be abused and exploited. On the other hand, is autonomous decision-making really possible? When one considers the power differential between physician and patient (Bhogal and Brunger, 2010), social and political structures which oppress women (Lippman, 1999), and possibly coercive or constraining contextual factors specific to prenatal screening (García et al., 2008; Hunt and deVoogd, 2003), it is unclear if wholly autonomous decisions are possible. The question of societal constraints to autonomous choice is particularly relevant when considering the relationship between prenatal screening and disability.

Of course, there are many other extra-discursive influences on choice, such as past and present reproductive history; experience and familiarity with disability; and available social and financial resources. Further consideration of the influence of these factors on choice is beyond the scope of this chapter, but interested readers may refer to a plethora of work on these topics (see Hunt et al., 2005; Rapp, 2000).

Methods

The illustrative samples of data presented in this chapter were collected and analysed as part of a larger inquiry into the process of informed decision-making about prenatal screening (Vanstone, 2012). Examples are presented here to illustrate different types of language and discursive tensions that emerged in the data. Further information about the broader study, including more detailed methodology, reflections on the process of the research, and authorial decisions about word choices, is available elsewhere (Vanstone, 2012; Vanstone and Kinsella, 2010; Vanstone et al., 2012).

Three sources of data were collected separately: 13 English language Canadian prenatal screening patient education pamphlets were collected by searching online, and by asking interview participants what pamphlets they had reviewed in the course of making a decision about whether or not to participate in prenatal screening; two Canadian policy documents (Chitayat et al., 2011; Summers et al., 2007) addressing prenatal screening were collected after a search of academic and policy databases, and contact with professional colleges and governmental agencies; 16 pregnant women were interviewed after prenatal screening was introduced to them by their family doctor, but before women who chose to participate had received their results. All women interviewed were under the age of 35, self-identified as low-risk, and were carrying their first pregnancy. Information on ethnicity was not an inclusion/exclusion criterion and was not collected.

Participants were purposively recruited through the London-Middlesex Public Health Unit's Prenatal Fair (10 women), advertisements on pregnancy and classified ad websites (four women), and through snowball sampling (two women). Women were sampled to represent a diversity of perspectives about participation in prenatal screening (see Table 3.1).

Sampling was completed when theoretical saturation was thought to be achieved, that is, when no new categories were seen to emerge in further interviews. This study received research ethics approval from the University of Western Ontario. In audio-taped interviews lasting between 25 and 69 minutes, women were asked about the process of being offered prenatal screening, how they came to make a decision about whether or not to participate, and their thoughts and feelings about the test. For the purposes of this analysis, a discursive analytic perspective (Fairclough, 1995; Foucault, 1972) was adopted. This involved an analytical sensitivity to language, which in practice meant a careful comparative consideration of word choice, figurative language, and metaphors across all three data sources.

Illustrative samples of discursive tensions: biomedical and everyday discourses

Through a comparison of the language choices from the three different sources, two distinct types of discourse were identified: a biomedical

Table 3.1 Participant demographics

Pseudonym	Participated in screening?	Age	Urban/rural	Number of years of education	Recruitment method
Abby	No	27	Urban	17	Snowball
Bridget	Yes	30	Urban	21	Online
Carrie	No	29	Urban	16	Snowball
Danielle	Yes	31	Urban	14	Prenatal Fair
Eva	Yes	20	Rural	13	Prenatal Fair
Farah	Yes	27	Rural	16	Prenatal Fair
Gail	Yes	24	Urban	10	Online
Holly	Yes	26	Urban	15	Online
Isobel	Yes	24	Urban	20	Prenatal Fair
Jade	Yes	28	Rural	16	Prenatal Fair
Kyla	Yes	29	Urban	17	Online
Lucy	Yes	29	Urban	12	Prenatal Fair
Madelaine	No	30	Urban	21	Prenatal Fair
Nadia	Yes	30	Urban	16	Prenatal Fair
Olivia	No	29	Urban	18	Prenatal Fair
Penny	Yes	28	Rural	12	Prenatal Fair

discourse used mainly in the policy documents and some pamphlets, however also present in some of the language adopted by women; and an everyday discourse used mainly by women, but which was also identified in some phrases used in the biomedical pamphlets, and in the earlier policy document (Summers et al., 2007). Because the intent of this chapter is to examine the way in which these discourses function and to consider the possible implications for informed decision-making, we describe our analytic findings by drawing on examples from each discourse. To describe the findings, we consider how the different discourses might be used to answer two apparently straightforward questions: 'Who' (or 'What') does prenatal screening test? 'What' (or 'Who') does prenatal screening detect?

'Who' (or 'what') does prenatal screening test?

Markers of biomedical and everyday discourses

Analysis of the three data sources revealed that biomedical discourse about who or what is tested by prenatal screening was frequently marked by depersonalised words such as *foetus, pregnancy*, and *embryo*. When words that imply personhood such as *infant* or *child* were used, they tended to be used to specifically mark the transition after birth. Sometimes the woman was named as the subject of the test (rather than the pregnancy). Biomedical discourse used the terms *pregnant woman* or *patient* to refer to women as the subjects of the test, or as the locus of biomedical risk. Everyday discourses revealed in women's accounts of who or what is tested by prenatal screening often included terms which implied personhood, such as *baby* or *child*, many times without distinguishing between whether or not birth had taken place. Women sometimes referred to themselves or others in a similar situation as *mothers* or *parents*, even before birth had taken place.

Policy documents

The language in the Society of Gynaecologists and Obstetricians of Canada (SOGC) policy documents on prenatal screening (Chitayat et al., 2011; Summers et al., 2007) is intended for health professionals, and therefore a predominant emphasis on biomedical discourse is not surprising. In these documents, the most prevalent word used to refer to who or what is tested by prenatal screening was *pregnancy*; *foetus* was the second most common term. *Pregnancy* was sometimes used in place of a more specific term such as foetus or embryo: 'the chance of identifying a pregnancy with a specific chromosomal abnormality' (Chitayat et al., 2011, p.738); 'the practice of using [technique] to identify at-risk pregnancies' (p.739). The differentiation between *pregnancy* and *foetus* or *embryo* was sometimes ambiguous. For example, Chitayat et al. (2011, p. 737) refer to 'pregnancies' as what is conceived, whereas Summers et al. (2007, p. 149) refer to 'foetus' as the product of conception. Ambiguity was also found in the language used to describe who was at risk (of having a condition); sometimes the pregnant woman was at risk (Summers et al., 2007, p.152), sometimes the pregnancy. The most recent SOGC policy document uses the word *woman* most frequently, and occasionally refers to the woman as a *patient*, a word that may be viewed by some as medicalising pregnancy, for example, 'patients undergoing first trimester screening' (Chitayat et al., 2011, p. 742). Neither policy document participated widely in the everyday discourse. The term *baby* is used only twice in both documents, once referring to a child who was born with trisomy18 and once referring to age-related risk of having an affected pregnancy (p.739).

Patient education pamphlets

Thirteen prenatal screening patient education pamphlets were analysed extensively elsewhere (Vanstone and Kinsella, 2010); these pamphlets used a wide variety of terms to describe the subject of prenatal screening, and drew on both biomedical and everyday discourses. Seven pamphlets were consistent in the language used throughout the publication, using *pregnancy* (1) or *baby* (6), with the same word used before and after birth and for affected and unaffected pregnancies; other pamphlets differentiated between *foetus* and *baby* at the point of birth. Several pamphlets participated simultaneously in both discourses, using the terms *foetus, embryo,* or *pregnancy* when referring to an affected pregnancy or decisions about pregnancy termination; and the terms *baby* or *infant* or *child* when describing the reassurance that screening tests may provide, or the desire to have an unaffected child.

Pregnant women

Discourses can occur in both talk and text, and examining the words women use in speech revealed tensions between discourses. Women overwhelmingly used everyday language when referring to 'who' or 'what' would be tested, however the occurrence of biomedical language was often notable. For instance, the word *foetus* was used by only three women and only used one time by each. In each case, the woman used *foetus* when explaining medical information she had received, such as the probability of detecting Down syndrome, or what the ultrasound measured.

Pregnancy was a term that was used by every woman, mostly to describe the process of being pregnant, getting pregnant, or feeling pregnant, except, notably, when it was used to describe decisions to *terminate, end,* or *continue* the pregnancy. In contrast to the policy documents, most women did not talk about the pregnancy as being at risk or as being the subject of the screening tests. In almost every woman's speech, *pregnancy* referred only to the state of being pregnant, unless they were talking about abortion, in which case they used language such as *terminate the pregnancy,* a phrase which reflects a biomedical discourse.

There were clear patterns of everyday discourse in the interviews with pregnant women; women commonly used the words *baby, kid,* or *child* to refer to the born and unborn, potentially affected or unaffected. *Baby* was by far the most common term used by women, usually prefaced by *a* or *your* when speaking hypothetically, and *my* or *the* when speaking personally. *Child* was also a common term, used in a similar way to *baby*. Seven women used the word *kid*, usually to describe what life would be like after they gave birth: '*I think I would love that kid no matter what*', or as a way of ascribing personhood to the foetus, such as one woman did when describing how it would be difficult to make a decision about whether or not to terminate an affected pregnancy because '*it is still your kid, right*'? Women's tendency to use personal terms such as *baby, child,* or *kid* may implicitly reflect their adoption of the idea of motherhood.

The use of medical discourse may be seen to distance this relationship, and is perhaps a necessary tactic to make possible the consideration of pregnancy termination.

What does prenatal screening detect?
Markers of biomedical and everyday discourses

The distinction between biomedical and everyday language was clear in text and talk about the conditions detected by prenatal screening; biomedical language about conditions was marked by precise terms for specific conditions, rather than the everyday language of *problems* or *issues*. Terms such as *disorder* or *disease* participate in a general biomedical discourse, although they are not used in the policy documents to describe the most common conditions tested for by prenatal screening (Down syndrome and incomplete neural tube closures). These terms were occasionally identified in the pamphlets and in the women's descriptions, perhaps suggesting a desire to borrow from biomedical discourse. However, given that such terms are not regularly used in authoritative biomedical sources in relation to prenatal screening they instead served to mark the outsider status of the user. All three data sources predominantly framed unaffected pregnancies as *normal*, with many women and patient education pamphlets also using the word *healthy* to describe a foetus or person unaffected by one of the conditions screened for.

Policy documents

In general, the language used in the clinical policy guidelines is specific and scientific. For example, a commonly used term that did not occur in the other sources was *aneuploidy*, which refers to the possession of an unusual number of chromosomes and refers to conditions such as trisomy conditions (e.g. Down syndrome) which manifest when three chromosomes exist (triploid) where two are typical (diploid). Both policy documents also refer to *open neural tube defects*, a class of conditions such as *spina bifida*, which result from incomplete neural tube closures, and *open foetal defects*, including *gastroschisis* and *omphalocele*. Neither policy document uses the word *defect* to refer generally to *birth defects*, language that was present in a few pamphlets. The word *disorder* is prevalent in both policy documents, referring to single-gene disorders, autosomal recessive disorders, and 'rare disorders of cholesterol and estriol biosynthesis', 'common and mild disorder, X-linked steroid sulfatase deficiency' (Summers et al., 2007, p.151). Chitayat et al. (2011) use the word more generally, stating that 'screening for a disorder should be undertaken only when the disorder is considered to be serious enough to warrant intervention' (p.738). *Disability* was not used frequently in Chitayat et al. (2011) or Summers et al. (2007), only to state that screening programmes should respect the needs and quality of life of people with disabilities (Summers et al., 2007), and to describe that particular conditions are associated with 'intellectual disability' (Chitayat et al., 2011, p.744), a move towards more politically correct biomedical language than the alternatives used by Summers et al. (2007) including 'mental handicap' (p.148) and 'mental retardation' (p.153). The word *condition* is used four times in each document, referring to 'chromosome conditions' (p.148) and 'genetic conditions' (p.153). Both Chitayat et al. (2011) and Summers et al. (2007) commonly use the terms *affected* and *unaffected*, but both also equate *normal* to *unaffected*, stating that prenatal screening has the 'benefit of reducing the numbers of normal pregnancies lost because of complications of invasive procedures' (p. 146).

Patient education pamphlets

The patient education pamphlets used many different terms to describe what the screening test detected, including *disease, disorder, defect, abnormality, anomaly,* and the names of the conditions, such as *Trisomy 18,* or *Down syndrome.* While most pamphlets had an internal consistency, there were few constant trends between all pamphlets in the types of words used to describe conditions. Some used neutral language that participated in biomedical discourse, such as *birth anomaly,* one used the neutral, imprecise word *difference.* The majority of the pamphlets used valued terms that were more general, such as *problem* or *birth defect.* Every pamphlet named at least one specific condition. Almost all the pamphlets placed these terms for disability in opposition to words or, in one case, *perfect,* for the purpose of stating that a negative result will lead to the birth of a healthy, normal, or perfect baby.

Pregnant women

Biomedical words were occasionally used by women; for example, *disorder* was used once each by three women and *defect* was used by six women, once or twice each. Women did tend to adopt biomedical language at times. For instance, four women used the term *chromosomal abnormality.* One woman used the terms *characteristics* and *disposition* to refer to the conditions tested for, but this was not common. The most common words used by women were *problem* or *issue,* describing 'chromosomal abnormality or some other problem', 'developmental issue', 'genetic issue' or just generally used to state that the test would provide information about whether there was a *'problem with my baby's health'* or the baby *'has some kind of issue'*. *Disability* was also a common word, used repeatedly by four women and a few times by three others. *Condition* was the word used by the interviewer, and many women adopted that term and used it once or twice after hearing it; only two women used the word *condition* before it was introduced. At times it was apparent that women were struggling to find the right words. In the interview transcripts there are many indications of this, for instance, 'I think it will tell you if there's a chance of some sort of problem with the baby, like some sort of defect, I don't even know if that's the right word, some sort of genetic issue with the baby' or queries such as 'Is Down syndrome really a disease or is it just a defect?'

Discussion

This chapter has described the presence of two distinct discourses in prenatal screening patient education pamphlets, policy documents, and pregnant women's speech. Language relating to a biomedical discourse was identified as strongly present in clinical policy guidelines, and language related to an everyday discourse was identified as strongly present in women's speech. These broad discourses were found to overlap in many of the prenatal screening educational pamphlets, which may be seen to act as a bridge between the two broad discourses, and serve as a means of 'educating' pregnant women into particular ways of understanding prenatal screening (Dixon-Woods, 2001). Congruent with Tom Shakespeare's (1999, 2006) writings on the subject we do not wish to imply the presence of an intentional eugenic conspiracy, 'abetted by science, to eliminate all disabled people' (Shakespeare, 2006, p. 87), or to suggest that individuals are somehow 'brainwashed or coerced' (p. 101) by discourse. Rather, we wish to highlight the contexts in which decisions are made, and the potential influence of those contexts on the options which may be available or appealing.

Drawing on the ideas of medicalisation and normalisation, it is important to consider the ways in which medical discourses structure social relations and the institutional processes of prenatal screening. Disability has long been constructed as a medical issue and the ways in which it is understood and spoken about are constituted by medical discourse and action, where decisions of whom to save, treat, or abandon are 'acts and omissions serving to continually reinforce and re-create medical notions of disabled' (Shildrick and Price, 1998, p. 227).

Medicalisation, intentionally or unintentionally, expands the domain of medical jurisdiction (Conrad, 1992). Morgan (1998) explains that the expansion of medical jurisdiction requires social acceptance or acquiescence, legitimising the use of medical concepts to describe and treat life phenomena. Normalisation is related to medicine's inclination to measure and compare bodies in order to govern them, which Foucault (1990) names bio-power. Bio-power functions by encouraging women to act, to participate in the discourse of the body as a knowable, measurable, standardised object (Foucault, 1990). Such practices have the effect of identifying and recategorising bodies which do not fit the medicalised ideal of normal. Through this process of identification and categorisation, individuals are encouraged to exercise their power to intervene and transform these potential people into something which would be identified as normal or acceptable (Foucault, 2003a). In this conception, 'normal' is a social construct, aided by the ability of medicine to measure, count, and calculate, in order to appraise or judge ways of being as either acceptable or unacceptable.

These practices of classification and codification are central to the emergence of the concept of disability, dividing some people from others and objectifying them (Tremain, 2005). The process of classifying or coding some types of people as 'other' can be seen in the biomedical discourse of measuring, identifying, finding, detecting, and reporting, so that labels of disorder, defect, disability, and abnormality can be applied. The interview participants used words such as *problem* or *issue*, which participate in the categorisation of an atypical person as 'other', however this language is less quick to label, and participates in a much larger category than the more heavily loaded words of *disorder*, *defect*, or *abnormality* (Vanstone, 2012).

We posit that informed decision-making participates in the process of bio-power by enabling women to choose to act; when participation in prenatal screening is discursively framed as enabling choice or providing reassurance it becomes more difficult to resist this process of normalisation (Seavilleklein, 2009). Belief that one is acting autonomously and making choices supports domination and normalisation by hiding the workings of power (Dreyfus and Rabinow, 1982); a mechanism that indicates the success of power (Foucault, 1990). By cloaking prenatal screening in language of 'choice' and autonomy, women are encouraged to participate in the workings of this powerful apparatus, to measure and identify anomalies so they can be reported and extinguished (Lippman, 1991; Seavilleklein, 2009). Some forms of counselling to support informed decision-making may promote this emphasis; when the valued nature of prenatal screening is camouflaged under the guise of non-directive counselling, it may be more difficult for women to identify the potential for a disjuncture between their own values and the values which inform the institution of prenatal screening (Vanstone et al., 2012).

When considering discursive imperatives to participate in prenatal testing and subscribe to medicalised ideas of 'normal', it is interesting to consider the cases of interview participants who chose to participate in prenatal screening but stated that they would not terminate their pregnancies if a condition was found (four women in the study) (Vanstone, 2012). Women who make this choice may be seen as participating in the medicalised discourse of measuring and classifying, yet simultaneously they may be seen as resisting discourses of normalisation, by stating that they would not choose to terminate their pregnancy if a condition was found. The four women who stated they would choose this approach explained their choice by stating that prenatal screening would give them a chance to educate themselves and prepare to raise a child with a disability, acknowledging that they would face additional challenges as a parent and that their child would require additional support (Vanstone, 2012). This reflects a simultaneous acknowledgement of and resistance to the medicalised discourse of normalisation. The imperative towards normalisation was acknowledged by these women as they expressed awareness that people with disabilities may face many challenges living within a society primarily constructed to accommodate typical people, an insight well explained by Wendell (1996). Women's unwillingness to act to ensure that their child conforms to this normalised standard may also be seen as an example of resistance with respect to the imperatives of normalisation.

The acknowledgement that people with disabilities face additional challenges was expressed by all women in the study (Vanstone, 2012). Interestingly, each woman paired this acknowledgement with a statement about how she personally would act to circumvent or mitigate these challenges. Some women stated that they would take action by terminating an affected pregnancy, while other women described different courses of action such as educating themselves and securing the necessary resources to support an affected child to their full potential. This unquestioned adoption of personal responsibility speaks to the transfer of responsibility for health from society to the individual, a hallmark of medicalisation (Morgan, 1998). Prenatal screening presents a clear example of this principle: the onus is placed on women to detect and abort foetuses with disabilities or to provide care for children with disabilities, obscuring the responsibility of society to help all people live to their full potential. This emphasis on individual responsibility as enacted through individual choice obscures the social context within which that choice is made, and which shapes the possibilities individuals can or are willing to consider (Shakespeare, 2005).

Autonomous informed choice is the mechanism by which modern day genetics is separated from eugenics (Duster, 2003). Granting women the choice to participate in prenatal screening empowers them to participate in governing their own bodies and families (Foucault, 1990) and disguises the workings of medicalisation and normalisation. However, what does choice really mean when the choices and the context within which those choices will be enacted are constructed by others (see Beaulieu and Lippman, 1995; Lippman, 1991; Lippman and Wilfond, 1992, Shakespeare 1999, 2005, 2006)? For instance, can the choice to raise a disabled child be considered informed and autonomous when parents are not offered sufficient information (Williams et al., 2002) to combat cultural representations of people with disabilities as 'pathetic, medical tragedies, dependent, and unfulfilled' (Shakespeare, 2005, p. 226)? Is it truly a choice to raise a child with a disability within a society that does not provide the necessary resources and support to ensure that child is given the opportunities of all other children? Is the choice to terminate a pregnancy after a condition is found because you cannot afford to care for that child truly autonomous (Sherwin, 1998)? When considering questions of informed choice, we must examine the context in which those choices are constructed.

Conclusion

Informed decision-making is built upon the concept of autonomous choice; when considering the social context of informed decision-making, specifically, the ways in which disability and normality have been constructed in western society, a significant tension emerges. When disability is so widely culturally represented as something negative that must be avoided, it becomes much more difficult to refuse prenatal screening and assume the uncertainty of not knowing or decline to terminate an affected pregnancy. The discursive creation of a dichotomy between normal/disabled perpetuates the construction of a society which does not accommodate or support the needs of people with disabilities. The discursive tensions observed in the speech of women suggest that some women may resist this particular construction. Without explicit acknowledgement of the values and assumptions that underpin prenatal screening, individual women may find it difficult to locate a foothold from which to make a decision that is not unduly influenced by dominant discourses. This may be particularly challenging for those that might resist the social and discursive pressures to participate in prenatal screening, and the unquestioned action to terminate an anomalous pregnancy.

4
The Limits of 'Choice': Abortion and Entrepreneurialism

Kate Gleeson

At the Cairns District Court in October 2010 a young couple was found not guilty of all charges relating to the importation and use of the abortion drug RU486. Police had charged Tegan Leach and her partner Sergei Brennan under sections of the Queensland Criminal Code concerning the procuring of miscarriage. It was alleged Brennan had arranged for a relative in the Ukraine to supply him tablets of the drug mifolian (a Chinese version of RU486) which he provided to Leach, who ingested them, procuring an abortion (Australian Broadcasting Corporation (ABC), 2010). Since 2006, Australian doctors have used RU486 selectively for the purpose of medical abortion, but its legality had not been tested prior to the 2010 trial, known as 'the Cairns Case'. Leach was the first woman to be tried in Queensland for procuring her own miscarriage, and possibly the first woman to face such a charge in Australia. Although the Cairns Case resulted in all charges being overturned, the legal status of RU486 remains arguably unclear. Most significantly, the case produced, rather than clarified, a number of questions about the social and legal status of self-abortion in the age of relatively 'safe' and accessible drugs like RU486.[1]

In this chapter, I examine the role of the state in the Cairns Case to suggest that the example of self-administered RU486 exposes some of the limits of neoliberal discourses applied to reproduction. Increasingly, neoliberal tropes such as 'patient as client' and 'individual choice' have shaped public discourses and policies concerning reproductive health, such as in the arenas of assisted reproduction, birthing practice, and to a lesser extent, selective (genetic) abortion in Australia. This agenda, while reflecting and perpetuating an erosion of the social welfare state, has provided some (debated) benefits for women in the provision of heightened 'choices'. However, abortion performed for 'social indications' stands alone in the reproductive policy milieu, in its contested social and ethical status and its historical regulation by the criminal law. In particular, the capacity for women to self-administer abortion poses challenges to medical, political, and legal establishments that patrol the practice. Historically, the autonomous, private act of self-abortion has sat uneasily in relationship to medical and legal establishments, posing a populist threat to hegemonic authority. This threat was thought to be neutralised by legal reforms enshrining medical control of abortion in modern criminal and health laws from the 1960s (Baird, 1998). In an unanticipated turn, the medicalisation of abortion by way of chemical abortifacients like RU486 has opened the practice once again to populist administration by individual women. I argue that the resurgent phenomenon of self-abortion exposes some of the potent limits of neoliberal framings of reproductive praxis. To illustrate this point, in this chapter, I shall examine the Cairns Case in relation to Foucault's observations about the nature of medicine made in *The Birth of the Clinic* (2003b) and Angela McRobbie's (2009) analysis of post-feminism under neoliberalism. I do this to provide a sociological analysis of contemporary abortion regulation within the deregulated social and economic environment in which young Australian women reside.

Children of the revolutions

Leach and Brennan were arrested in 2009 after police searched their house in regard to an unrelated matter and discovered empty blister packets of mifolian tablets and accompanying material written in Ukranian. When questioned, the couple told the police that the drugs were taken by Leach after she became pregnant, and that she was no longer pregnant. Local doctors had wrongly told Leach that she could not access a medical abortion in Cairns. Perturbed by the idea of a surgical procedure, she sought alternative treatment. Brennan admitted to arranging for the importation of the drugs, which he knew to be legal in his native Ukraine, believing that Customs would have confiscated them had they been illegal to import (Zlotkowski, 2009). On this basis, Leach and Brennan were charged respectively with offences related to procuring a miscarriage and supplying abortifacient drugs, with maximum penalties of between three and seven years gaol. There is every indication that the couple believed their activities to be uncontroversial, involving a private decision unrelated to the criminal law. Both declined legal representation when questioned by the police (ABC, 2010).

On her arrest Leach was 19 years old, having grown up in the post legislative-settlement period of Queensland abortion governance that commenced with the 1986 trial of *R v Bayliss and Cullen*. In that case, in which two doctors were acquitted of all abortion-related offences, Judge McGuire clarified that the 1899 Criminal Code governing abortion provided for the performance of 'a surgical operation upon an unborn child to preserve the mother's life' (*R v Bayliss and Cullen* (1986) 9 Qld Lawyer Reps 8). On these grounds, abortion performed by doctors is understood to be legally protected in Queensland. The McGuire Ruling is one outcome of the medicalisation of abortion governance commenced in Britain in the nineteenth century, when the exceptional occurrence of 'therapeutic' abortion performed by doctors was legitimised in jurisprudence (Gleeson, 2011). Jurisprudential medicalisation was completed in Australia in the twentieth century on the relocation of pregnancy and birth from the home and local midwives, to doctors and hospitals, as part of a modern campaign of 'staking of professional terrain' for the new specialisations of obstetrics and gynaecology (Baird, 1996, p. 9). This relocation formed part of a process of 'rearticulating' pregnancy as the 'nurturing of a human being' – an intervention into working-class social practices by which the woman within her family was 'truly being constructed, not as wife, but as mother' (Finch, 1993, p. 123). In making his judgement protecting doctors, Judge McGuire stated that it applied only to 'exceptional cases' as determined by the medical profession:

> This must be clearly understood. The law in this State has not abdicated its responsibility as a guardian of the silent innocence of the unborn. It should rightly use its authority to see that abortion on whim or caprice does not insidiously filter into our society. There is no legal justification for abortion on demand. (*R v Bayliss and Cullen* (1986) 9 Qld Lawyer Reps 8)

In all Australian States and Territories, abortion performed by doctors is legally protected by a rationale more or less in keeping with the Queensland direction. Leach and Brennan were targeted by police alleging their actions contravened Section 225 of the Criminal Code, stating that 'Any woman who, with intent to procure her own miscarriage, whether she is or is not with child, unlawfully administers to herself any *poison or other noxious thing*, or uses any force of any kind, or uses any other means whatever, or permits any such thing or means to be administered or used to her, is guilty of a crime' (ABC, 2010, emphasis added). It was argued the protection the McGuire Ruling affords doctors not to apply to women undertaking self-abortion, even in the age of big-pharmaceutically authorised RU486. In court the prosecution focused on the alleged 'noxious' nature of RU486, stating, 'if you accept that Tegan Leach got her miscarriage, she got what she intended. Were the drugs injurious to the woman? We say they were' (ABC, 2010).

Leach resided in a time and place in which, understandably, she understood abortion to be a matter of personal 'choice' and deliberation. Despite caveats made about 'exceptional cases' in judgements such as McGuire's, abortion is the most common surgical procedure for women of reproductive age in Australia. In all States except South Australia, abortion services are delivered typically by doctors in private practice in free-standing clinics, with the Commonwealth Medical Benefits Schedule providing patients partial rebates for abortion since 1974.[2] National abortion data are not recorded, but it is estimated that doctors perform at least 80,000 abortions each year and around one in three Australian women will have an abortion in her lifetime, the majority performed for 'social' indications (Pratt et al., 2005). Based on South Australian data, in 2009 the abortion rate was 15.6 per 1000 women aged 15–44 years; approximately 21 per cent of reported pregnancies ended in termination (Chan et al., 2011). Although in wide use in Europe since the late 1980s, RU486 was unavailable in Australia until 2006 when special cross-bench legislation was passed in the Australian parliament to allow its use by doctors with authorised prescriber status. Concerned that the McGuire Ruling addressed only surgical procedures, in 2006 the Queensland branch of the Australian Medical Association obtained advice from the State Attorney General, to the effect that the ruling would also protect the authorised procurement of medical abortion by doctors (Walker, 2006, p. 4). Still, doctors' use of RU486 remained limited at this time, with no drug company applying to register the drug and there is evidence that general practitioners remained confused about its legal status. The legacy of this situation continues. While in the UK medical abortions account for around one-third of all doctors' abortion procedures, in Australia, surgical abortion performed under light sedation remains the norm (Gleeson, 2010). In August 2012, the Australian Therapeutic Goods Administration approved an application to import RU486 by a company established by the reproductive health organisation Marie Stopes International (Peatling, 2012). In June 2013, the drug was listed on the Commonwealth Pharmaceutical Benefits Scheme, making it significantly more affordable (ABC, 2013).

Although the authority to 'permit' an abortion rests with the doctor performing the procedure in accordance with the law (a factor subject to feminist critique, see Petersen, 2000), in Queensland, as in most Australian jurisdictions, no medical or psychiatric referral is required for women to attend clinics and request the termination of their pregnancies. This situation, which appears to defy McGuire's direction, is the outcome of lobbying – medical, civil libertarian, and feminist – that commenced in the 1960s. While historically, medical lobbyists arguing for reform tended to characterise abortion as a grave medical need, feminists portrayed it as a matter of women's human rights and/or a 'choice' made by individual women (Gleeson, 2014). Despite the medicalisation of reproduction, it is the feminist treatment of abortion that solidified in the popular mind 'a woman's right to choose' as one of the most successful and 'ubiquitous' discourses of second-wave feminism (Ruhl in Gleeson, 2014). The feminist rhetorical framing of abortion as 'choice' has been appropriated to align with emergent neoliberal discourses which exult 'choice' as applied to healthcare, including women's sexual and reproductive praxis. Typically, and perhaps paradoxically, abortion is characterised as both a medical procedure and a 'choice' – a situation resented by those who oppose abortion. Most women report surprise on being informed that abortion is a matter for the criminal law (Rosenthal et al., 2009) and it is estimated that one-third of general practitioners do not know if abortion is listed in the Crimes Act or Criminal Code in their State or Territory (Rouse, 2010).

The neoliberal revolution in healthcare commenced in earnest in Australia on the election of the Howard coalition government in 1996. It was characterised by a rhetorical focus on the patient as client, a belief in the self-evident efficiency of market forces dictating healthcare delivery, and a coterminous shrinking of public services to further the productivity of the private sector (the substantial subsidising of which revealed the ideological significance of neoliberalism to the government). Neoliberalism understands citizens as rational consumers of public goods, including healthcare (Horton, 2007, p. 3) and promotes an individual's right to choose apparently almost any commodity or service as a fundamental modern freedom. While evident in an array of policies, the Howard government's agenda on health and education, in particular, demonstrated the 'life long commitment to the politics of choice, social conservatism and individual responsibility' of Prime Minister Howard who presented the 'choice between the public and private' systems in these areas as a 'universal right, not the privilege of a wealthy few' (Maiden, 2006, p. 113). Throughout the 1990s maternity and reproductive policies were reformed in keeping with a bipartisan neoliberal health reform agenda evident in the programmes of various

State and Territory governments. While not radical, these reforms capitalised on a spirit of entrepreneurialism to provide women and their partners with heightened maternity choices (while falling short, for example, of providing for feminist policies such as state-supported homebirth) (Reiger, 2006, p. 334). For the most part, abortion practice had already been relinquished to the private sector, reflecting the historical legal ambiguity of the practice and its ongoing political contention. While this has resulted in unequal abortion access between rural and metropolitan women, and problems associated with a shrinking pool of abortion practitioners, it has allowed doctors unrestrained by statehealth to make entrepreneurial developments in practice, which benefit women (Ripper, 2001).

Increasingly, abortion practitioners have striven to have abortion viewed and treated as 'just like any medical procedure' governed by the ethos of 'informed consent' dictating all elective medical procedures. But still it remains governed by the criminal law in most jurisdictions. In Queensland, like all day surgeries, abortion clinics are subject to expensive and laborious licensing regulations (which arbitrarily prevent the performance of abortions after 20 weeks gestation). Such regulations arguably inflate costs, while aiming to provide a standardised, safe medical environment for women as abortion 'consumers' (Gleeson, 2010). As a child of both feminist and neoliberal revolutions, it is little wonder that Leach characterised her decision to have an abortion as a personal 'choice' made in regard to her own capacities, in consultation with her partner and her family. Such a procedure is arguably the norm for the majority of abortions in Australia. Along with her existence amid the prevailing health policy landscape, Leach appears to have understood herself as the self-constituting subject of neoliberalism. In other words, she applied the types of vigilant, independent self-regulating 'techniques of the self' that Foucault (in Horton, 2007) has identified as appropriate to neoliberal governance. This is a form of governance that seeks to integrate the 'self-conduct of the governed into the practices of government' by constructing the ways in which individuals are 'required to assume the status of the entrepreneurial self' (Horton, 2007, p. 1).

The entrepreneurial female self

The entrepreneurial self facilitates neoliberal governance in its expectations of self-reliance, which protects the neoliberal state from citizens' claims made for redistributive equality and public service provision in the form of a welfare state. The entrepreneurial *female* self is particularly concerned with the performance of corporeal and reproductive self-discipline, facilitated in the contemporary age by what McRobbie (2009) has identified as 'postfeminist' discourse. Postfeminism is a political strategy aimed at facilitating a programme central to neoliberalism of undoing the 'anti-hierarchical struggles' of past social movements through the 'simultaneous incorporation, revision, and de-politicisation of many of the central goals of second-wave feminism' (p. 130). It does this, in part, to annihilate the threat to liberal democratic regimes of feminist strategies that attempt to 'subvert the dominant political and economic systems' by enabling women's equal political representation and participation in public and private decision-making (Hawkesworth, 2004, p. 978). Postfeminism is a 'kind of anti-feminism' reliant, paradoxically, on an assumption that feminism has been taken into account in the neoliberal world order and 'transformed into a form of Gramscian common sense', while it is also 'fiercely repudiated, indeed almost hated' (McRobbie, 2009, p. 255). In particular, the apparent 'fusing' of 'palatable elements' of liberal feminism with concepts of 'self production and rational choice so fundamental to neoliberalism' is deployed as a 'decoy for domination' of all women, and young women especially (Baker, 2008, p. 6).

Foremost among McRobbie's (2009) claims about postfeminism is the prevalence of a new sexual contract entered into by young women, which facilitates them existing as individual consumers and workers so that they may perform as 'economically active female citizens' of neoliberalism. Contraception and 'responsible' (non-procreative) sexual practices are central to this contract, which is supported by a widespread vilification of young mothers by state and society. The new sexual contract of postfeminism holds that

> So long as she does not procreate while enjoying casual and recreational sex, the young woman is entitled to pursue sexual desire seemingly without punishment. Indeed the appropriate uses of sexual pleasure are prescribed within the many manuals and forms of instruction which constitute the terms and conditions of this new sexual contract (McRobbie, 2009, p. 85).

McRobbie's theories are supported by qualitative research that reveals the readiness with which young Australian women (Leach's peers) identify themselves as procreationally *'responsible'*, individualistic, self-governing agents. In keeping with the new sexual contract, in Australia, young motherhood has generally been cast as counter to the discourses of neoliberalism and postfeminism which 'reject welfare dependency and encourage young women to delay motherhood until they have attained an education, career and some sexual and romantic experience' (Simic, 2010, p. 430). In government documents and policies, teenage pregnancy is construed as a potential welfare burden, with young sexually active women identified as being potentially 'at risk' of 'long-term economic disadvantage' in motherhood (ibid.). Teenage pregnancy is targeted as in need of remedy and prevention, a message that has generally been internalised by its subjects. Since the early 1970s, the number of teenage pregnancies has mostly declined in tandem with a heightened stigmatisation of teenage motherhood arising in relationship to the de-stigmatisation of 'out of wedlock' births in the broader adult community. According to South Australian data, the teenage pregnancy rate (per 1000 women aged 15–19) declined in the 1970s and 1980s along with a decline in the teenage birth rate, but increased in the 1990s until 1996, when it declined again. From 2003, this decline was associated with a decline in the teenage abortion rate. The teenage pregnancy rate in 2009 was 32.7 per 1000 women, the lowest rate recorded since 1970 (Chan et al., 2011). In 2009, teenagers accounted for 18.1 per cent of all abortions and 4.1 per cent of all live births (ibid.). The proportion of 'known' pregnancies terminated was 52.9 per cent for teenagers compared with 20 per cent for women of all ages (ibid.).

Accordingly, Joanne Baker's 2008 study of 55 young women (aged 18–25) of a large regional city in North Queensland indicated a pervasive adoption of identities as 'hyper responsible' selves having internalised the perceived need to 'live up to' the 'widely canvassed new freedoms for women' promulgated by postfeminism (p. 57). The young women strove to identify themselves as responsible, so as not to *put themselves* 'at risk' of economic disadvantage. Their accounts of their lives appeared overwhelmingly 'influenced by an obligation to represent themselves as free-willed, uncoerced and experiencing equality. This was achieved in a number of ways, the most prevalent of which was the extensive use of the concept of 'choice' (Baker, 2008, p. 59). Baker observed a 'strong tendency for young women to demonstrate responsibility and resilience in order to distance themselves from notions of weakness or the dreaded "victim" status' (ibid.). To this end, in keeping with prevailing hegemonic, women-centred 'self-help' discourses, the subjects redefined disadvantage as 'offering them opportunities to make "the right choices"; to demonstrate strength, to improve and develop resilience' including, in some circumstances, the choice to have an abortion (ibid.).

They particularly valued the ability to make decisions, even difficult decisions about reproduction, *on their own*, as self-actualising responsible individuals. In keeping with Baker's observations, in response to police questioning in 2009, Leach and Brennan explained their decisions to undertake an abortion as *personal* decisions made as a couple in regard to their youth and perceived capacity as parents. Brennan represented himself in the responsible terms of the entrepreneurial self: 'I'm a young fella, you know, I want to give my kid the best when I have a kid but at the moment I don't feel I can give them the best and that's the way she [Leach] feels too'. Leach reflected in a similar, responsible, individualistic vein: 'I suppose the reason I don't want to have the child is 'cause I can't even look after myself' (Zlotkowski, 2009).

Abortion 'choices'

Considerations by which women weigh up abortion choices have been documented at least since Carol Gilligan's groundbreaking *In a Different Voice* (1982), which characterised women's morality as having a 'responsibility orientation' (as opposed to men's 'justice orientation'). Data from a recent Melbourne School of Population Health study reveal a similar pattern in how women experience their decisions made about abortion. The 2009 study (of 60 women aged 16–38 considering an abortion) indicated that the women's decisions to seek an abortion were 'rarely motivated by a single factor' (Kirkman et al., 2010, p. 150). The fact that a pregnancy was simply unplanned (although some of the pregnancies were planned) was not the deciding factor in decisions, which were made in consideration of the 'ramifications of the pregnancy' (p. 152). In every case, the women described making decisions about their pregnancy that took into account their life circumstances. Each woman described 'complex lives and social contexts' within which she made decisions, assessing her 'capacity to be a good mother and to provide adequately for the potential child; they thought about their relationships and the man concerned; those with children considered their needs; and many women assessed their readiness for motherhood or another child' (ibid.). The women commonly felt 'too young or otherwise unready for the commitment of motherhood; others were unwell or not prepared to accept the physical changes of pregnancy' (p. 150). The most common reasons for seeking abortion were family completion or the desire to delay pregnancy.

In their long-standing focus of their perceived capacity as individual mothers, it might be argued that women have consistently acted as 'entrepreneurial selves' when deliberating reproductive decisions. As I have noted, this persona is generally rewarded in neoliberal political regimes, however it also opens women's abortion choices to accusations of frivolity, and makes aborting women vulnerable to their portrayal as threatening the existing social order. Indeed, Leach and Brennan's lawyer strove to distance his clients from entrepreneurial, responsible (threatening, autonomous) personas, instead characterising them as 'just two kids who got scared being pregnant' (Walker and Hyde, 2009, p. 1). When acting in their capacity as entrepreneurial selves women have been portrayed as 'selfish' individuals, rather than the procreationally 'responsible' individuals they aim to construe themselves as. In particular, when pitted against pronatalist and anti-abortion discourses, the entrepreneurial neoliberal woman who chooses abortion is marginalised, such as in responses made to the publication of 2006 South Australian teenage abortion data. With apparently no knowledge of their individual circumstances, local Family First Party politician Dennis Hood characterised terminations undertaken by teenagers as lost opportunities for the gratification of others, stating, 'My wife and I took a long time to have a child and we were considering adoption but we looked into it and found it was incredibly difficult. It'd be terrific to see those babies, instead of being aborted, being given up for adoption' (News Limited, 2006).

In contrast, some women interviewed in the Melbourne School of Population Health study thought that abortion was 'more responsible than adoption' (Rosenthal et al., 2009). The 'selfishness' of women is a long-standing trope of pronatalist discourse, deployed since at least the time of the 1904 Royal Commission into the Decline of the Birth Rate (Gregory, 2007, p. 63). For example, contemporary Australian (male) politicians have attempted to damn abortion choices as frivolous, by relating them to the desire to be able to 'wear a bikini', for example (Brankovich, 2001, p. 90). As Health Minister, current Prime Minister Tony Abbott described abortion as having been reduced to a 'question of the mother's convenience' and experienced 'almost by some as a badge of liberation from old oppressions' (Gleeson, 2014). In response to the Cairns Case, anti-abortionists justified abortion prohibitions as 'protection against young people or older people not standing by their offspring. They have a duty of care to that offspring' (ABC, 2010).

Health consumerism in the age of the clinic

The Cairns Case exposes potent limits of neoliberal framings of reproductive praxis by illustrating acutely that the state welcomes the entrepreneurial (female) self in so far as she does not disrupt authoritarian medical and political regimes. More broadly, the case may help to illustrate a phenomenon long noted by health policy specialists: a disconnection between neoliberal rhetoric and the reality of health services as they are delivered. While policies of health consumerism suggest an open and accountable system for patients, some have argued that the model of 'patient as consumer' may actually be 'a myth' because the 'consumer model does not fit well into health care' (Hogg, in Horton, 2007, p. 3). The inequalities of power maintained between patient and medical expert suggest that 'consumerism' as a discourse is not generally appropriate to health, 'because it fails to accurately describe the actual behaviour of patients in the concrete setting of the hospital, the clinic or the practitioner's surgery' (Horton, 2007, p. 3). Abortion as consumerism, performed with drugs prescribed and purchased by oneself, further disrupts the 'proper' hierarchical relationships within this setting identified by Foucault and others.

In *The Birth of the Clinic* (2003b) Foucault observed the birth of modern medicine in the final years of the eighteenth century, which came to be characterised by modern diagnostic methods hitherto unarticulated. From the nineteenth century, in the process of diagnosis doctors began to describe 'what for centuries had remained below the threshold of the visible and the expressible' (Foucault, 2003b, p. xxi). In this process, the relation between the 'visible and invisible – which is necessary to all concrete knowledge – changed its structure' (ibid.). For Foucault,

> This new structure is indicated – but not of course, exhausted – by the minute but decisive change where by the question 'what is the matter with you' with which the eighteenth century dialogue between doctor and patient began (a dialogue possessing its own grammar and style) was replaced by that other question 'where does it hurt?' in which we recognise the operation of the clinic and the principle of its entire discourse. From then on, the whole relationship of signifier to signified, at every level of medical experience, is redistributed: between the symptoms that signify and the disease that is signified, between the description and what is described, between the event and what it prognosticates, between the lesion and the pain that it indicates, etc. The clinic – constantly praised for its empiricism, the modesty of its attention, and the care with which it silently lets things surface to the observing gaze without disturbing them with discourse. (ibid.)

At this time, the modern medical 'gaze' became productive rather than reductive, contributing to the production of *individuals* as observed and defined by medical science. It was this formal scientific recognition of the individual, writes Foucault (2003b, p. xv), that 'made clinical experience possible; it lifted the old Aristotelian prohibition: one could at last hold a scientifically structured discourse about an individual'. While productive of individual personas as patients, the medical treatment of the nascent individual was also productive of a power relationship evident in the 'establishment of a "unique dialogue"' between patient and physician (ibid.). This dialogue reflected the 'special contract' that Foucault identified as being revived and invigorated in the twentieth century 'in the interests of an open market, so-called "liberal" medicine' (ibid.). The physician's enquiring of the patient 'where does it hurt', produced the patient as a subject requiring interpretation, divination and treatment. But in the case of abortion for social indications, women typically make *entrepreneurial* decisions of a social, not medical, 'responsibility orientation' (Gilligan, 1982). They do so in defiance of a concerted campaign staged over two centuries by the state and medicine to control the practice of abortion. Where women's abortion decision-making arguably represents a clear example of the entrepreneurial female self in action it serves to expose the limits of neoliberal health discourse. Legally, and according to best medical practice, the woman wanting an abortion must present to a physician explaining 'where it hurts' to seek *diagnosis* and *permission* to have her pregnancy terminated by a third party in the orthodox surrounds of the clinic. A woman who self-aborts subverts this orthodoxy by embodying entrepreneurialism. Hence the practice of self-abortion has consistently defied juridical and medical hegemony, posing a significant threat targeted by the state and caricatured in medical and legal discourse as having left a 'horrible trail of morbidity and mortality' prior to the widespread legalisation of surgical abortion throughout the West (Solinger, 1998, p. 4).

In response to the arrests of Leach and Brennan Queensland Labor Premier Anna Bligh resisted repeat appeals to her to intervene in the case. Although Bligh identifies herself as 'pro-choice' and as having 'very liberal views on abortion', she refused to take this opportunity to act on the State platform of her political party, which seeks the abolition of all criminal abortion offences (Greer, 2009). Bligh deferred to Westminster convention and stated that although she found the arrests to be 'very disturbing' it would be inappropriate as Premier to intervene in a criminal case prosecuted by the Director of Public Prosecutions (Walker and Hyde, 2009, p. 1). In discussing the case, Bligh was at pains to emphasise that self-administration of abortion in the absence of the assent of an Australian doctor posed a threat to the criminal law and society. When questioned on national television about the case she stated,

> I just really make the point that this is a – the charge has been about self-procurement. I don't know of any legislation or any circumstance in Australia where self medication like that would be legal and even if Queensland considerably relaxed its own legislation you certainly wouldn't be, you know, condoning that sort of activity. (ABC, 2009)

In a further reinforcement of the power and authority of the medical establishment, Bligh's sole intervention in abortion policy and law constituted a government bill to clarify the legal protection of abortion *doctors*, at the time of the Cairn's Case. Alarmed by the arrests and implications for the legal status of RU486, obstetricians working in Queensland hospitals ceased performing medical abortions, which had typically been performed with drugs in maternity units on women found in mid-pregnancy to be carrying a foetus with severe abnormalities. After numerous women were reported to have travelled interstate to access such procedures, and in response to demands made by doctors, Bligh introduced the Criminal Code (Medical Amendment) Bill, stipulating that the Queensland Criminal Code provided for the performance of 'a surgical operation *or medical treatment*' upon an 'unborn child to preserve the mother's life'. The Bill received bipartisan support on the guarantee that it provided only a clarification for medical doctors, 'would not "liberalise" the law' (Australian Associated Press, 2009), and had no discernable impact on the criminal charges involved in the Cairn's Case.

Conclusion

Leach and Brennan experienced the wrath of some who objected to their self-regarding entrepreneurialism. During the 18-month period of their arrest and after their names were published in the press, it is alleged that Brennan's car was damaged and the couple's home firebombed (Betts, 2009, p. 25). But in court, the direction of the judge and the findings of the jury were interpreted by pro-choice campaigners as a vindication of the rights of women to exercise their personal judgement about abortion, a 'clear message that not even a jury is willing to convict on these charges' (Marsh in ABC, 2010). The prosecution's case rested on the alleged *noxious* nature of RU486 tablets contravening the law – 'They are intended in one sense of the word, to be injurious, hurtful and harmful' (Byrne in ABC, 2010). But Judge Everson directed the jury to consider expert evidence that 'there are virtually no complications' with using RU486, which is 'not harmful to the person taking it'. Defence counsel also argued that RU486 was not noxious to women and implored the jury to consider 'life as it really actually occurs today. Ladies and gentlemen, I ask you on behalf of these two young people that – put an end to the nightmare that they've had to go through and return in each instance a verdict of not guilty' (McCreanor in ABC, 2010). Within an hour of deliberation the jury returned its verdicts acquitting the couple of all charges (ABC, 2010). Bligh resisted calls for the decriminalisation of all abortion offences, stating that a majority of MPs in parliament did not support such change (Schwarten and Agius, 2010), while Cairns obstetrician Caroline deCosta optimistically declared that the Cairns judgement 'effectively decrminalises RU486 in Queensland' (deCosta, 2010).

As documented in this chapter, for women of the feminist and neoliberal revolutions, life as it 'really actually occurs today' tends to involve self-regarding decisions about reproduction made on the basis of their perceived capacities as responsible mothers. In recent years, doctors have reported a burgeoning international black-market trade in RU486 and other abortifacients (Walker, 2009). This points to women's continued or heightened autonomous decision-making about abortion, and perhaps a failure in accessible abortion service delivery.

Women's abortion entrepreneurialism complements the individualistic ethos of neoliberalism internalised by young women raised to pride themselves in their independence and responsible, self-sufficient behaviour. It thus reflects one of the core functions of neoliberalism, whereby individuals are 'persuaded to make meaning of their life as if it were the outcome of individual choices made in furtherance of self-interest and self-actualisation' (Baker, 2008, p. 54). Nonetheless, women's reproductive entrepreneurialism may prove to be at odds with medical, legal, and political authority that has fought tenaciously to circumscribe and control women's abortion choices. Hence a sociological view of abortion as it 'really actually occurs today' may help expose some of the limits of neoliberal discourse applied to health policy.

Acknowledgements

The author would like to acknowledge that the research for this chapter was supported by an Australian Research Council Australian Postdoctoral Fellowship (DP0986934).

Notes

1. Like most strong pharmaceuticals, the 'safety' of RU486 is contested, but I suggest that its clinical trials and measured dose mean that it is generally accepted as 'safer' than traditional means of self-abortion.
2. In South Australia most abortions take place in the public health system.

5
Gaps in Post-Birth Care in Neoliberal Times: Evidence from Canada

Cecilia Benoit, Camille Stengel, Rachel Phillips, Maria Zadoroznyj, and Sarah Berry

Introduction

Research in a number of countries shows an increasing trend for maternal services to focus on infant health, and on the surveillance of parents and their parenting skills, rather than on the provision of broad ranging support for mothers (Dennis et al., 2007; Zadoroznyj, 2006). A wide range of reports produced in the United Kingdom (UK) over the past several years, for example, have centred on the notion of 'early years' or 'foundation years' interventions, with the aim to reduce child poverty and inequality in life chances, and to ultimately forestall persistent social problems linked to early parental neglect (Allen, 2011; Field, 2010). Within these reports, the importance of mothers' mental and physical health is cited in relation to childhood health and well-being, but the proposed solutions are most often short-lived, rather superficial interventions such as brief visits by 'health visitors', often aimed at screening for risks. Additionally, such visits are framed almost exclusively in terms of expected improvements in mother–child bonding ('attachment') and/or breastfeeding rates, and their alleged consequences for early brain development and immunity, rather than any substantive improvements in the health and overall well-being of mothers themselves (Marmot et al., 2010).

These developments have been accompanied by major changes in the delivery of maternity services. One change is the steady increase in Caesarean-section (CS) rates. In the UK, CS rates increased from 12 per cent in 1990–1991 to 23 per cent in 2005–2006. CS rates are even higher in the United States (US), at 30 per cent in 2005 (Hamilton et al., 2007), and Australia, where 31.5 per cent of women gave birth by CS in 2009 (Australian Institute of Health and Welfare (AIHW), 2011; Einarsdottir et al., 2012). As detailed below, Canadian CS rates are also relatively high. In contrast, Finland has seen no increase in its CS rates of 16 to 17 per cent since 1994; indeed, the Finnish rate has declined slightly since 2005–2006 (Hamilton et al., 2007).

The World Health Organization (WHO) (1985) recommends rates of 10 to 15 per cent. High CS rates suggest elective (non-medically necessary) intervention, and while CS surgeries represent an important birthing option, they are also linked to a range of risks for mothers and babies. Such risks include higher rates of maternal and infant morbidity (Public Health Agency of Canada (PHAC), 2008). Consequently, health professionals advise against practising CS surgeries in excess of evidence-based guidelines (PHAC, 2008).

Another important change in the delivery of maternity services over the past several decades has been an overall reduction in maternal length of hospital stay following birth. Such reductions have followed the erosion of cultural norms about the need for a 'lying in' period (Brown et al., 2002). Standard hospital lying-in periods have been reduced from between one to two weeks for an uncomplicated vaginal birth in the 1950s, to stays of two to three days (or fewer) in Canada, UK, US, Sweden, and Australia (Donnellan-Fernandez, 2011). A striking example of the contraction of in-hospital post-birth stays is in the US, where by the early 1990s, the lying-in period had taken on a 'drive-through' character; hospital stays of 12 to 24 hours for uncomplicated vaginal births, and 48 to 72 hours for uncomplicated CS births have become standard (Declerq and Simmes, 1997). As with rising CS rates, there has been considerable controversy surrounding the question of whether earlier discharge of mothers and babies is safe, and uncertainty about the degree to which shortened post-birth stays promote positive health outcomes (Brown et al., 2002).

These changes in both perceptions and structuring of post-birth care are taking place at the same time as many families are experiencing barriers to privately organised care (Zadoroznyj, 2006). Fiscal policies of cost containment in recent decades, coupled with neoliberalisation policies stressing individual responsibility and reliance on market forces, have resulted in the contraction of state-provided care services in a range of sectors and states. Reliance on market mechanisms commodifies care arrangements, transforming care into 'products' for purchase, and the means of care provision into specialised jobs and occupations (Zimmerman et al., 2006, p. 20). The result of which is a 'care deficit' – a situation in which the demand for care exceeds its supply for those individuals who are unable to afford care services (Hochschild, 1995).

To date, the implications of these changes in care provision, including the possibility of a care deficit, for new mothers have received limited attention. In this chapter, we examine post-birth care provision in one Canadian city. As a matter of context, we provide a brief background of recent changes in post-birth care in Canada.

Post-birth care in Canada

Similar to many other countries, Canada has reorganised the content and delivery of post-birth care services in recent decades (Zadoroznyj et al., 2012). The length of time Canadian women spend in hospital following childbirth has decreased dramatically, from five to seven days in the 1960s, to between 24 to 48 hours after vaginal delivery in the current decade (Canadian Institute of Health Information (CIHI), 2004).

As noted earlier, this trend has taken place at the same time that the national caesarean delivery rate has steadily increased, with total caesarean rates increasing from 17.6 per cent in 1995, to 21.1 per cent in 2000 and 25.6 per cent in 2004 (PHAC, 2008, p. 29). This increase of 45 per cent over a decade is likely due to an increase in both elective and emergent CS births (CIHI, 2007). In 2005, the Vancouver Island Health Authority in British Columbia (BC), where our study took place, had the highest provincial CS rate – 32 per cent (British Columbia Perinatal Health Program, 2008). Although the CS rate for mothers aged 40 or older is currently double (42 per cent) the rate for mothers aged 20–24 (21 per cent), there is little evidence that this variation is based on mothers' demand – the so-called 'too posh to push' argument (Bourgeault et al., 2008). Rather, variations in obstetric practice and government cost-saving measures are among the main reasons for the CS rate increases for Canadian mothers across all age groups (CIHI, 2007).

In addition to women being discharged earlier to private home environments, Canada's public health care system currently covers a narrow range of post-birth care services. Typically, where post-birth care services do exist, they are low-intensity interventions that tend towards mechanisms of surveillance, guidance, and referral, rather than more intensive, home-based supports for new mothers (Shaw et al., 2006). At the federal level, this is restricted to the provision of informational supports for the provinces and the publication of national guidelines for maternity and newborn care (Health Canada (HC), 2003). At the provincial/territorial level, publicly covered post-birth care services following discharge from hospital involve a single home visit from a public health nurse; in some regions a telephone call from a public health nurse constitutes the extent of available support (HC, 2003).

On the positive side, publicly funded midwifery services have become available for care throughout pregnancy, birth, and post-birth. After considerable public debate and advocacy by consumer organisations, in the mid-1990s midwifery became institutionalised and publicly funded initially in the province of Ontario, with BC following soon thereafter. Today, the midwifery option is available in seven regions in roughly half of the provinces in Canada (Benoit et al., 2010).

Yet the impact of this midwifery expansion to date has been small. In fact, less than 5 per cent of births in Canada are currently attended by a certified midwife (Canadian Association of Midwives (CAM), 2010). While the percentage is higher in some provinces, including Ontario, with a rate of 10 per cent (Ontario Association of Midwives (OAM), 2013), and in our research site of BC, where 11 per cent of deliveries were midwife-attended in 2011 (CAM, 2011), a substantial proportion of women in all parts of the country who want to see a midwife are currently unable to find one (CAM, 2010). Many pregnant women and their families must pay for their care out of pocket for both pre- and post-birth care from privately practicing midwives. Even in jurisdictions where midwifery services are publically funded, research shows that less educated women, younger mothers, those without a partner, Indigenous women, and those living in rural and remote areas or socio-economically disadvantaged communities are less likely to have access to midwifery care during pregnancy, labour, and delivery and in the post-birth period (CAM, 2011; PHAC, 2009).

As the role of the state in providing care to post-birth women has declined, a wide range of services for purchase on the market has grown to fill the care gap. While these services have garnered media attention in Canada (Whittaker, 2007), there are currently no published research studies on the for-profit post-natal services. Post-birth doulas[1] – who advertise online – often propose tangible, high-intensity supports such as newborn care, breastfeeding and bottle feeding support, child-minding services, meal preparation, household chores and management (including laundry, plant, and pet care services), errand-running, and peer support/counselling. Many of these post-natal care providers hold degrees in nursing and midwifery, and/or have completed specialised training as post-birth doulas and lactation consultants.

Unfortunately, the relatively high costs of hiring private post-natal care providers make these forms of support accessible only to those who are able to pay for them. Providers who advertise online generally charge around CAD$25 on a per-hour basis, or anywhere from CAD$100 to CAD$1000 for overnight to week-long package deals, respectively. There is currently no information available on user demographics, patterns of use, or outcomes associated with these forms of commodified care, though such information would offer insight into the types and levels of unmet needs that exist. In sum, the provision of post-birth care in Canada is stratified by geographical location, social status factors, and capacity to pay for services on the market (Christie and Bunting, 2011; Kornelsen, 2003; van Teijlingen et al., 2009). We now explore these challenges for new mothers in one Canadian city.

Study design and methods

The data analysed in this chapter are drawn from a larger study of social determinants of pregnant and new mothers' health in one urban region of BC. A key objective of the project was to comparatively examine the experiences of mothers under physicians' and midwives' care. Our sample selection was theoretically informed and based on two overarching criteria: (i) diversity of backgrounds, and (ii) choice of maternity care provider. Our purposive sample included pregnant women who represented a range of ages, ethnicities, educational levels, parity, and economic status, and had chosen either a certified midwife or physician for their primary attendant. The proportion of women choosing to have the care of a midwife is thus artificially high compared to the general population, which estimates put at 25 per cent in the local region, an utilisation rate higher than that of any other city in Canada (Pope, 2012). We attained our sample by distributing posters and flyers to places that pregnant women frequent in the Victoria Census Metropolitan Area (CMA), including physicians' and midwives' offices, pre-natal classes, single-parent resource centres, and low-income outreach programs. While our non-random sampling technique precludes us from knowing if our findings can be generalised to the broader regional population, we believe that our sample reflects the diversity of social and economic backgrounds and the style of maternity care available through the public health care system in the area (see Table 5.1). In total 106 women responded to our research postings, with an estimated population-based recruitment rate of 3.5 per cent. We completed interviews with 93 women any time during their third trimester of pregnancy (Wave 1), 89 at four to six weeks post-birth (Wave 2), and 83 during the four to six months post-birth period (Wave 3). Thirteen participants were lost at each stage because they became unavailable for

Table 5.1 Selected characteristics of study population compared to the Victoria Census Metropolitan Area (CMA), 2006 (Income in Canadian dollars (CAD))

Measure	Sample population	Victoria CMA
Aboriginal background	4.5%	2.8% (females)
Visible minorities	11.2%	9.0% (females)
Less than high school graduation	7.2%	9.9% (female age 20–34)
Average (mean) annual income	$52,565	$66,594
Gross household income	$53,500 (median)	$59,015 (median)
Own home	40.4%	61.8% (by household, not specific for age or gender)

an interview or had scheduling conflicts, had a therapeutic abortion or stillbirth, or moved outside the region and were not accessible by telephone. Our final participant retention rate for Wave 3 was 89 per cent (n=83). Our analysis of the 13 participants who were not included in the study due to attrition or missing/incomplete data shows that they had a lower mean income and education level than the others. Just under half of Wave 1 participants (n=42) were under the care of a certified midwife and just over half received care from either a maternity physician or obstetrician (n=51). The four interviewers involved in data collection collaborated during pre-testing of the instruments and interviewer training to ensure consistency of delivery.

One of the purposes of qualitative inquiry is to report the perspectives of the people interviewed, in their own words (Denzin and Lincoln, 1998). During the interviews, participants described their experiences of pre- and post-natal care, and the interviewers probed for detail where necessary in order to clarify meaning and promote in-depth responses. Our interviewing strategy sought to understand the birth and post-care experiences as described and interpreted by the participants themselves (Flood, 2010, p.13). The perspectives of our participants were our paramount focus, as these lived experiences expose 'taken-for-granted assumptions' of pre- and post-natal care in Canada (Moustakas, 1994). To facilitate triangulation of data on topics of central interest within the study (birth, parenting, care, and health experiences) the interviews included both closed and open-ended questions, a variation of the mixed-methods approach (Small, 2011).

A feature of our mixed-methods design was to follow many of our closed-ended questions with a probe asking participants to explain their response, and then an additional open-ended follow-up question. The two questions relevant to our present analysis are: How satisfied were you with the post-birth care you received from paid care providers from the time of your baby's birth, up to the present? and Would you have wanted any of these types of care, but they were not accessible to you? If so, what were the barriers to your accessing this care? We reasoned that our research design, which involved both quantitative and qualitative data collection methods, would afford richer, more nuanced findings than using only one method. Specifically, we felt that it would enable us to quantify and qualify key variables, giving us an opportunity to consider both types of data on topics of interest and to elaborate on the meaning and experiences subsumed within survey statistics. In addition, after having completed key survey questions, participants were keen to elaborate on what factors they had considered when choosing their survey answer and to narrate stories that exemplified their experience. We delivered more sensitive questions and demographic questions, including those on income level and depression, using a self-administered, written questionnaire completed by the participant at the end of the face-to-face interview. In the case of this chapter, responses to relevant closed-ended questions were entered and analysed using SPSS 12.0 software. The qualitative data were analysed by the second author, with the first and third authors reviewing the coding categories for consistency, validity, and preliminary interpretation. Our study was approved by the Human Research Ethics Boards at the University of Victoria, Canada.

Results

Sample characteristics of participants

Our descriptive characteristics are based on the subsample of Wave 2 participants (n = 89) (interviewed four to six weeks post-birth) who gave complete answers to questions regarding their access to support and care in the post-birth care period. As shown in Table 5.1, participants were slightly more likely to identify being of Aboriginal or visible minority background, and to have lower income and lower home ownership than the population in the study region (Statistics Canada, 2006). Participants had somewhat higher levels of high school completion compared to the local population. The lower income and home ownership of the sample are likely a reflection of their younger, childbearing age.

Participants' median number of days in hospital was 2.4 days overall, and 1.9 days for persons reporting a vaginal birth; these data support the literature noted earlier regarding a reduction in length of hospital stays following the birth of a child compared to earlier generations. While the demedicalisation of birth, as evidenced by short-term hospital stays, may not in and of itself signal a lack of care in the post-birth period, coupled with the finding that approximately one-third of participants (33.8 per cent) reported that they desired post-birth care services that were not available, it would seem that some new mothers experience a care deficit upon discharge. Further, the closed-ended data suggest that a greater proportion of participants with incomes below the median for the census region reported a gap in post-birth services (39.5 per cent) compared to participants with an income above the median (27.3 per cent). The most commonly cited barriers to obtaining post-birth services included 'cost' (42.8 per cent) and 'unavailability of home care supports' (38 per cent).

As noted above, we asked participants to expand on the topic of access to post-birth care services, and barriers to receiving the care they wanted. The next section of the chapter focuses on themes that emerged, both in terms of participants' satisfaction with access to post-birth care, as well as lack of access to the services they wanted to help them through the early stages of being a new mother.

Qualitative analysis

Satisfaction with post-birth public services

Several participants either did not identify any additional post-birth care needs, or specifically described aspects of Canadian post-care services that contributed to a positive experience during the early postpartum period. Most of these participants noted practitioners and services that contributed to quality continuity of care. Sarah[2], aged 27 and seven weeks postpartum with her third child, described her satisfaction with her maternity doctor's team: 'Just wonderful people, they're just, they're awesome. They really listen; they never treat me like an overactive, over-reactive mother. You know they always take my opinion very seriously and uh, they're just really caring and really great.' Another participant, Annie, expressed gratitude for the public health nurse who visited her post-birth:

> The nurse contacted me right away when I got out of the hospital and she came to check up on me. [I]t's kinda nice to know that

they'll come to you. [Because], you know, when you first get out of the hospital and especially after a C-section [you are] sore. You don't really wanna go anywhere; you just wanna be home. So it's nice to have that for them to come to you. (aged 34; first child, five weeks postpartum)

Annie's response addresses the importance of accessibility to health care during the post-birth period and having services available that are flexible to the needs and physical capacity of the new mother. Many participants who mentioned positive aspects of post-birth care also highlighted easily accessible, helpful information. This included the 24-hour nurse hotline, which they saw as a beneficial support system for answering questions. Participants praised the workers of the hotline service as 'knowledgeable' and 'helpful' and easily accessible by phone. Other participants discussed the benefit of having a midwife for their most recent child. Kathy explained:

I went with the midwife this time round [and] I just felt that I was, I felt really, really well looked after [...] it's just so different having, you know, two women midwives where that's all they do, versus the GP who does all kinds of things and doesn't specialize in [...] the amount of time, I think that's a huge thing that they, the midwives, offer. (aged 35; second child, 19 weeks postpartum)

Another participant, Myra, highlighted that in addition to the more lengthy visits offered by midwives in comparison to physicians, another benefit was the availability of in-home postpartum care and flexible appointment times:

I can call them [the midwives] anytime and they will come over. Like it's not even, it's never a question like – if I ever needed to get in to see them now, like it was – they came here in the first two weeks and then I've been going there and if I needed to get in, I know they would just squeeze me in and – so that's why I like it. (aged 25; second child, seven weeks postpartum)

The dedicated time that the midwives' spent on these participants as well as the flexibility with regard to service time and location were greatly valued in the post-birth period. The qualities that made up a positive labour, delivery, and post-birth experience were the same qualities addressed as absent from other participants' post-birth experiences.

Dissatisfaction with post-birth services covered under the health care system
A major theme articulated by participants' who expressed dissatisfaction with their interaction with the health care system was the perception of inattentiveness during the labour and delivery period, and a lack of follow-up during the post-birth period. Concerning her hospital experience, Jessica stated:

> You really are like a number in the hospital. They're, you don't, they're not a lot of caring women and I'm not sure whether it's because they're tired of their job, you know and they're not happy that way or whether they're just you know, but there were a couple of kind women there, you know [...] you just, you don't really get that attention that I really believe that you deserve and you need. (aged 36; third child, 13 weeks postpartum)

A second-time mother four weeks postpartum, Lena, aged 31, stated 'I kinda [sic] fell through the cracks a bit with this second baby [...] I don't know what but nobody called me afterwards to remind me of things like shots and so I actually went, I went far too long before I got his shots.' Several respondents commented on a feeling that health care providers 'did not pay attention' and had a 'million other things to do', and that it would be beneficial if they could 'spend more time with the patient'. Some participants identified a need for more emotional and social support services, in particular supports that were not connected to risk assessment activities. The desire for this type of support came from a variety of avenues, including health care professionals, mental health services, and a space to informally socialise with other mothers. Kelsey, aged 22 and a first-time mother at six weeks postpartum, expressed a desire to have a health care professional to confide in while her son was in the hospital that would not

> Potentially deem me 'unfit' to have my baby because I'm depressed or something [...] that just scared me so I, I guess I had access to somebody to talk to but I didn't use it because I felt it probably caused more trouble than I would have wanted.

Women in Canada, including in the study area, are routinely screened for postpartum depression as part of public health care services (Stewart et al., 2004). Using the Beck Depression Inventory, we found that 15 per cent of participants at Wave 2 and 21 per cent of participants at Wave 3 reported moderate depression symptoms (Benoit et al., 2007).

Some new mothers, such as Kelsey, who identified the need for mental health services during the post-birth period did not access the desired services because they were concerned that doing so would undermine the perception of competency and capacity to care for a child (Benoit et al., 2007).

What women wanted and was not publicly available

As noted above, one-third (33.8 per cent) of the women wanted additional services than were provided via the health care system but were not available due to varying factors. Many of these women lacked a strong informal support system and the income to purchase post-care services out-of-pocket. Theresa, a first-time mother of twins, remarked:

> Well since I've been home like there have been times when having the two has just been really, really intense and really hard. I'm getting a grip on it now, but there were times within that first six weeks that I just felt like I was gonna lose my mind. It would be really nice if there was somehow just a number you could call, in the community, just to, I don't know, like listen to you for a minute or, or, I don't know, rush over and hold one of your babies! (aged 30; first child, six weeks postpartum)

Sabina wished that the public health nurse would provide more continuous care during the initial weeks after the birth. But this is not a covered public care service in the study region, and as a result she had to go without:

> I would have liked the public nurse to come back again because she said she was going to and she didn't because I had breastfeeding questions and you know he had that acne, that um, from breastfeeding and I didn't, I didn't, wasn't sure whether that's what it was or not. Just questions that you have. (aged 30; first child, six weeks postpartum)

Three-quarters (74.7 per cent) of the new mothers in our study reported that someone, primarily friends and family members, came forward to help them during the post-birth period; mothers of participants were noted most commonly as the source of familial support. Without such care, they would have had to purchase it on the market, as Becky, aged 33 and six weeks postpartum with her second child, stated: 'If, if I didn't have my, my sister and mum, I definitely probably would consider a doula'. Other participants who wanted more post-birth care sought out a private doula, a service which is not publicly funded. Tina was able to pay for her doula, but saw it as an expense that she should not have to pay for:

> Like, we paid quite a bit for the doula. [Y]ou know it's expensive uh, anyway it would be nice if there was um public health care. I think they're worth it, it just like, it was something that we paid for and we knew wanted the support so but definitely it was expensive. (aged 24; first child, 18 weeks postpartum)

Other participants who did not have access to family care or the needed economic resources to pay for a doula went without. As Barb, aged 23 and a first-time mom six weeks postpartum, noted: 'I would have liked a doula but they're very expensive'.

Breastfeeding advice has become more commonplace in Canadian hospitals and all midwives are trained to routinely give such advice. During Wave 2 data collection, 64.8 per cent of new mothers reported breastfeeding exclusively, but by Wave 3 only 50 per cent were breastfeeding exclusively. In addition, 89 per cent of the women reported problems with breastfeeding their infants during Wave 2 while 35 per cent reported problems in Wave 3.

We asked the women who responded 'yes' to having breastfeeding difficulties what sorts of difficulties they had. The most prominent of these were: sore nipples (the most common source of difficulty), difficulty latching on, sleepy baby, and milk undersupply. Karli expressed her concerns about breastfeeding, stating:

> When we left the hospital I kinda worried because I still wasn't, you know, breastfeeding that [...] My milk hadn't come in and there was things like that so um. [I]t was fine in the hospital cuz I had that support but once you go home you don't have that support any longer. (aged 31; first child, four weeks postpartum)

Other respondents similarly commented that 'breastfeeding was the hardest' and that it would have been beneficial to have a lactation consultant who had time and expertise to devote specifically to the task. As these participants noted, lactation consultation is a very important service as new mothers have a short window of time to establish breastfeeding before they may turn to bottle/formula feeding because they are worried that the baby is not being adequately nourished. The public health nurse visit offered to residents in the study region may not be sufficiently timely or intensive enough to meet this important need. In fact, a lactation consultant is arguably a valuable addition to midwifery care in the first two weeks as the midwives help with breastfeeding but some people, particularly first-time mothers, require more intensive support in early post-birth period. In our study only one-quarter (25.8 per cent) of participants mentioned they had used the services of a lactation consultant in the post-birth care period.

Discussion and conclusion

In this chapter we examined the provision of post-birth care for mothers in one Canadian city. The findings presented here indicate that the majority of participants were happy with their post-birth care services, noting that they contributed to a positive experience during the first six weeks of the birth of their baby. Most of these participants noted practitioners and services that contributed to quality continuity of care. However, a minority of the women in our study expressed a desire for post-birth care services that were not available to them locally. Many of these women reported incomes below the median for the census region. Unavailability of home care supports was at the top of their list of barriers to access the post-birth care services they desired.

The adoption of neoliberal reform ideologies in countries such as Canada and Australia has reduced government responsibility for the care of post-natal women and increased private responsibility (Dennis et al., 2007; Zadoroznyj et al., 2012). This shift has resulted in a care deficit, particularly for women without the financial or support networks to secure appropriate post-birth care. The cultural politics of post-birth care are premised on the often mistaken assumption that women need little, if any, care after their discharge from hospital, an understanding which is, in turn, based on very narrow conceptions of care itself (Forster et al., 2008). Where government-based home care is provided within these contexts, it is generally limited in scope, fragmented in terms of its provision across multiple carers, and differs in type and quality between hospitals and geographic regions (Fenwick et al., 2010; Schmied et al., 2010). To the extent that community-based services exist, they tend to focus on the infant rather than the mother, or on a form of surveillance aimed at identifying 'at risk' groups. In some developed welfare states, such as Sweden and the Netherlands, the two branches of the welfare state have cooperated and integrated to such an extent that they are able to meet the care needs of women in the post-birth period (Benoit et al., 2005; Hochschild, 1995). In other countries, including the US, Australia, the UK, and Canada, these branches do not closely overlap. It is within the latter contexts that a gap in care exists. In some of these countries, midwives and family networks provide adequate social care to some new mothers but these tend to be women who are more advantaged. The market – through the private services of doulas, lactation consultants, etc. – is variously called upon to fill the care gap (Leitner, 2003). To the extent that individual mothers or families rely on the market for care provision, issues of equity and quality of care are pivotal (Folbre and Nelson, 2000; Pocock, 2006).

Parents' and children's welfare in the post-natal period is often closely connected (McCoy et al., 2010; United Nations Millennium Project, 2005). Appropriate post-birth care from government services that attend to the needs of new mothers in turn can aid with the creation of a stable, reduced-stress environment for children during an important phase of their life (Zadoroznyj, 2006). Currently, the potential of the post-natal period as a period of health promotion opportunity is not being fully realised. Additional supports for new parents, which reflect the importance and synergism of both parent and infant health promotion, should be considered.

Acknowledgements

This research was supported by a research grant from the National Network on Environments and Women's Health, Canada, and postdoctoral fellowships from the Michael Smith Foundation for Health Research and the Canadian Institutes of Health Research. A deep thank you to the women who took part in the interviews. Thanks as well to the Centre for Addictions Research British Columbia (CARBC) for the research assistantship, the office space, and the supportive work environment, and to Marie Marlo-Barski for helping to edit the manuscript.

Notes

1. 'Doula' comes from the Greek word meaning 'a woman who serves'. Doulas are trained (non-medical) professionals who support women before, during, or after birth (DONA International, 2014).
2. Pseudonyms are used to protect the identity of the participants.

Part II
Reproductive Bodies and Identities

6
Unborn Assemblages: Shifting Configurations of Embryonic and Foetal Embodiment

Deborah Lupton

Introduction

When the pregnancy of Kate Middleton, the Duchess of Cambridge, was announced in early December 2012, the news received high attention in the news media and social media outlets. What was immediately noticeable about this coverage was the immediate configuring of a new personage: that of the 'royal foetus'. Spoof Twitter accounts were set up on behalf of the 'royal foetus' purporting to be tweeting from 'inside the royal womb'. Various comments were made by others on Twitter concerning the wealth and social standing that the 'royal foetus' already enjoyed. A commemorative plate in the style of royal souvenirs celebrating events such as births, weddings, and coronations was even mocked up, using a generic ultrasound image to denote this new individual in lieu of the traditional photograph.[1]

This response to Middleton's pregnancy is the culmination of a series of cultural changes that have occurred since the emergence of scientific medicine and its associated technologies and surgical techniques, as well as photojournalistic techniques added by magnification, endoscopic cameras and computer imaging and the advent of social media. All of these have contributed to new ways of portraying, conceptualising, and dealing with the unborn.[2] In previous eras, apart from medical students and researchers and those who viewed preserved embryos and foetuses in jars at fairs or museums, the appearance of the unborn for most people remained largely a mystery. In contrast, visual images of the embryo and the foetus have now become so prevalent in popular culture that their presence has become taken-for-granted. We have reached a stage at which embryos and foetuses may have their own Facebook profiles and, like the 'royal foetus', Twitter accounts, and post messages from the womb, where the ultrasound, originally a technology designed for medical purposes, has now become yet another means by which the 'person within the womb' can be visualised and positioned as an individual and these images shared with up to thousands of others on social media sites and where unborn development is commonly represented as occurring without the assistance of or connection to the maternal body.

I would contend that partly as a result of this increased visibility of the unborn and its visual conventions, a number of changes have occurred in configurations of the unborn, particularly over the past half century or so. At the same time as the appearance and developmental stages of the unborn have reached the public domain, they have also become more valued, to the point that they are fetishised in contemporary western culture, represented both as highly precious and as vulnerable to harm. The concept of infancy is now routinely extended back into the gestational phase of development. Unborn entities are commonly portrayed as proto-babies, deserving of full personhood status and the same kinds of rights and privileges that are accorded infants. The common convention of the visual portrayal of the unborn body as separate from and seemingly unconnected to the maternal body also contributes to this concept.

In this chapter, I examine changes in ideas about the unborn since the introduction of these technologies, practices, and visual representations. The chapter explores how the pre-modern concept of the hidden, mysterious organism that was inextricably part of the maternal body changed to encompass the notion of the unborn as mini-humans, individuated from the maternal body and even in conflict with or at risk from their mothers. Following discussion of the ways in which discourses and representations of and practices around the unborn are constantly shifting, I introduce the concepts of the unborn and maternal assemblages. I then provide an overview of histories of the unborn and discuss various key dimensions of contemporary configurations of the unborn and maternal assemblages: the role played by visualising technologies, the commodification of the unborn and risk and the unborn-maternal assemblage.

Contingencies of unborn assemblages

Scholars pondering the nature of humanity and personhood in relation to unborn bodies argue that, like other human bodies, they are social objects which are always already invested with cultural meaning.

From the time an unborn organism is conceived from the union of a human oocyte and sperm, and even before (Karpin, 2010), its meaning is shaped via cultural and social understandings. There has long been a debate in religious and legal circles concerning the point of development at which the unborn entity is considered a person. Definitions of personhood, and by extension, notions of individual embodiment, are constructed via social and cultural understandings. They are dynamic and shifting, open to change and contestation, and therefore highly contingent.

Given the debate over whether or at what point in their development the unborn are fully 'human' or 'persons' and the integral role played non-human actors such as technologies in visualising, representing, measuring, and monitoring these entities, such a perspective offers much to the understanding of the social worlds in and across which the unborn move and are configured. In the contemporary era such social worlds include not only the private and intensely intimate spaces and relationships of the women and men who provide the gametes that result in the conception of the unborn but also a range of other places, spaces, and social actors, including the medical clinic, the in vitro fertilisation (IVF) clinic or human embryonic stem cell (hESC) laboratory, the pro- or anti-abortion organisation or activist group, and the mass media and digital media outlets, in and between which the unborn assemblage may move and achieve meaning.

The unborn have for some decades been the focus of highly emotive and political contestations concerning their position in relation to the human/non-human binary opposition, particularly in relation to abortion politics (see Ginsburg, 1990; Hartouni, 1992; Hopkins et al., 2005). This debate has intensified recently in the wake of the development of hESC science, where embryos are used for research and therapeutic medical purposes (Franklin, 2006a; Hogle, 2010). Although foetuses are further along in their development than embryos and thus closer to the point at which personhood and human status are attributed, as not fully developed bodies and as still unborn and contained within the maternal body they provoke uncertainty and therefore contention about their status. Unborn entities are placed on the margins of the boundaries between human and non-human or in the spaces between these categories, and thus are located on a continuum rather than a binary definition of humanness (Casper, 1994a, 1994b).

Depending on the perspective one takes, an unborn assemblage may be considered already human/not yet human, part of/individuated from the maternal body. These positions may indeed fluctuate for the one person, depending on the context. A wanted pregnancy may be considered 'my baby' from the time of first realisation of pregnancy; an unwanted pregnancy may be considered 'just a bundle of cells' (Pfeffer, 2008). Different countries and even regions within a country have different ways of distinguishing legally between a 'foetus' and an 'infant', based on such attributes as gestational age, size, and whether the organism in question is considered 'viable' outside the uterus.

Anthropologists have demonstrated that different cultures have differing ways of deciding when unborn and even post-born entities are deemed to become 'infants' or designated as 'human' (Conklin and Morgan, 1996; Gottleib, 2000; James, 2000; Kaufman and Morgan, 2005). There are 'processes of coming-into-social being' (Kaufman and Morgan, 2005, p. 321) which are related not to biological attributes but rather to accepted understandings within a social group or culture. In any cultural context the ontological distinctions between the 'embryo', 'foetus', and 'baby/infant' are constantly shifting, open to change and contestation. Some cultures locate the beginning of personhood while the unborn body is still *in utero*. In many other cultures, however, newborn infants are considered to be not fully human until they have demonstrated certain behaviours or lived for a defined time period (Conklin and Morgan, 1996).

Unborn entities may be understood as 'assemblages', or constantly changing configurations produced by the interaction of an individual's body with a range of heterogeneous elements, including other bodies, non-human living organisms, material objects, discourses, practices, space, and place. The term 'assemblage' has emerged from the philosophy of Deleuze and Guattari in conjunction with actor network theory, a major theoretical approach in science and technology studies (Marcus, 2006). The concept of the assemblage acknowledges the contingent nature of embodiment and subjectivity and the importance of recognising the interaction of bodies with others' bodies and with non-human agents. It highlights the components from which social phenomena – including human bodies and subjects – are comprised and with which they relate. It goes beyond a social constructionist perspective, which tends to assume that once a phenomenon has been constructed its meanings are fixed, to a perspective which allows for constant change and the interplay of meaning between social actors and objects, for the making and remaking of phenomena.

Unborn assemblages are sometimes individuated from and sometimes joined with the maternal assemblage. These assemblages may come together and merge at some points and at other times come apart.

Birth or pregnancy loss events are the pivotal moment at which the unborn-maternal assemblage comes apart in material, fleshly terms. As I explain in further detail below, visualising and medical technologies that depict the unborn as separate from the maternal body also work to disentangle these assemblages. When unborn entities are destined never to enter a woman's body, as is the case with surplus embryos created for IVF that may be then discarded or used for hESC research, they never enter into a merged physical assemblage with the maternal body, but may still be thought of as 'my baby' by the woman who has contributed her ova. There is no single 'unborn assemblage', therefore. Just as any other human body is open to change and contestation in its meaning, unborn assemblages are mutable, changing in form not only as they physically grow and develop but also as the social worlds in which they are located shift and change.

Histories of the unborn

It is important to emphasise that concepts of the unborn in western societies have changed considerably in recent times. Despite their potent contemporary position as cultural icons and fetishes, embryos and foetuses as they are represented and conceptualised today are relatively recent figures in western cultures. One major difference is that in earlier times the embryo or foetus was never referred to as such and was viewed as inextricably part of a woman's body. While medical drawings and anatomical wax figures from the early modern period commonly represented the unborn body as individuated from that of the maternal body, the vast majority of lay people were not exposed to these representations and did not think of the unborn in these terms (Duden, 1993; Newman, 1996). Indeed the foetus was not routinely treated legally or clinically as if it were a separate body until the 1960s (Featherstone, 2008; Weir, 1998).

Before the invention of medical and scientific technologies that could conduct anatomical dissection or autopsies on unborn corpses, test for pregnancy hormones, visualise the unborn body, or hear its heartbeat, the unborn, particularly in their earlier stages of development, were enigmatic, hidden creatures. For pregnant women the unborn assemblage was experienced as an invisible constellation of bodily sensations that for the first several months of the pregnancy only they could feel, and others relied upon the women to recount these haptic sensations. The concept of 'quickening', in particular, was used to describe the first movements of the foetal body felt by the pregnant woman, and these movements confirmed to her that she was pregnant (Duden, 1993; Featherstone, 2008). Thus before the advent of scientific medicine the unborn entity may have given signs and intimations of its presence, but there was never certainty as to its existence until it actually passed out of the pregnant woman's body.

The emergence of scientific medicine in the late eighteenth century resulted in major shifts in the ways in which the unborn were understood and treated. Once scientific medicine had established, in the nineteenth century, that fertilisation involved the meeting of ovum and sperm, and had developed greater knowledge about embryonic and foetal development from autopsies and vivisection, the unborn organism became increasingly viewed as human and alive from an earlier stage (Dubow, 2011; Featherstone, 2008). By the late nineteenth century, doctors began advising pregnant women to protect their unborn's health by engaging in certain protective behaviours. A sense of the unborn as an individual with its own identity and embodiment – and indeed, interests – separate from that of the pregnant woman began to develop (Dubow, 2011; Duden, 1993; Featherstone, 2008).

New surgical and therapeutic techniques developing in the twentieth century contributed to the new notion of the foetus as a patient in its own right, with separate needs and treatment from the maternal body. The introduction of foetal surgery in the 1990s, in which foetuses are operated upon while *in utero*, is one such example. Medical writings about foetal surgery portrayed the foetus as a living human being, an embodied actor within the pregnant woman's uterus. Practitioners working in the field routinely referred to the foetus as 'the kid' or 'the baby', while the pregnant woman was represented as the 'organic host' for the foetus in some medical writings (Casper, 1994a; Williams, 2005). The conceptual boundaries between the foetus and infant were also challenged and rendered unstable by medical technologies which were able to facilitate the survival of premature infants at gestational ages which previously were too early to support survival, pushing back the medical and legal definition of viable life to earlier stages of development. In the UK, for example, the definition of a 'foetus' has been reduced from 28 to 24 weeks as survival rates for prematurely born infants have increased at earlier gestational ages, with some (albeit a tiny minority and with a high risk of disability) now surviving even at 22 weeks (Williams, 2005).

Even more important in developing new ways of thinking about unborn embodiment and in configuring the foetus as a subject was the invention of visualising technologies that were able to penetrate the recesses of the uterus and expose its insides. From the middle of last century, visualising devices became integral to new imaginings of the unborn assemblage, reconfiguring it as the novel icon of the 'public foetus', or more recently, the 'public embryo' (Duden, 1993; Franklin, 2006b; Morgan, 2009; Petchesky, 1987). In the 1960s, apart from X-rays, which were rarely used because of the harmful effects of radiation upon the foetus, no technologies were available to image, scan, test, or screen the unborn. Obstetric ultrasound became routinely offered to pregnant women from the late 1970s. In many countries it is now offered to all pregnant women at least once during their pregnancy, and more often than this for many: in Germany, for example, the majority of pregnant women undergo an ultrasound at each prenatal check-up (Erikson, 2007).

Visualising the unborn: the role of the obstetric ultrasound

Once ultrasound became a commonly used technology, the once opaque and secret environment of the uterus, perceptible in the early months of pregnancy only to the pregnant woman herself via her physical sensations, was opened to public observation. The unborn could now be seen *in utero* moving about, with recognisable features and limbs, even before the pregnant woman could feel these movements. As a result the haptic perceptions, or those generated by the senses of touch, taste, the sense of space and atmosphere that were once so important to give evidence of the unborn's existence and well-being, became largely replaced by optic interpretations (Duden, 1993; Morgan, 2009).

These representations have changed the ways in which pregnant women think about the body growing within them and how they are treated in the medical setting. The visual evidence of the ultrasound is privileged by both medical professionals and pregnant women over the woman's own haptic experiences of the presence of the unborn body developing within them. While for medical staff the ultrasound image is regarded both as a medical, technical image for diagnostic purposes and as a means of promoting 'bonding' between prospective parents and their unborn, for the pregnant woman or couples involved, it is above all a social image, the first photograph of 'our baby'. As many scholars have contended, quite apart from any medical reason for ultrasound it has come to serve an important social function in assisting prospective parents to forge an emotional bond with their unborn (for example, Mitchell and Georges, 1997; Nash, 2007; Palmer, 2009a; Petchesky,

1987; Roberts, 2012; Taylor, 2008). Ultrasound images may also serve to confirm what until that moment has been a 'tentative pregnancy' (Rothman, 1993), in which the couples may withhold developing an emotional bond with the unborn until they know for sure that it is normal and healthy (Mitchell and Georges, 1997; Williams et al., 2005).

Showing ultrasound images to others is often part of the process of making the pregnancy public, used to demonstrate to others what the unborn 'looks like' and proving that it actually exists and is 'normal' in its morphology. Sonographers routinely make reference to the foetus being 'shy', 'not wanting his picture taken', 'athletic', 'smart', 'just like his Dad', 'very good', or 'cooperative' and may speak to it, telling it to 'Smile for the camera' or 'Say hello to Mama' (Kroløkke, 2010; Mitchell and Georges, 1997; Palmer, 2009b; Roberts, 2012). The introduction of 3/4D ultrasound imaging has contributed further to this infantilising of the foetus because this technology is able to show greater detail of the foetus's face, expressions, and therefore supposed emotional state, interpreted as 'smiling', 'fear', or 'frowning' for example. Sonographers undertaking 3/4D ultrasounds and the prospective parents looking on often seek to identify parts of the foetal body or face as looking like features of the parents – their nose, eyes, or long legs – in an attempt to personalise the foetus, give it an identity, and render it as part of the family (Nash, 2007; Palmer, 2009b; Roberts, 2012).

As this suggests, where once the unborn became a child at birth, ultrasound images serve to render it as an infant pre-birth and bring it into being as an assemblage with which observers have a particular social and therefore ethical relationship. These images have therefore resulted in the 'social birth' of the new human to shift from the moment of physical separation from the maternal assemblage to earlier phases of unborn development, so that the bestowing of such social attributes as gender and name often takes place before physical birth (Mills, 2008; Morgan, 1996). Here the liminality of concepts of the foetus and the infant is exposed, with these images promoting a blurring of boundaries between the two.

Once the focus of the ultrasound image is directed at the unborn, the woman's body becomes merely a conduit by which the 'pictures' can be taken of this unborn individual. She may be asked by the sonographer to move her body in certain ways so that 'better images' can be taken of the centre of attention: the unborn. All eyes, including those of the pregnant woman, are turned away from her body once the imaging process begins to a focus on the unborn images on the screen (Palmer, 2009b). The images produced tend to show the foetus in a dark, undifferentiated space that is not easily recognisable as a part of the pregnant woman's body. The umbilical cord and placenta that are clear visual reminders of the interconnectedness of the unborn and maternal assemblages may be considered to be extraneous to the image and even as obscuring a clear view of the unborn. They may even be digitally edited from printed ultrasound images to provide a 'better' picture of the unborn for the parents to take home (Nash, 2007; Palmer, 2009b), an image in which the convention of the portrayal of the unborn assemblage as individuated from the maternal assemblages is preserved.

The unborn as commodities

We are now in the midst of an era in which unborn assemblages have become increasingly commodified, their visual representation and form integral features of popular culture. Photographic portrayals of the unborn have routinely been used in advertisements, television documentaries, and coffee table books and anti-abortion material, as well as information texts produced specifically for pregnant women such as handbooks, websites, and medical brochures (Hartouni, 1992; Stormer, 2008; Taylor, 2008).

The production of visual images of the unborn for the consumption of the public began with the photojournalist Lennart Nilsson's photographs, appearing first on the cover of the American *Life* magazine in 1965 and in a colour photographic essay within, and then later in coffee table books, a 1990 issue of *Life* and several television documentaries, has been particularly influential in reconfiguring the visual portrayal of the unborn body. Nilsson's images, as well as those by Alexander Tsiaras in his anatomical atlas of unborn development made using computer-enhanced magnetic resonance imaging (MRI) first published in 2002, represent the unborn body as sublime, glowing like a jewel in bright colours, a figure of beauty and wonder. The embryonic and foetal bodies in these images float in their own space, with little indication of the maternal body from which they came. The blackness in which these unborn entities float and against which they are contrasted appears instead as the infinity of the universe, the unborn body and this space together representing the awesome power and mystery of life (Stabile, 1992; Stormer, 2008).

Like photographic portraits of the unborn, ultrasound images have also made their way into public culture and have contributed to the commodification of the unborn image. A thriving industry now exists in the commodification and marketing of 3/4D obstetric ultrasound images of the unborn. Clinics have been established that encourage clients to attend to make a 'bonding' ultrasound to have a 'sneak peek at bub'. Packages are marketed for special 'entertainment' for baby showers: showing off views of the foetus for family and friends ('showcase your new little star', as one Sydney clinic puts it). Some clinics even provide mobile services, bringing their technology to people's homes so that the imaging can be performed in a relaxed and private environment. Packages may include the session itself, DVDs and glossy photos. When they attend such a session, expectant couples and their 'guests' (other family members or friends attending the session) thus are offered a chance both to 'see' and to 'consume' the unborn assemblage *qua* baby (Kroløkke, 2010, 2011).

Ultrasound images have been employed in many different commercial products, including 'ultrasound art' (canvases featuring the enlarged and cropped ultrasound image that has been colourised), jewellery containing the images, t-shirts for pregnant women, custom photograph frames to show off the image (bearing 'Love at first sight' or 'A miracle is born' legends), baby shower cup-cake decorations, invitations and thank-you cards, Christmas ornaments, and photographs of pregnant women displaying their naked bellies with a transposed foetal ultrasound image placed over their bared flesh to demonstrate what lies within.[3] Other commodities featuring the unborn include foetus-shaped cakes for baby showers, jewellery shaped in the image of a foetus, embryo and foetus dolls dressed in baby clothes, foetus cookies, and foetus soaps.[4] On websites such as 'The Visible Embryo' (www.visembryo.com/baby), in addition to viewing images and information about embryos and foetuses for each week of gestation, one can purchase products online showing these images such as posters, t-shirts, caps, and mugs.

In the use of the image of the unborn as a commodity, here again the tendencies towards the infantilising of the unborn and the representation of them as precious entities individuated from the maternal body are evident. Images of the unborn are aestheticised in these representations. Part of this process is removing these images from the fleshly and messy reality of the womb: physical evidence of the maternal body is again erased from view. Such images contribute to a 'biotourism' approach to the unborn assemblage, in which awe and wonder are inspired by the glorious vistas of inside the uterus, a mysterious and foreign space to which viewers are able to have visual access via these images (Kroløkke, 2010).

Risk and the unborn/maternal assemblage

Over the past few decades, a discourse of risk has grown around the pregnant body, in which the subject 'at risk' is not the woman herself but the unborn entity growing inside her (Kukla, 2010; Lupton, 1999, 2012; Lyerly et al., 2009). The pregnant woman, by monitoring and regulating her own actions, is expected to create a shield of safety around her unborn by preventing any potentially polluting substances to pass into the uterus. She is expected to avoid certain kinds of 'risky' consumption such as eating the 'wrong' kinds of foods, drinking alcohol, or smoking, and to take up other types of consumption deemed to contribute to the unborn's health and development, such as the ingestion of prenatal vitamins. All of these behaviours are part of how a 'good', 'responsible', and 'loving' mother is constructed (Lupton, 1999, 2011, 2012; Taylor, 2000).

In such discourses and practices, there is evidence of a continual slippage between the concept of the 'embryo' or 'foetus' and that of the 'baby'. The unborn are positioned as vulnerable and frequently as in need of protection from the risky habits of their mothers, even as potentially 'abused' by the maternal body in which they are contained (Bell et al., 2009; Hartouni, 1991; Karpin, 1992; Kukla, 2005, 2010; Lupton, 1999, 2012; Petchesky, 1987). In legal discourse foetuses may be rendered as independent 'parentless minors' requiring protection (Hartouni, 1991, p. 28). The concept of the 'foetal citizen', bolstered by the emergence of foetal medicine and the subsequent portrayal of the unborn as patients in their own right, has been used to construct a legal entity that allows foetal advocates to promote such causes as antiabortion and the regulation of pregnant women's behaviour, including forcing medical care upon them against their wishes in the interests of this citizen (Dubow, 2011). In the US, pregnant women have been prosecuted for child neglect, foetal abuse, and even foeticide because they used drugs such as cocaine, amphetamines, or heroin while pregnant. Such cases rely upon the discourse of child protection and therefore constitute yet another way in which the unborn are positioned as already children (Dubow, 2011; Hartouni, 1991; Karpin, 1992, 2010; Ruddick, 2007).

Here the unborn and the maternal bodies are positioned as interconnected, as a joint rather than individual assemblage. However this interembodiment is often positioned as detrimental to the precious unborn body. There is ambiguity and a paradox evident in these approaches to the maternal/fetal body. While some laws position the unborn body as a separate entity from the mother, thus allowing a woman to be sued following birth for injury experienced by the foetus while *in utero*, the very supposition that the actions of the mother caused damage to the unborn assumes that their bodies are connected and inseparable. Legal and biomedical discourses represent this inseparability as a point of vulnerability for the unborn (Karpin, 1992, 2010).

The unborn body in this context is conceptualised as permeable, open to the porous maternal body, shaped and influenced by the mother's consumption of substances such as food, vitamins, or drugs, her exercise habits, the viruses or bacteria she may harbour, and even her moods. Due to the intense focus on the health and well-being of the unborn, the pregnant woman again disappears from view: this time in terms of her needs or desires, which are expected to be subsumed to those of the unborn she is carrying. Pregnant women are represented as the carriers of the precious unborn organism, vessels rather than as individuals in their own right with their own needs and priorities that may not always coincide with those of the unborn.

Conclusion

This discussion has highlighted the contingent nature of unborn assemblages: the ways in which the cultural meanings, discourses, and practices surrounding and directed at these bodies are subject to change. Since the 1980s, feminist critics have critiqued the disappearance of the pregnant woman from visual representations of the unborn (see, for example, Hartouni, 1991; Maher, 2002; Petchesky, 1987; Rapp, 2000; Stabile, 1992). In the contemporary era, as part of the bringing together of a diverse set of practices, technologies, and social actors the unborn assemblage is positioned more than ever as independent and autonomous of the maternal body, and as increasingly infant-like even from the earliest days of development. The deeply relational, interembodied, and liminal nature of the unborn-maternal assemblage (Maher, 2002; Young, 1990) appears only to be highlighted when pregnant women's responsibilities for the health and development of their unborn are emphasised.

Positioning the unborn as already 'babies' has broad implications for how they are conceptualised and treated. Unborn assemblages become humans who are as adorable, cute, innocent, precious, and equally as worthy of protection as are infants. This configuration represents a powerful counter to those who seek to represent the unborn as not having achieved full moral personhood, as do pro-abortion activists or hESC researchers. Is it any coincidence that countries such as the US and the UK have experienced an apparent resurging in conservative political moves to limit abortion choice for pregnant women and in anti-abortion groups protesting outside clinics? Since 2008 various bills have been introduced into some US state legislatures by members of the 'personhood' movement seeking to redefine personhood as beginning at the moment of fertilisation of human gametes and thus contesting abortion as well as some forms of contraception, IVF treatment involving the freezing or disposal of embryos and the surgical removal of ectopic pregnancies (Collins and Crockin, 2012). In the UK, various conservative politicians, including the Prime Minister David Cameron and the Health Secretary Jeremy Hunt, have questioned the current legal limit for abortions and suggested the reduction of the current cut-off gestational age from 24 to 22, 20 or even as low as 12 weeks (Childs and Evans, 2012).

The implications for reframing reproduction are profound. The visual disappearance of the pregnant woman's body from the space occupied by the unborn in medical and popular cultural representations has become ever more obvious a phenomenon in the past decade. Paradoxically, this visual disappearance is occurring at the same time as medical and legal discourses insist on the primacy of the pregnant woman's actions in affecting the health and development of the unborn within her, and indeed may even suggest that the unborn entity is trapped within the maternal body. Pregnant women must deal with intense scrutiny from others and a great sense of responsibility for the health and well-being of the precious unborn they harbour within them at the same time as their embodied connection to the unborn is constantly erased from view in medical and popular imagery. Perhaps this explains the emergence of the genre of professional pregnancy photography that shows an ultrasound image superimposed on a pregnant woman's belly, or depicts her holding an ultrasound photograph in front of it. These images may be interpreted as attempts to bring the pregnant body 'back in' to the gestational scene for the woman involved, making her feel as if she is 'part of the picture' and reconnecting her to her unborn in a benign way.

Acknowledgements

A portion of the material appearing in this chapter has also been published in my book *The Social Worlds of the Unborn* (2013, Palgrave Macmillan).

Notes

1. See my Storify on initial social media and news reactions to the pregnancy announcement: http://storify.com/DALupton/cultural-portrayals-of-the-royal-foetus.
2. I use the term 'unborn' here to denote any type of human organism produced from the union of human gametes at any stage of gestation. The term 'unborn' may suggest that this organism is destined for birth eventually. However in some cases, such as in abortion or pregnancy loss, and especially in relation to the multitudes of *ex vivo* embryos that are produced for IVF purposes but which may be surplus to requirements, embryos and foetuses do not survive to reach birth and become infants. Any descriptor for denoting such organisms will inevitably carry with it some ideological baggage, but I have decided upon this term in the absence of any less-charged term and in the interests of avoiding continual use of the more cumbersome terminology of 'embryos and foetuses'.
3. See my Pinterest board 'The Ultrasound as Cultural Artefact' (http://pinterest.com/dalupton/the-ultrasound-as-cultural-artefact) for examples of these uses of obstetric ultrasound images.
4. See my Pinterest board 'The Sociology of the Unborn' (http://pinterest.com/dalupton/sociology-of-the-unborn) for examples.

7
Picturing Postpartum Body Image: A Photovoice Study

Meredith Nash

Introduction

A primary concern for pregnant women often centres on whether they will 'bounce back' to their pre-pregnancy weight and body shape (Nash, 2012a). This is unsurprising given the current expectation in western cultures that women should regain their pre-pregnancy bodies quickly in line with the dictates of normative femininity. These cultural norms are targeted at women during pregnancy and have emanated from moral panics around maternal 'obesity' and the circulation of images of celebrity mothers who have 'bounced back' from pregnancy with little evidence of childbearing left on their bodies (Nash, 2012b). Supermodel Heidi Klum is often named as having mastered the art of 'bouncing back' after she strutted down the catwalk in lingerie looking slender less than six weeks following the birth of her fourth baby (People, 2009). The cultural expectation to 'get your body back' also means that women are expected to be *active* in reclaiming their pre-pregnancy selves. In doing so, they are 'celebrated as successful, powerful women—women to be emulated, admired, and envied' (para 4). The celebration of slender postpartum bodies in popular visual culture has encouraged a post-feminist view that regaining control of the body following childbirth is 'empowering' and that it should be a goal for all mothers (Nash, 2012a). As I have written elsewhere, the language of empowerment surrounding postpartum body norms is problematic because it presents women with a 'third shift' of work that further entrenches them in feminine body projects (Nash, 2011).

This chapter presents an opportunity to contribute to and extend feminist and visual sociological scholarship by examining how a sample of pregnant women in Australia documented their postpartum embodied experiences through digital photographs. I argue that photography is powerful in helping women to articulate the ways in which subjectivities and bodily boundaries are reframed in the postpartum period. One aim is to identify to what extent women's self-produced photographs accord with dominant cultural ideologies surrounding postpartum body norms and how we may read this, using feminist perspectives. Throughout this chapter, women's individual embodied experiences of post-pregnancy are reflected in the production and viewing of their own photographic images.

Postpartum embodiment

There is a significant body of interdisciplinary scholarship that examines women's body image over the life course (Grogan, 2008). Several studies have addressed postpartum body image in the West using quantitative methods (e.g. Clark et al., 2009; Jenkin and Tiggemann, 1997; Jordan et al., 2005) and focussing on affluent women. These studies have suggested that body image dissatisfaction increases postnatally and tends to heighten between six and nine months postpartum, and it is often correlated with weight following birth. As Rallis et al. (2007) have argued, bodily dissatisfaction may also be due to the fact that women feel like they no longer have an 'excuse' to be 'fat' following childbirth (Pauls et al., 2008). Several studies have identified factors that may influence postnatal body image such as poor mental health including depressive symptoms, dietary restraint, and physical comparison tendencies (Gjerdingen et al., 2009). Additionally, poor postpartum body image has been associated with decreased sexual satisfaction and disrupted marital relationships (Ogle et al., 2011).

Qualitative studies of postnatal embodiment tend to provide richer accounts of women's *lived* experiences (Ogle et al., 2011). Using in-depth interviews, these studies have provided a glimpse into women's anxieties around managing their bodies that feel unfamiliar. Qualitative studies have also flagged the pressure placed on women to 'bounce back' postpartum. Interviews with US women revealed that they were engaged in dieting and exercise regimens in order to regain control of their postnatal bodies (Upton and Han, 2003). A Swedish study noted that women often felt 'like they could hardly cope' with postpartum bodily changes and that they were most dissatisfied with the looseness of their vaginas and alterations to the size of their breasts (Olsson et al., 2005, p. 383). Some women in this study viewed plastic surgery as a means of returning to a more familiar body (ibid.). The 'mommy makeover' (breast augmentation, liposuction, and abdominoplasty) has been discussed by feminists as a disturbing but growing trend (Abate, 2010).

Postpartum bodies in visual culture

Pictures of postpartum bodies in western visual culture are largely confined to celebrities. This genre of imagery is predicated on cultural discourses that implore that women's bodies must always be slim and disciplined, even during childbearing (Nash, 2012a). The fascination with celebrity mothers in the West has been linked with the 1991 *Vanity Fair* cover photograph of US actress, Demi Moore, naked and pregnant (Tyler, 2011). Moore appeared on the cover of *Vanity Fair* again in a 1992 feature entitled, 'Demi's Birthday Suit'. Moore appears to be wearing a man's suit but upon closer inspection the viewer realises the 'suit' has been painted on her taut postpartum body. Moore's postpartum picture communicated that 'the pregnant belly has here become a fashion accessory, to be donned for a certain time and then taken off' (Matthews and Wexler, 2000, p. 204). These photographs made Demi Moore the figurehead for a movement centred around pregnant/postpartum beauty, facilitating the more recent rise of the 'yummy mummy' (Nash, 2012a).

A 'yummy mummy' is an affluent woman who 'can squeeze into size six jeans a couple of weeks after giving birth, with the help of a personal trainer' (McRobbie, 2006, para 1). 'Yummy mummies' are regularly featured on women's magazine covers showing off dramatic postpartum weight losses. Yet it is no longer acceptable for celebrities to 'just' have 'fit' postpartum bodies – they must engage in a range of neoliberal body projects (exercising, dieting, breastfeeding, etc.) in order to '"bounce back" more quickly to a new, even better body' (O'Brien Hallstein, 2011, p. 117). Photographs typically accompany media articles about celebrity postpartum weight loss, reinforcing the view that there is an 'ideal' postpartum body type/shape (Gow et al., 2012; Us, 2012). Images of celebrity mothers 'debuting' their postpartum bodies in bikinis communicates a message not only about slenderness and discipline, but also of 'sexiness' (O'Brien Hallstein, 2011, p. 123). Most media articles that document celebrity postpartum weight loss do not engage with celebrities' feelings of body dissatisfaction which further 'idealises' celebrity motherhood (Gow et al., 2012). There is rarely any discussion of the entourage of staff (e.g. gym trainers, chefs, nutritionists, chauffeurs, and nannies) that support a celebrity's postpartum body work.

When celebrity mothers do not attempt to regain pre-pregnancy beauty, they are vilified in the media for failing at performing femininity 'correctly' and their postpartum bodies are labelled as 'fat'. One blogger counted 109 news articles scrutinising US actress Jessica Simpson's attempts at postnatal weight loss (Stewart, 2012). In conjunction with viewing images of slender postnatal celebrities, everyday mothers are encouraged to act on their aspirations to 'get their body back' using the advice found in pregnancy fitness magazines which are premised on controlling the negative effects of childbearing on the body (Nash, 2011, 2012c). In line with this 'third shift' of body work, products such as postnatal corsets and stretch marks creams, and diet and weight loss programmes, are sold to women in order to help them 'manage' their unruly postpartum bodies.

Although pregnant bodies (Matthews and Wexler, 2000) and breasts (Ayalah and Weinstock, 1979) have been documented photographically since second wave feminist movements, the same attention has not been paid to postnatal bodies. However, images of post-pregnancy bodies *do* exist outside of the realms of the media. 'Mommy' blogs have become an important space for women to critique the normalising tendencies of post-pregnancy visual culture using their own photographs (Lopez, 2009). Started in 2006 by US mother Bonnie Crowder, 'The Shape of a Mother' (SOAM) (www.theshapeofamother.com) was one of the first 'mommy blogs' to focus on postpartum embodiment. Crowder began the blog because she wanted to put more 'realistic' images of post-pregnancy bodies in public. The site features mainly middle-class, white women's self-produced photographs of their postpartum bodies along with descriptions of their feelings about their bodies and other statistics (e.g. age, number of pregnancies, pregnancy weight gain, and number of weeks postpartum in photograph). The site is categorised into posts about the body (bellies, breasts, plus sized, etc.) and reproductive experiences (caesareans, infertility, child loss, etc.). The posts from women on the site demonstrate ambivalent feelings about their postpartum bodies ranging from disappointment ('I wish we could afford a tummy tuck') to pride ('My body surprised me ... now I'm very happy about my looks'). Reader comments are moderated to maintain a supportive space (Husbands, 2008).

The diversity of postpartum bodies featured on SOAM is amazing and confronting. The photographs powerfully communicate the realities of postpartum embodiment for everyday women. However, most photographs on SOAM have been taken in a mirror with women's heads cropped out of the frame.[1] The headless images make it difficult to 'see the whole woman and her experience of pregnancy'. The accompanying stories and narratives that women write are essential in bringing the individual self back into focus on SOAM (Serfaty, 2003). In revealing details about their postpartum bodies, the women featured on the site imagine/assume that they are 'speaking' to a like-minded community of women (De Laat, 2008). SOAM has become popular as result of the processes of identification that occur between site contributors and the readers – readers learn about themselves and their own postpartum bodies as they examine the photographs and narratives of other 'real' women.

Theorising photography and selfhood

This chapter is grounded conceptually and theoretically in the belief that photographs represent socially and culturally specific ways of seeing individual experiences of pregnancy. The intersection between 'memory' and photography has been the subject of interdisciplinary inquiry (Barthes, 1972; Bourdieu, 1990; Sontag, 1977). This scholarship has enabled critical reflection on the power of photographs in modern life; how images capture aspects of the social world; how they regulate visibility; how they add new dimensions to our memories; and also how we define memories. Such work has also illuminated the relationship between images, spectators, and power (Berger, 1972). What we deem to be 'photographable' in our everyday lives is predicated on collective moral and aesthetic values and inflected by social structure (Bourdieu, 1990, pp. 6–8).

The notion of 'everyday' photography is important in this research because digital photographs have become a regular feature of life in the West (Sarvas and Froelich, 2011). Individuals can now easily capture mundane moments in addition to the 'special' moments in their lives that typically defined 'everyday' photography in the era of the film camera (Bourdieu, 1990). Yet the social regulation of photographs described by Bourdieu (1990) still implies that some subjects are more 'worthy' of being photographed than others as photographs are now subject to even greater scrutiny as they are shared online.

Feminist scholars have focussed on problematising the claim that photography has on the 'real'. Sexual difference in photographs often appears to be based on what is 'seen' and this is problematic because visible differences between male and female bodies appear to be grounded in 'nature' (Evans, 2001). Thus early feminist scholarship centred on how women's experiences have been culturally remembered, how women's bodies have been represented, and how this can be disempowering. Feminists have raised questions about spectatorship and masculinist 'looking' and controlling the gaze through photographic production of images (Mulvey, 1975). Feminist critics have also argued that family photographs taken by women often reflect an attempt to manage appearances and identities, submerging conflict and unhappiness in an attempt to make 'happy' memories of family life and are often mechanisms for *forgetting* the past (Evans, 2001, p. 113). However, Rose (2010) has suggested that women's everyday photographic practices often closely reflect women's domestic lives as they photograph their families and friends. In this way, for feminists, photography is 'an ambivalent and complex field of cultural practice' (p. 9).

Methodology

Data in this chapter are derived from a project in which I used a modified version of the 'photovoice' method to explore pregnant embodiment in Hobart, Australia (2011–2012). Photovoice is a powerful social research method for empowering marginalised groups as it 'enables them to act as recorders, and potential catalysts for change, in their own communities' (Wang and Burris. 1997, p. 369). This method has been applied less often in the context of women's health (Wang, 1999; Wang and Pies, 2004), or in an Australian context (Wilkin and Liamputtong, 2010). With exception to my own research, No studies have described the use of photovoice with pregnant women (see Nash 2013, 2014).

Photovoice comprises three stages that are supposed to 'empower' participants and lead to social change: (1) the creation of participant-produced images; (2) critical reflection by the participants on the process of taking photographs and the photographs themselves using the SHOWeD method of inquiry (a set of specific questions to help with the reflective process)[2]; (3) sharing participant-produced photographs with the community (Wang and Burris, 1997).

Methods

Twelve participants were self-selected using multiple points of contact. In-depth, semi-structured interviews were conducted in 10-week intervals from between 10 and 20 weeks pregnant to after birth: in total, four interviews with each participant. Interviews were audio-/video-recorded with consent. This study was approved by the University of Tasmania's Human Research Ethics Committee. Pseudonyms are used throughout this chapter.

The first meeting was an introductory interview in which details about the pregnancy and the participant's background were discussed. Participants were given digital cameras and they were asked to photograph themselves and their lives during pregnancy and postpartum

(20–24 pictures per month). The only instruction given to participants was to take photographs that reflect their experiences. Participants were also given a prompt with framing questions for taking photographs, if required. During subsequent interviews, I downloaded participants' photographs on to a laptop and we viewed them together. Through a process of photo-elicitation (Harper, 2002) and employing a feminist phenomenological interviewing style (Levesque-Lopman, 2000), I conducted interviews that explored the photographs in relation to the participants' individual experiences, how they chose what to photograph, and how they felt about the experience of taking pictures. I also asked questions about pregnant/postpartum body image, exercise, and clothing. The final stage of this project involved participants in selecting 'significant' photographs to display in a public exhibition.

Data analysis

Analysis of the interviews and the photographs employed thematic analytic techniques (Braun and Clark, 2006). I closely read the transcripts, summarising key points raised by the participants and in relation to their photographs to understand the data in context. I reviewed the transcripts as a set in order to identify common themes. Once a set of themes had been identified, I reviewed the transcripts again to identify material related to each theme. The themes continually evolved and were refined throughout this process. My analysis in this chapter focuses on photographs that represented issues of importance to the participants in relation to postpartum body image.

Participants' characteristics

Participants were between the ages of 24 and 42, with a mean age of 32.5 years. Most women described themselves as Anglo-Celtic, middle-class, and tertiary educated. All of the participants lived with a male partner and half of the cohort lived with children. The majority of women in the study worked in paid employment (42 per cent in full-time work) in professional roles.

Participant-produced photographs

Participants produced nearly 1,500 photographs, with most participants taking between 20 and 30 photographs per trimester of pregnancy and postpartum. The participant-produced images captured the everyday context of women's feelings about their bodies as well as events and emotions associated with their pregnant/postpartum lives. All photographs in this chapter are published with the consent of the participants.

This body is 'not me': experiences of postnatal embodiment

In the fourth interview, all participants expressed that they were less satisfied with their postpartum bodies compared to their pregnant bodies. Most women expressed the view that their bodies felt unfamiliar or 'not me' and this was coupled with a fear that they would never return to a 'normal' embodiment. Participants understood that returning to a more comfortable former embodiment was critical in reclaiming their pre-pregnancy identities.

The six first-time mothers were surprised and anxious about postpartum bodily changes, particularly in relation to body size/weight, 'leakage' (of breast milk), and physical tiredness[3]:

> I don't know what I expected. They all say it takes months and months for [your body] to get back to normal but I don't know, maybe I just think that I'm immune to these things. (Priscilla, eight weeks postpartum)

> I haven't lost the [pregnancy] weight and it's just terrible and I feel like – why does pregnancy have to make you gain weight? ... It's really, really hard and you're not meant to go on a diet when you're breastfeeding ... I thought that with breastfeeding I could lose weight ... I don't know, maybe I should breastfeed more. (Julie, 10 weeks postpartum)

> I felt at one point that I had zero control over what my body was doing. I was trying to think of something I could take a photo of and the only thing I could think of that came remotely close was like a 'for lease' house sign. I didn't expect it – it was a really full on feeling. (Kirsten, 12 weeks postpartum)

> I'm still not where I was before I was pregnant and I just wonder when that's going to happen. Maybe I'm being a little impatient but I want my body back! A lot of people say that breastfeeding will help you lose weight. But when is it going to happen for me? (Kaz, six weeks postpartum)

These extracts reflect the collective ambivalence surrounding postpartum embodiment among first-time mothers. Most women did not realise the extent to which childbearing enacts *profound* bodily changes, both physical and emotional. First-time mothers were frustrated that they could not pinpoint when they would regain control of their bodies. When their bodies did not meet cultural expectations, they expressed guilt about their inability to 'get their bodies back' quickly. Julie and Kaz drew on wider cultural discourses and public health campaigns that emphasise that women should breastfeed because it will help them to lose weight postpartum (Department of Health and Ageing, 2009). They were struggling with competing 'normalising pressures' for new mothers to breastfeed and also to maintain a slender body (Crossley, 2009, p. 85).

Given these feelings of uncertainty, for the most part, first-time mothers did not photograph their postpartum bodies. Only one new mother photographed her postpartum belly. Photographs trigger memories of former selves and such processes of remembering appeared to be painful for women who did not want to confront these bodily changes (Harrison, 2002). However, the quotes do seem to suggest that new mothers did think that it was *possible* to reclaim their former selves, but that the process was taking too long.

The six experienced mothers had more 'realistic' expectations of how their postpartum bodies would look.[4] They expected weight gain and changes to body shape as well as periods of feeling out of control. These participants also had a sense of the bodily discipline associated with 'bouncing back' and how difficult it was to undertake this body work while looking after children. Most of these participants were aware that their bodies would never be the 'same'. However, this awareness did not necessarily mean that they were *satisfied* with the appearance of their bodies. Nevertheless, they were more willing to photograph their bodies during the postpartum period compared to the new mothers. Five of the experienced mothers photographed their postpartum bodies, generating a total of 14 photographs.

> I turned 43 [when the baby was born]. I can't expect my body to be fabulous at this time. I feel – whereas probably when I was younger I did expect to be able to do more. But I didn't – I wish I hadn't let myself go. (Marjorie, mother of five children, 10 weeks postpartum)

> I hate my belly obviously because it's all floppitty like rice paper but [my son] is only three months old. I have to just remember that. (Candice, mother of three children, 12 weeks postpartum)

> I'm well aware I've got at least five or six kilos I need to move. But, yeah, I'm feeling pretty good [postpartum]. But I know I am a bit flabby in the legs and the bum. (Lena, mother of two children, eight weeks postpartum)

These extracts speak of the discomfort that the group of experienced mothers felt about their changed appearances. However, their accounts reflect less frustration about the uncertainty associated with 'getting their bodies back'. These women had a different relationship to their bodies than the new mothers and the photographs that they took of themselves became part of 'a shifting collage [of pictures and selves] which is produced by and within the activities of the present' (Slater, 1995, p. 139). Their willingness to photograph themselves also suggests that the experienced mothers might have been actively challenging the cultural view that postpartum bodies should be hidden from view. Self-photographing may have 'freed' these women of posing in front of the camera with the intention of 'being pretty' (Lee, 2005).

Zoe took five photographs of her postpartum body. Twelve weeks after the birth of her third child, Zoe related that she was more distressed about her postpartum body than she had been previously:

> This time I'm really upset about my body. My belly button's moved, everything's changed, everything's in a different spot to what it was and it's not going away like it did with the other two [pregnancies] so I feel a lot more negative. I'm never going to wear a bikini, or I never plan to, but it's just thinking everything's all twisted out of shape. I suppose I should be grateful for [my body], it's a small price to pay for another child, isn't it?

Given the proliferation of negative medical and cultural metaphors that are used to characterise the experience of childbearing (Martin, 1992), it is unsurprising that Zoe felt like her body no longer 'fit in' with norms for femininity. Yet Zoe was careful to give me the impression that she was 'obsessed' about her appearance. When she said, 'I suppose I should be grateful for [my body]', she seemed to imply that having a changed appearance was acceptable given the functional purpose of a maternal body. By invoking essentialist discourses, Zoe was able to absolve herself of responsibility for her changed postpartum body. However, as Ussher (2006, p. 91) has argued, the adoption of such discourses can rob women of agency – it makes them feel as though they have little control over their bodies and, in turn, they can feel alienated from them. Alternatively, it is possible that Zoe thought that 'pride' in her postnatal body was the 'correct' reaction to perform for me because, in Australian culture, 'good' mothers are represented as being focussed on their children and not themselves (Nash, 2012a). Given the current western cultural climate in which pregnant and post-pregnant bodies are subject to public scrutiny, Zoe's narrative suggests that she was attuned to this but that she also wanted to contest the cultural view that pregnancy 'ruins' a woman's body forever.

Zoe's comments about her postpartum body were made in light of Figure 7.1, an image that she wanted to send to SOAM following our conversation about the website in a previous interview:

> I thought I could do my own photo for SOAM ... [The website] makes you feel a little bit better about yourself, not from looking at other people's [postpartum bodies] but I found from writing about my post-baby body. It just made me appreciate it a little bit more rather than the everyday thoughts you have [that] are very negative.

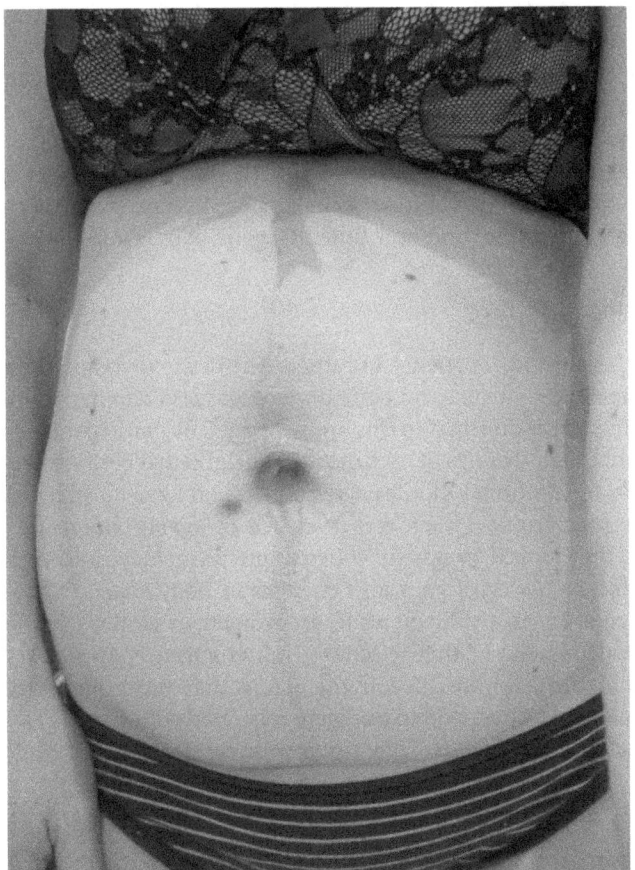

Figure 7.1 Zoe

Zoe observed the 'magical' abilities of celebrities to 'bounce back' from pregnancy with little (visible) effort: 'You just see the end product and think how did they do that? How come I can't do that?' Taking photographs of her own body appeared to be a way of resisting discourses that implore that *all* women look slender and toned post-birth. By photographing her caesarean scar, stretch marks, and puckered skin, Zoe made herself appear in direct contrast to contemporary popular representations of the slender and often air-brushed postpartum female body – an act of social power through visibility. Importantly, in this moment, Zoe was in control of the production and dissemination of an image of her body and its accompanying narrative.

However, in line with the photographs featured on the SOAM website, Zoe did not include her face in her photograph (Figure 7.1). When one includes their face and head in a photograph as a subject, one opens themselves up to scrutiny by a viewer. A photograph of a postnatal belly without an accompanying head transforms into a picture of a postnatal belly, and not a woman. The picture becomes 'not me'. Aside from concerns about anonymity associated with posting images on a public website, Zoe's 'not me' photo appeared to be deliberate given her concerns about her appearance and how unfamiliar her body felt following birth.

Dressing

For both new and experienced mothers, the desire to 'bounce back' to a more stable (pre-pregnancy) embodiment was also an effect of practical concerns such as dressing. In this liminal stage, my participants explained to me that their feelings of 'not me' were heightened because they could not wear their 'normal' clothes. Dressing was no longer a straightforward exercise – maternity clothes were described as 'too big' and made women feel like they looked 'pregnant', whereas pre-pregnancy clothing was still too tight given weight gain and changes in body shape (Nash, 2008). Post-pregnancy is a difficult time for women to clothe themselves as more is demanded of their clothes given changed body size as well as the requirements of breastfeeding (e.g. tops that make the breasts easily accessible). All of the participants were frustrated with the lack of choices they had in terms of what they could wear comfortably.

> I only have four outfits now that I can wear out. (Julie)

> That's what my wardrobe is whittled down to [referring to Figure 7.2]. I like I feel I'm a dag all the time because I've got so little clothes that I feel like I can comfortably breastfeed in. (Judy)

Picturing Postpartum Body Image 127

Figure 7.2 Judy's closet

> I reckon 75 per cent of that stuff [in my closet] I can't wear. That box in the corner is ... what I wear every day to hide my tummy. (Zoe)

Clothing is an important social symbol in which information about identity is communicated to others (Tseelon, 1997). For my participants, the inability to fit into their 'normal' pre-pregnancy clothing signalled a failure of the postpartum body and reinforced a loss of identity. Whereas most women in the study wore tightly fitting maternity clothes, post-pregnancy, the women wore looser clothes that accommodated their larger bodies and also covered them up.

Certain types of post-pregnancy clothing were viewed to visually express motherhood, specifically comfortable trousers (e.g. leggings, track pants) and slippers.

> Tracky daks[5] and slippers seem to be the story of my life. When I'm at home I wear trackies but I tend to wear jeans if I'm going out. I want to look presentable when I go out. (Lena)

> I've never owned so many pairs of track pants ... That was my wardrobe for probably six or eight weeks ... I have never owned slippers and now I own them and I wear them ... I don't feel like I've made an effort when I put them on and I don't feel like I've gotten dressed for the day so that makes me feel a bit yucky then. If you had a day at home where you weren't doing much, you would feel it would be really nice to put on your track pants and comfy tops and just slouch around, but wait until it's all the time – it's not such a relief. (Zoe)

Figure 7.3 Zoe's track pants

Track pants and slippers presented images that were discontinuous with their pre-pregnancy identities and these items of clothing made women feel less confident. These sentiments correspond with Guy and Banim's (2000, p. 318) understanding of the discourse 'The Woman I Fear I Could Be', which draws together recollections and current anxieties that women reported about their clothes. This discourse is relevant here because it reflects my participants concerns over not being able to use clothing in the same ways to manage identity as in pre-pregnancy.

Lena and Zoe seemed to be conflicted about what they wanted from their clothing during the postpartum period. From one perspective, comfortable clothing was essential to accommodate the new demands being placed on their bodies as mothers (e.g. weight gain, breastfeeding, and sleepless nights). From another perspective, comfortable clothing was worn by women in the study when they 'felt fat' (Kwon and Parham, 1994). Comfortable clothing is generally loosely fitted and does not highlight the contours of the body nor does it enhance sexual attractiveness. Thus such clothing contradicts norms for feminine dress and accords with an image of an asexual maternal body.

However, track pants and slippers are items of clothing that are not typically worn outside of the home. Several scholars have acknowledged that clothing reflects one's occupation and social status and it appeared that Zoe and Lena felt that this clothing did not appropriately reflect their social statuses as middle-class working women. Zoe implied that track pants and slippers embodied 'lazy' dressing or a negative image of a stay-at-home mother. There was a class dimension to these comments – wearing track pants and slippers seemed to embody the image of a British working-class 'chav mum' described by Tyler (2008).

Women were also conflicted about wearing larger sized clothing postpartum. Living in a culture that prizes thinness (even during childbearing), my participants felt threatened by the prospect of wearing larger sized clothing because of what it meant for their social identities. 'Fat' female bodies in the West are often associated with laziness, asexuality, and a lack of discipline (Murray, 2005). Having a larger body places women on the margins of femininity.

Throughout their pregnancies, both Lena, aged 34, and Kaz, aged 36, had relatively positive views of gaining weight. However, by the postpartum interview, Lena and Kaz spoke to me about their bodily dissatisfaction. This dissatisfaction appeared to be located, in part, in the realisation that both women now had to wear larger sized trousers.

> These are my $9 Big W jeans. Perfect for those people that hope not to be in that size for ages. I've never been a size 16.[6] I don't think

Figure 7.4 Lena's jeans

> I look like a size 16 but obviously I am a size 16 ... obviously I'm not looking as good as I thought around the tummy. (Lena, eight weeks postpartum)

> Suddenly the fashion that I'm used to wearing, I just can't wear. [It] makes me feel like it's somebody else's body. I went to the op shop yesterday and bought these jeans – size 14.[7] My first pair of size 14 jeans. I put on the jeans but then I put on my belt and my belt didn't do up! (Kaz, six weeks postpartum)

The purchase of a larger sized pair of jeans proved to be a confronting moment for these women as they realised that their images of themselves did not match their physical selves. The larger sized jeans served as material evidence of time passing and of their bodies changing. In Figure 7.5, Lena uses a mirror to reflect upon her multiple selves and the image became a powerful retrospective memory aid. Lena used the mirror in her photograph to 'see' her body in new ways. Lena's memory work, on this occasion, called into question her previous memories of herself and instead produced 'counter-memories' of who she was and how she appeared to herself and others – an unexpected outcome. Both women felt that they had transformed into unrecognisable selves and the larger sized jeans represented a physical and mental challenge in reclaiming former selves or 'bouncing back'. Kaz and Lena had not invested in their 'new' bodies – both women went to discount stores and spent less than $10 on their new jeans, which indicates that they saw their 'new' larger bodies as 'temporary' (see Figure 7.5).

Figure 7.5 Lena wearing her size 16 jeans

But why were my participants worried about wearing larger sized trousers if they knew that it was only 'temporary'? A large female body is a 'reminder of motherhood' (Adam, 2001, p. 41). Kaz drew on this cultural stereotype when I asked her about her size 14 jeans:

> I feel like my body makes me look like just a mum ... there's something about being heavy and frumpy and not able to fit into my usual clothes that makes me feel like I'm not me anymore. I'm scared about losing that identity ... I associate a big body with ... losing the lightness of being a small person. That's how I was. I've had a pretty interesting intellectual life and I always have been a smaller person. So that life is not there anymore. I've had to give up that part of my life to make sure that [my baby] is nurtured. Day in and day out its nappy changes and feeding, washing nappies, changing clothes, bathing ... It gets a bit boring and dreary.

Wearing larger sized jeans was material evidence that Kaz had transitioned into a different phase of life – motherhood. She was coping with not only the 'loss' of her 'normal' body but also the 'loss' of her professional identity. The 'loss' of her 'old' body pushed Kaz to contemplate the conflict between her current/future self as a mother and her former identity/embodiment as a slender working woman. Kaz's comments resonate with the narratives of the US participants in Upton and Han's (2003, pp. 683–684) study. They found that postpartum women struggled to regain not just their own body but an 'ideal' body. This desire competed with the cultural message that pregnancy changes a woman's body forever. Upton and Han point out that 'getting *the* body back' should really be read as 'getting *a* body back' (2003, p. 686). As Kaz's extract reveals, women push themselves to achieve an idealised role or body type that is not limited to the actual appearance of their body but that also reflects how they feel about themselves and how they cope with the demands of motherhood and everyday life.

Discussion

The narratives discussed in this chapter demonstrate that post-pregnancy was a time of uncertainty and ambivalence for my participants. Both first-time and experienced mothers were, for the most part, dissatisfied with the appearance of their postpartum bodies and their photographs and narratives reveal important information about how they struggled to cope with their bodily changes. As noted, the women spoke about their post-pregnancy embodiment through a discourse of 'not me' to emphasise the unfamiliarity and uncertainty they felt in their 'new' bodies. 'Bouncing back' was viewed as a way to resolve these feelings. The photographs that they took also aligned with this view. Dressing was a key moment that brought the reality of their 'not me' embodiment to the surface in their daily lives. Being able to fit into their pre-pregnancy clothing was viewed to be a sign of bodily success – that they had reclaimed their pre-pregnancy selves. Wearing larger sized clothing was difficult for women who did not see themselves as 'large' and they had to reconcile themselves to a temporary wardrobe of 'not me' clothing.

This chapter has also demonstrated that digital cameras were tools that allowed women to portray themselves and their postpartum experiences uniquely. Several participants used photographs to create self-portraits using specific poses and framing in order to define the photograph and its meaning in relation to their embodied experiences and in contrast to representations of post-pregnancy in popular visual culture. One of the unresolved tensions surrounding the participant-produced photographs was the tendency for some participants to 1) represent their postnatal bodies without a head in photographs, or 2) to avoid taking photographs of their postpartum bodies. The headless photos and the avoidance of taking photos may be attributed to the public nature of this project and the possibility of having photographs exhibited at a later date. Alternatively, as my participants were feeling ambivalent about their postnatal bodies, it is unsurprising that some women would be reluctant to be portrayed in this way. A photograph of a belly without a head becomes 'not me'. For Zoe, the 'not me' photo was deliberate given her anxieties about feeling like she had an unfamiliar body. In this way, the headless photo is an example of how women brought their embodied experiences into a Cartesian dualist framework when they were unsure of how to cope with bodily changes.

My participants' ongoing concerns over appearance and 'getting their bodies back' and wearing the 'right' clothing also reflected changing norms for post-pregnant embodiment. In postmodernity, women are becoming mothers in the midst of a changing cultural landscape (described in the Introduction to this book; see also Nash, 2012a). While my participants discussed their inability to conform to such high standards of postpartum bodily maintenance, undeniably they also felt pressure to conform to contemporary models of postnatal femininity that emphasise slenderness. My group's visual and narrative accounts of trying to manage their postnatal bodies point to a 'third shift' of bodywork in which women must labour (literally) in order to maintain a body that is in line with normative feminine ideals (Dworkin and Wachs, 2004, p. 114; Nash, 2011, 2012c).

Participant-produced photographs invite us into a world of experiences and emotions in post-pregnancy that is normally concealed from view. The photographs that my participants took of themselves are valuable because they stand in contrast to popular representations of post-pregnancy. It was often the mundane images of track pants, barren wardrobes, and self-portraits in a bathroom mirror that provided the most unexpected insights into women's transitions to motherhood.

Acknowledgements

Thanks to Eliza Burke and anonymous reviewers for feedback on this chapter.

Notes

1. The headless photos may be attributed to the public nature of the website and the need for anonymity.
2. The questions that spell the acronym **SHOWeD** include: 'What do you See here? What is really Happening here? How does this relate to Our lives? Why does this situation, concern, or strength exist? What can we Do about it?' (Wang, 1999, p. 189)
3. 'First-time mother' or 'new mother' refers to participants who gave birth for the first time during the course of the study.
4. 'Experienced mother' refers to participants who already had children.
5. Australian colloquial term for tracksuit pants.
6. AU size 16 is equivalent to UK size 14 or US size 12.
7. AU size 14 is equivalent to UK size 12 or US size 10.

8
'My Doctor Told Me I Can Still Have Children But ... ': Contradictions in Women's Reproductive Health Experiences after Spinal Cord Injury

Heather Dillaway and Catherine Lysack

Introduction

Women who have sustained a traumatic spinal cord injury (SCI) are sexually active, menstruate, become pregnant, carry healthy babies to term, have a range of childbirth experiences, use birth control, experience menopause, and have the same risks for reproductive diseases, illnesses, and conditions as able-bodied women. In other words, disabled women can have full reproductive lives. Yet we know relatively little about how women with SCI think about and experience 'normal' reproductive processes within the context of a permanently impaired body. Using data from 20 in-depth interviews with US women with SCI, this chapter describes the impairment-related contexts and social barriers that women with SCI confront in relation to their reproductive capacities. As the data show, knowing only that disabled women are diagnosed as 'capable of getting pregnant' or that they have unique health conditions after SCI tells us little of their lived experiences of reproduction. Our data reveal that the everyday experience of reproduction for a woman with an SCI is much more complex and uncertain than her 'normal' reproductive capacity might suggest.

Women and SCI

The focus on women's reproductive issues after SCI is relatively new, in part, because SCI primarily happens to men. Approximately 270,000 individuals in the US currently have a SCI and an estimated 12,000 new injuries occur each year; however, 81 per cent of all injuries occur to males (National Spinal Cord Injury Statistics Center [NSCISC], 2013).

Approximately 70 per cent of SCIs occur to males worldwide, so women have a greater percentage of SCIs outside of the US (Rutberg et al., 2008). SCI is a devastating injury (Dijkers, 2005); some persons lose the ability to use their legs and lower body (paraplegia) while others lose this ability from the neck down (tetraplegia). Complete or partial motor paralysis necessitates lifelong wheelchair use and coping with a range of serious medical complications (Jensen et al., 2007). In effect, they must deal with the impairment and the physical effects of the impairment at the same time as the social barriers created by an ableist society that defines their bodies as 'abnormal' (Hughes and Patterson, 1997; Morris, 2001; Shildrick, 2002; Thomas, 1999).

Furthermore, women with SCI must confront their 'double disadvantage' of being women *and* disabled (Deegan and Brooks, 1985; Lloyd, 2001; Shildrick, 2002; Thomas, 1999; Wendell, 1996), facing prejudice and discrimination because of both social locations. For instance, disabled women are often stereotyped as 'asexual', 'celibate', 'non-reproductive', or 'infertile', and not mothers (Becker et al., 1997; Nosek, 2000; Welner, 1999; Wendell, 1996). Sexual beings and mothers are assumed to be able-bodied while disabled bodies are assumed to be broken, childlike, and incapable of 'normal' activities (Wendell, 1996; see also Shildrick, 2002). This has serious consequences for disabled women's health. McColl (2002) describes the health challenges facing women with SCI as a precarious 'house of cards'. The most serious problems include a lack of primary and preventative health care in basic areas of reproductive health (e.g. pelvic exams, Pap smears, mammograms). The reasons for this include physicians' lack of knowledge about the impact of impairments on reproductive health (Reitz et al., 2004), and a lack of sensitivity to women's health needs (Becker et al., 1997; Brandeis, 2003). Kaplan (2006) details how even a simple gynaecologic exam may be difficult to obtain because many physicians will never have (or, rather, assume they will never have) patients with disabilities in their practice and lack the appropriate office equipment or staff.

Unique reproductive health experiences?

Many questions remain about how disabled women make decisions about reproductive health within a context of limited medical knowledge and public stereotypes. There is some evidence that women with SCI face higher risks in pregnancy and delivery because of their higher risk of diabetes and high blood pressure (Nosek et al., 2001); thus some impairment-related contexts may shape reproductive experiences. The rates of forceps use, vacuum extraction, and caesarean section are also higher for disabled women, but data are largely anecdotal (Jackson and Wadley, 1999). It is unclear why these interventions are utilised more often with disabled women, but the reasons may be related to doctors' lack of knowledge about how to deal with disabled birthing bodies. In addition, a temporary period of amenorrhea is probable immediately after SCI, but we do not know how this amenorrhea affects disabled women's fertility (DeForge et al., 2005; Dillaway et al., 2013). DeForge et al. (2005) published the largest systematic search of the literature on fertility to date. They found that no study between 1996 and 2003 focused on the fertility of women with SCI – all focused on men. Furthermore, there is a dearth of research on disabled women's everyday experiences of fertility, and women's actual contact with doctors, doctor's offices, and lay individuals about their reproductive capacities and reproductive health care. We also rarely hear the voices of disabled women as they experience fertility/infertility.

Existing research on women's reproductive health after SCI is often written *in response* to myths about the 'non-reproductive' and 'asexual' nature of disabled women. For instance, much of the existing research has simply argued that women *can* still become pregnant after SCI (see American Association of Clinical Endocrinologists [AACE], 2006; Welner, 1999). Other research seems to be generated purposely to inform health care providers about special screening and treatment issues that arise from the effects of physical impairment (e.g. bladder management, labour management, osteoporosis, cardiac problems, or lack of sexual feeling) or the risks of particular birth control methods (DeForge et al., 2005; Estores & Sipski, 2004; Nosek, 2000; Welner, 1999). This research reinforces the definition of women with SCI as different and 'abnormal', and stands in contrast to research that documents the 'normality' of women with SCI (Shildrick, 2002). Thus, a tension exists in this small body of literature: Do women with SCI have 'normal' or 'abnormal' reproductive health experiences? Overall, we know little about the embodied experiences of reproduction among disabled women (Turner, 2001). We also know little about how a 'normal'/'abnormal' dichotomy might limit disabled women's reproductive options and experiences (Shildrick, 2002). By focusing on their 'difference' alone, we forget that disabled women are living full reproductive lives that are affected but not completely determined by their physical impairments (Shildrick, 2002). The purpose of this chapter is to address these gaps in knowledge, drawing on in-depth interviews with US women with SCI.

Conceptual framework

The literature on the sociology of the body and within disability studies can be useful in expanding how we conceptualise disabled women's reproductive experiences. Merleau-Ponty (1962, p. 206) has observed that the 'lived body' is a location of meaning and identity as well as a material entity: 'We are in the world through our body, and ... we perceive that world within our body.' While the objective, physical body can be an object of intellectual inquiry, so can the 'subjective body of personal experience' and also body image as individuals negotiate 'identities, experiences and social relationships' while living in/with particular bodies (ibid.). Turner (2001, p. 253) has also argued that it is necessary to examine the production of bodies within society – i.e. the social rules, constraints, and barriers that structure embodied experiences. In looking at the latter, Turner suggests that this is where we find that bodies are 'vulnerable', 'contingent', and socially created (not 'natural') (ibid.).

The disability studies literature expands on sociological approaches to embodiment by 'separating out "impairment" (that is, the functional limitations of our bodies and minds) from "disability" (that is, the disabling barriers of unequal access and negative attitudes)' (Morris, 2001, p. 2). Hughes and Patterson (1997) and Thomas (1999) further remind us that we need to pay attention to embodied experience of impairment while also paying attention to the social barriers (inequalities) associated with impairment. As Thomas (1999) explains, impairment is not a fixed property of a person but, rather, a 'social relational' entity, which means that how one lives with a physical impairment is more telling about what an impairment truly is for a person than any material reality of the body. Societal definitions of impairment matter in shaping disabled women's embodied experiences (Shildrick, 2002).

Shildrick (2002, p. 81) has proposed that physically impaired bodies represent 'an openness and vulnerability that western discourse insists on covering over' – a 'monster' that 'haunts' us but can never be completely 'expelled'. Disabled women's bodies, she argues, are liminal and exist at the nexus of 'normal' and 'abnormal' (ibid.). Women's reproductive bodies have been similarly theorised through a framework of monstrosity and liminality in their blurring of corporeal boundaries. Bodies that lack dexterity control, that use physical space differently than socially expected, that leak, or that grow and produce life are all suspicious in their ability to cross over socially constructed boundaries of what the self or subject is.

Method

This chapter is underpinned by a feminist disability perspective. A feminist disability perspective emerges from the premise that disabled women are *both* women *and* disabled persons and, therefore, both gender and disability shape their thoughts and experiences (Deegan and Brooks, 1985; Lloyd, 2001; Wendell, 1996). Disability can shape the way in which women might experience gender, and vice versa, in that gendered meanings and experiences are filtered through the context of having a disability as much as the meanings of and experiences of disability are filtered through their gendered social context. In line with Shildrick (2002), we steer away from a definition of the disabled gendered body as 'abnormal' or 'monstrous' (when compared to male, 'able', or non-reproductive bodies) and instead allow disabled women to define their own lived experiences of reproduction. Nonetheless, we acknowledge the importance of the 'normal'/'abnormal' dichotomy in defining women's embodied experiences of disability and reproduction, and seek to understand how and when disabled women's reproductive experiences both accommodate and disrupt this dichotomy.

Second, we align ourselves with feminist scholarship that has expanded conceptualisations of reproduction (Ginsburg and Rapp, 1991; Rich, 1977). In postmodernity, women's 'reproductive' experiences include *more* than just conception, contraception, pregnancy, and birth. Reproductive experiences also include: menstruation and menopause, contemplation of fertility, use of and problems with contraceptives, negotiating body image (e.g. asexual and non-reproductive images in the case of disabled women), and making informed choices and seeking health care in any of these instances. We are also of the view that reproductive experiences must be viewed as embodied (Shildrick, 2002; Turner, 2001) and social (Ginsburg and Rapp, 1991). In this chapter, we discuss disabled women's interactions with doctors, their feelings about being able to get pregnant, and their contact with asexual, non-reproductive, and non-mother images associated with disability, to continue broadening definitions of 'reproductive' experiences.

Participants

The sample was recruited from an earlier study on independent living after SCI in Detroit, Michigan, USA ('Community living after spinal cord injury: Models and outcomes'; R01#1HD43378, funded by the National Institutes of Health; PI: Lysack). All participants granted permission to be contacted for future research and, therefore, the list of participants from the earlier study became the sampling frame for the current study. After multiple attempts, we contacted 27 of the 44 women (61 per cent) who participated in the original study by telephone. Twenty-four (88 per cent) agreed to participate in our study on 'reproductive health and SCI', and we completed in-depth interviews with 20 women (83 per cent).[1] Ethics approval was secured for the follow-up study before any data were collected and informed consent procedures were followed according to university regulations. Participants received USD$35 at the completion of the interview.

Sample characteristics

Participants' average age was 46 years (range: 27 to 66 years), and they had been injured for 19.5 years on average (range: 3 to 41 years). Eight of the women (40 per cent) never married, seven (35 per cent) were married, with the remainder divorced or separated (5 or 25 per cent). All women described themselves as heterosexual. Eleven women (55 per cent) had paraplegia, and nine women (45 per cent) had tetraplegia. Automobile accidents were the cause of injury for half of the sample (n = 10), followed by violence (n = 4), pedestrian-automobile crashes (n = 2), falls (n = 2), a sporting accident (n = 1), and an acute spinal tumour (n = 1).

The women in this study had a broad range of reproductive health experiences. Overall, 17 of the 20 women (85 per cent) reported having menstrual experiences at some point after their injuries, although only seven were still menstruating at the time of the interview (because of the ages of women in the sample). Fifteen of the women (75 per cent) reported a pregnancy at some point in their lives, and six women had healthy babies after their injury. Three women lost pregnancies at the time of injury. Six women reported using some form of birth control post-injury. Most women (65 per cent) were in their late reproductive years because they were aged 35–45 at the time of interview, or at the age when the endocrine, biological, and clinical features of perimenopause commence (Freeman et al., 2001). Thus most of the women were looking back over decades of reproductive health experiences pre- and post-injury (see Dillaway et al., 2013).

Data collection

Each woman participated in one face-to-face interview, lasting two to three hours on average. All interviews took place in the women's homes. We employed a feminist phenomenological interviewing style (Brown et al., 2006; Creswell, 1998; Reinharz, 1992; Rubin and Rubin, 1995), meaning that we kept our interviews conversational and 'loose'. A feature of feminist phenomenological interviewing is that the interviews are guided by the participant. However, probing questions arose from interviewees' narratives after general questions were asked (e.g. 'Tell me about X') (Reinharz, 1992, p. 21; see also Creswell, 1998). All interviews included questions about overall health and physical functioning (to explore the effects of impairment), reproductive health attitudes and experiences, contact with health care providers, and reproductive health-seeking behaviours post-injury.

Two research staff conducted the interviews (Lysack and a research assistant). Together, they had seven years of experience conducting interviews with women with disabilities and they received additional training about women's health interviewing from the first author. Training ensured that key topics were recognised, probed, and handled sensitively. Sometimes the best data came at the end of the interviews, when women felt comfortable enough to speak unprompted about reproductive health issues that were important to them (e.g. desires for children, sexual relationships, childbirth experiences, patient-doctor interactions, etc.).

Data analysis

A computerised database was used to track data collection, transcription, coding, and analysis. The interviews were transcribed verbatim and qualitative software (QSR NUD*IST) facilitated coding and data analysis. The two interviewers randomly checked 5 of the 20 transcriptions each against the tape-recordings for accuracy and found a nearly perfect correspondence. To analyse the data, we followed the constant-comparison method (Glaser, 1978).[2] Analysis began with a close reading of all 20 transcripts by both authors. In a first round of coding the authors reached 100 per cent agreement in defining which interview conversations were about 'reproductive' experience. In the second stage of coding, we organised the data into three general categories: (a) 'the ability to get pregnant'; (b) 'feelings about reproductive capacity'; and (c) 'unique reproductive contexts'. We also reached 100 per cent agreement in this second stage. Data within these three categories were compared and contrasted to identify similarities and differences among interviewees, and to find nuances within the grouped data. Early findings were shared with one woman in the study; discussions with her helped us to finalise our analysis.[3] While our systematic coding and analysis procedures help to support our interpretations of the data, the ultimate validity and trustworthiness of the data must be judged within the wider context of the rich descriptions offered (Creswell, 1998; Glaser, 1978).

The results reported below are the interpretation of the women's experiences by the research team, although every effort has been made to provide first-hand accounts and verbatim descriptions of women's lived experiences. Throughout this chapter, we use pseudonyms to protect participant anonymity and confidentiality. Furthermore, we focus only on a selection of participants in our findings in order to provide continuity in the stories that we share.

Results

In this section we illustrate how women had to respond to varying assessments of how 'reproductive' they were and whether they, in turn, deserved a full range of reproductive health care opportunities in light of their impairments. Sometimes they also faced the effects of impairment but, for the most part, reproductive opportunities were shaped by social and not impairment-related barriers. Ultimately these women found themselves in a contradictory space as they led (or even contemplated) their reproductive lives.

'You can still get pregnant'

Women in our sample were not stereotyped as non-reproductive by their physicians; in fact most were told by doctors that 'they could still get pregnant'. Terry (two children before injury) discussed what she learned from doctors post-injury:

> Th[is] was what I learned in the hospital: "(a) You can get pregnant again; (b) If you take birth control you will die; [and] (c) You'll never walk again." They told me [this] 3 times a day, all day long, and they wondered why I was crying.

Damita (one child after injury) reported a similar conversation with a doctor.

> I know one of my greatest concerns right after my injury wasn't so much about walking but whether I could be a mother or not ... I remember asking the doctor, "Could I still have children?" To me that was far more important than being able to get back on my legs ... The doctor told me that I could still have children but ... I was still in denial about that for 14 years ... Yeah, he said, "If you're going to engage in something like that (sex) you need to protect yourself" ... even with that, I figured [that] because I couldn't feel the part, it just

didn't work ... [W]hen I found out I was pregnant, that's when it really sunk in ... he was right!

Prompted or unprompted, doctors shared information about reproductive capacity very quickly after injury and characterised women as having the 'choice' to get pregnant. In general, impairment was not defined as limiting reproductive capacity; as material entities, their bodies were defined as 'normal'. There were exceptions to this, of course. Candace (two children after injury) talked about how her doctor never mentioned the possibility of pregnancy after injury. She attributed this to the fact that her injury happened 40 years prior to the interview (in 1968): 'It wasn't really talked about by the doctors at that time ... [Now women can say,] "You know what? I want to have a child, so I'm going to do this"'. On the whole, women reported that doctors were supportive of their desires to have children (or not) after injury.

Am I 'normal'?

As indicated by the extracts above, women sometimes thought that doctors were wrong about their reproductive capacity, or that they were not ready to hear information about reproductive health in the aftermath of their injury. Thus living a reproductive life was not always as easy as merely deciding to get pregnant and having a child. The embodiment of impairment involved thinking about oneself as 'abnormal' and questioning 'normal' reproductive capacity at least at the start; starting to think about oneself reproductively after injury (within the context of a new and permanent disability) took time.

Terry also discussed how health care providers defined the return of menstruation after her injury positively, but it was hard for her to see it positively because in the moment she did not see why others were so 'happy' about a 'normal woman's function'.

> When they [her periods] came back, [small laugh], the nurses at the rehab hospital cheered. ... They were so excited, [they said,] 'That's a normal woman function and that's fabulous!' I went, 'Okay, this was the only part about this whole accident that was actually seeming good. Will you stop cheering 'cause I'm crying!' They cheered up and down the hallway of this hospital.... I'm like, 'You guys need to get a life, because I'm not seein' what is happy about this'.

Despite reactions from medical providers, Terry felt that she was embarking on a different 'normal' than the 'normal' that able-bodied women experience, and she did not want to be compared to 'able-bodied' women. The 'normal' versus 'abnormal' dichotomy activated by the reactions of hospital staff to Terry's menstruation led to mixed emotions – she did not know how to think about menstruating as a disabled woman.

Kendra (one child before injury and pregnancy loss at time of injury) also found it difficult to think of herself as both reproductive and physically impaired, and decided she would rather have a hysterectomy than deal with the logistics of both.

> I'd love [a hysterectomy]. I said, 'Will you pull out my uterus?' They said: 'No'. [*Interviewer: Were they concerned that if they took your ovaries you'd get osteoporosis?*] Yep ... I'm taking Fosamax.[4] I don't like my periods, they're messy ... I don't like it, I want nothing to do with it, okay? But they won't let me have my uterus pulled out.

Kendra wanted to end the hassle associated with menstruation but she also wanted to limit her risk of pregnancy. Within the context of her disability she did not want to manage 'normal' reproductive processes anymore. Thus doctors were sometimes more positive about 'normal' reproductive function than the women themselves. Unprompted, nine women in our sample (45 per cent) specifically mentioned that they thought of themselves as 'normal' and 'healthy' in relation to their reproductive health, and others (n = 6) also acknowledged their ability to reproduce (even when they did not refer to themselves as 'normal'). Yet it still took time for women to contemplate how their experience of living with an impairment might shape their reproductive lives. While acknowledging themselves to be reproductive like any other woman (and activating the 'normal' versus 'abnormal' dichotomy as they talked about themselves), embodying reproduction in the face of impairment was complicated and confusing at times (see Dillaway et al., 2013). Having a body that was capable of 'normal' reproductive functions was less important than what it meant to live with disability every day and simultaneously engage in reproduction.

Unique reproductive contexts

Warnings often came with doctors' discussions of participants' abilities to get pregnant as well. For example, doctors wanted to make sure that Terry knew that she could not use hormone-based birth control (e.g. the pill) because of the risk of blood clots. Similarly, Damita's doctors wanted to make sure that she understood that a lack of feeling in her pelvic region did not preclude pregnancy. Kendra's doctors would not contemplate hysterectomy post-injury because of her higher risk of osteoporosis; instead they offered her Depo-Provera[5] (and eventually a tubal ligation) to meet her desires for non-reproductive status. Even though many women in our sample led full reproductive lives post-injury, their reproductive options and 'choices' were often constrained by impairment-related contexts.

Further, despite being told by doctors that they were 'normal' and capable of having children, our interviewees faced considerable obstacles when attempting to access reproductive health care. While some barriers were structural (e.g. exam tables that could not accommodate disabled bodies), a large proportion of barriers were attitudinal (e.g. doctors and office staff sometimes chose to not make the effort to examine disabled women's bodies). These social barriers left women struggling to make sense of their own identities and reproductive health care options: in theory they were 'normal' women in need of reproductive health care but they experienced significant problems of access. Kendra brought her daughters to her gynaecological appointments (to lift her out of her wheelchair and to hold her legs while she was on the exam table). She understood that doctors would *not* accommodate her impairment-related needs. Unfortunately, some women in the study opted out of regular gynaecological care because of these types of obstacles and others acquiesced to incomplete or substandard health care. In these cases, structural and attitudinal barriers interacted with the effects of impairment to simultaneously shape disabled women's reproductive 'choices' and challenges.

Questionable mothers?

Unprompted, four women in our study reported interactions that reaffirmed negative attitudes towards disabled mothers. Terry contemplated having a third child (what would have been her first child post-injury) but she was worried about the physical risks of pregnancy after her injury and therefore thought about adoption. Eventually, Terry decided to forgo additional childrearing altogether because disabled women are often seen as questionable (and unfit) mothers.

> [P]eople said, 'There is no way that you would ever get a [child] because you would not qualify'. Because I'm in a wheelchair they would not deem me as an appropriate parent. And so I'm not a decent parent for the two kids (born pre-injury) that I have? ... It didn't matter that financially we could have had [a third child]. We have the room ... [W]ho would ... help me be able to convince these idiots that I would be a good parent? ... I would have been desperate

if I would have thought that I no longer, you know, that somebody would say I couldn't be a mother.

Damita also talked about lay attitudes towards disabled motherhood.

> [M]y doctors have been pretty thorough in explaining to me, you know, about having children and that you can still do that even in spite of your injury. But it's not so much that, it's about how people have responded to the fact ... that I have a child ... That I gave birth ... I think the *hardest* thing (her emphasis) that I've had to come to grips with is that people look at me as an asexual being ... And I just think it's a common response to women with disabilities ... [F]or example, you and I could be walking down the street and my kid is with me and they'll probably think that my kid is your kid. You know, 'cause they think, 'How could she have sex?' *[Interviewer: Do people ever say that to you?]* It's implied ... I think sometimes we're viewed as half-people, not whole individuals. Or, as children. You know, as if we're not capable of doing things.

In this extract, Damita described her surprise and anger in response to individuals who view disabled women as asexual or unfit mothers. Like Damita, participants found themselves trying to negotiate contradictory attitudes towards their impairments in the context of reproduction. From one perspective, it was acknowledged by doctors that participants were sexual beings who were able to conceive like other 'normal' women. From another perspective, women were confronted with negative public attitudes in relation to their reproductive capacities. Finally, participants also faced impairment-related situations that made for frustrating interactions with doctors and hospital staff. For example, staff had to adapt their medical interventions to accommodate the physical impairments of women with SCI and the women themselves had to learn how to 'be' women with irrevocably altered bodies. While we do not have data on the reasons why the non-mothers in our sample chose not to have children post-injury, or why some women in our sample did not have more children after injury, we suspect that the simultaneous weight of impairment-related contexts and social barriers may have limited their reproductive options.

Discussion and conclusions

The majority of women with SCI in our study were told by doctors that 'they could still get pregnant'. Yet while doctors reminded women of their 'normal' reproductive status, they also warned them about the effects of their impairment on reproductive 'choices'. Further, when women in our study tried to seek reproductive health care, structural barriers in medical office settings and doctors' unwillingness to deal with their disabled bodies meant that accessing high quality reproductive health care was difficult. Thus medical providers left women in a contradictory space, effectively defining them as both 'normal' and 'abnormal' in their reproductive and disabled statuses. The women struggled with having a body capable of reproducing and living with a disability (that either limited reproductive options or created structural and attitudinal barriers to this reproduction). Women had to actively work through their own thoughts about their impairment and reproductive capacity, their embodied experiences of reproductive activities, others' attitudes towards their bodies and their potential motherhood, and the effects of impairment on their reproductive 'choices' and health care. While the impairment consequent to SCI might not limit reproductive opportunities and options, our data reveal that the everyday experience of being a woman with an SCI was much more complex and uncertain than their 'normal' reproductive capacity might suggest.

Additionally, some interviewees were exposed to asexual and non-reproductive images of themselves in other non-medical, public interactions. Women were aware of their physical impairment and what it meant to live with SCI but they were not always prepared to justify their existence to a (non-disabled) public that continues to view disability as not only 'difference', but also 'deficit' (Finkelstein, 1996; Hughes and Patterson, 1997; Shildrick, 2002; Thomas, 1999). Participants reported that their most significant struggles were in 'public' spaces, either around unsupportive medical staff or around lay individuals who repeatedly challenged their legitimacy as reproductive beings. Thus the barriers participants faced in leading reproductive lives were overwhelmingly attitudinal or social-relational in nature. Even their individual experiences of physical disability were socially created in part (and not completely determined by the physical impairment itself), because of medical providers' lack of understanding of disabled women's needs.

It is important to note that almost a third of our sample bore children post-injury and many women made 'choices' post-injury about menstruation, birth control, sexual relationships, hysterectomies, and other reproductive procedures. Thus limitations on their reproductive 'choices' did not prevent them from living out reproductive lives. Nonetheless, some interviewees never had children, and this may be because they were discouraged from realising new reproductive experiences after injury. Others were apparently finished having children before injury and may have felt the impact of disability on reproductive experiences less directly.

Reproductive experiences were not uniform across our sample and not all women experienced disability in the same way. Of the five women we quoted in this chapter, only Candace was tetraplegic. Terry, Damita, and Kendra were paraplegic. Paraplegic women may find it easier than tetraplegic women to navigate reproduction post-injury, because impairment effects are different; however, we cannot judge this from our current data. Moreover, it is beyond the scope of the chapter to discuss differences among interviewees in birth control use, respiratory problems, urinary tract problems, sexual function, motor control, etc. In this way, varying impairment-related conditions that women with SCI faced make it difficult to generalise our disability-related findings about reproductive experience. Some women in our sample were also more willing than others to battle with health care providers to access high quality reproductive health care and make possible the reproductive experiences they desired. Understanding why some interviewees felt a greater sense of 'empowerment' as they engaged in reproductive lives is extremely important if we aim to understand disabled women's reproductive lives in full. On this front, there is still much to explore.

Disabled women in our study are not *just* the sum total of their impairments. However, these women experienced unique reproductive conditions and accessibility issues that were related to both physical impairment and the inability of non-disabled individuals and social institutions to effectively deal with disabled bodies. Shildrick (2002) has argued that all bodies (especially when they enter the 'space of discourse') do not fit with the norm, and this idea applies well to women's reproductive bodies and disabled bodies. As the data show, a 'normal'/'abnormal' dichotomy cannot sufficiently explain embodied experiences of disabled women's reproduction (Shildrick, 2002). Knowing only that disabled women are diagnosed as 'capable of getting pregnant' or that they have unique health conditions after injury tells us nothing of their lived experiences of reproduction or the barriers they face.

On one hand, this study cannot be used to generalise about the reproductive experiences of disabled women (in that SCI is a very specific kind of injury and we interviewed a very small sample of women) or women in general. On the other hand, the lens we use in this chapter can help reframe how we think about all contemporary women's reproductive lives. We explored how we might separate out women's reproductive capacities and their related meanings from women's embodied experiences of reproduction. We also highlighted the social barriers that particular groups of women might face in their reproductive lives. Our findings allow us to see how women think and feel and experience their reproductive bodies, as they navigate complicated social-relational landscapes related to both reproduction and disability. As reproductive 'choices' and challenges change across time and place, and bodily experiences expand for some and contract for others, scholars of reproduction should continue to complicate notions of both reproductive capacity and lived reproductive experiences. As this chapter shows, the 'normal' reproductive body is an empty concept in postmodernity.

Notes

1. We do not know whether the women who were successfully contacted differed from those who did not respond. We acknowledge that this is a possible sampling bias.
2. This is an ethnographic technique and we found it useful in our study since many of the themes in the data were unexpected; we undertook cross-interview comparisons to understand the nuances in our coded data.
3. 'Member checking' is a common method of establishing validity in qualitative research (see Creswell, 1998).
4. A prescribed drug used to treat and/or prevent osteoporosis.
5. A prescribed hormonal contraceptive that women receive as an injection every three months.

9
Taking a Long View of the 'Right Time' for Fatherhood

Fiona Shirani

Introduction

This chapter explores the 'right time' for fatherhood, drawing on qualitative interviews with UK men in heterosexual relationships. The timing of parenthood has become a popular topic of debate, particularly the apparent trend towards 'delaying'. A 2010 release from the Office for National Statistics[1] detailing that births to older mothers have trebled in 20 years prompted media articles questioning 'when is the right time to have a baby?' (e.g. Atkins, 2010). However, such articles focussed on women's decision-making, have led to criticism from readers that men's crucial role in establishing the 'right time' for parenthood had been overlooked. The trend towards delaying has prompted concern about age-related infertility, as the number of in-vitro fertilisation patients rose between 1985 and 2005 from 3,717 to 32,626 (Human Fertilisation and Embryology Authority, 2009). Akin to the media discourse, there has been an evident bias towards a focus on women in the existing research on fertility decision-making. Yet as men's attitudes are as relevant as women's to understanding fertility behaviour (Jamieson et al., 2010), it is clear that the way in which men negotiate the timing of fatherhood is an issue which needs to be explored. Alongside wider concerns about delaying, teenage parenthood continues to be identified as troublesome (Edwards et al., 2010). The apparent problematisation of these divergent timing experiences has prompted some to describe age at first birth as a marker of social polarisation (Kneale et al., 2009), subsequently leaving a relatively small window of culturally determined 'appropriate' age to embark on first parenthood.

Life course theorists propose that the impact of an event has much to do with its timing (Macmillan and Copher, 2005), highlighting timing as an important area for research. Age is often applied as a standard to judge the 'proper' timing of life events or 'timeliness' of life transitions. The individual is then placed in a logical social temporal order of being 'on' time (or not) in their development (Mills, 2000). Consequently, people in certain age groups may be regarded as more suitable for parenthood than others. However, critiquing what may be seen as an overly deterministic life course approach, others have argued that decisions about the timing and sequencing of important transitions have become more individualistic, with individuals creating choice or 'do-it-yourself' biographies (Beck and Beck-Gernsheim, 2002). Whilst both approaches offer useful insights, Hunt (2005) contends that given the postmodern turn it is arduous to suggest that people go through clearly demarcated stages as in previous decades. However, society continues to shape and limit life experiences and chances. Subsequently, despite indications of individualisation processes, the idea of agency without limits seems inappropriate, as is the idea of destandardisation in the sense of dissolution of life course structures. Therefore, whilst we may be witnessing increasing disparity in age at first parenthood, it is clear that not all transitions are regarded equally.

In accounting for timing decisions, being able to provide the 'best start' for the child is relevant. Previous work has highlighted family life as an area where moral identities are at stake (McCarthy et al., 2000). In contemporary society, being moral does not suggest an unwavering commitment to being good, but actively and reflexively exercising choice with responsibility (ibid.). Much of the work in this area focuses on relationships between parents and children, yet with increasing expectations that parents should foreground the needs of even their unborn offspring, decisions about when to become a parent are also imbued with notions of morality. One aspect of this is the ability to provide the 'best start' for the child (Shirani et al., 2012). For example, both Townsend (2002) who has explored this question in relation to fathers and Currie (1988) regarding mothers, describe the importance of secure employment, stable relationship, and suitable housing in constituting the 'right time', thus highlighting 'right time' as related to the configuration of particular material circumstances.

Whilst fatherhood is an increasingly popular area for research, there has been reluctance amongst some feminist researchers to focus on men and fathering out of concern that it may yield men new confirmation of their importance to, or power over, women and children (Segal, 2007). However, feminists have also recognised the benefits of redefining fathering to disrupt traditional notions of masculinity and make for greater gender equality (Silverstein, 1996). A focus on men's timing decisions in relation to fatherhood is crucial, as continuing empirical neglect of men may reinforce a view of women as primarily reproductive, positioning them as responsible for what has been described as a 'crisis' of delayed childbearing (Simpson, 2010). In an attempt to redress this imbalance, this chapter draws on data from a qualitative longitudinal study to explore how men describe the 'right time' for fatherhood.

The study

Much of the data presented in this chapter comes from the Men-as-Fathers project at Cardiff University (led by Professor Karen Henwood) and part of the UK-wide qualitative longitudinal network 'Timescapes'. For the Men-as-Fathers project, interviews were conducted in 2000 with 30 men from East Anglia once before the birth of their first child and twice within the first year afterwards. When the study became part of the Timescapes network, 19 of these men were interviewed again in 2008–2009 (Sample A), the remaining 11 could not be traced (Sample B). A further sample of men from South Wales was also recruited in 2008 (Sample C) and interviewed once before and twice after the birth of their first child. Whilst working as a researcher on this project I also undertook a PhD on the 'right time' for fatherhood (Shirani, 2011) and recruited an additional sample of men who had become fathers age 40+ or under 22, in order to provide further detail on the experience of off-time transitions in terms of age (Sample D). This constitutes a total sample of 53 participants ranging in age from 15 to 54 at the time of the child's birth. All men were in heterosexual relationships; most were married (n = 37) or cohabiting (n = 13), and the majority of men were also in full-time employment (n = 43).[2] Whilst I collected a large proportion of the data, the involvement of multiple researchers in a qualitative longitudinal study has held implications for the research, an issue considered in-depth elsewhere (Shirani, 2010).

The interviews were qualitative and semi-structured, allowing for a balance of comparability and flexibility, and analysed thematically using Atlas ti software. Whilst qualitative longitudinal data sets provide opportunities for case study analysis, the large data set also created opportunities for synchronic analysis across the sample at a single point in time, giving a broad view of timing experiences. In their first interviews all participants were asked whether the pregnancy was planned and if they felt this was the right time or not. This simplistic dualism failed to capture the complexity of the men's experiences in relation to timing and responses fell into five categories; unplanned, mistimed, right time, semi-planned, and delayed, indicating a continuum of planning. This chapter focuses predominantly on 'right time' in order to provide an insight into the way men defined the preferred circumstances for embarking on parenthood. Whilst discussion of what it means to make an off-time transition is not included here (and warrants further consideration in its own right) the concept of right time was meaningful for men across the sample, as even those who did not feel that they had achieved an on-time transition often described the different circumstances in which the timing would have been 'right'. Consequently this chapter aims to provide a detailed exploration of right time in light of the current gap in the literature on men's fertility decision-making.

Analysis

Chronological and biological time

For the majority of participants there was a strong sense of a culturally approved 'good' age to have children, ranging from late twenties to late thirties. Anyone younger was seen to lack the necessary maturity and stability, whilst older fathers would lack energy to keep up and be too far removed from the experience of childhood today. The men related chronological age to maturity – something they viewed as a prerequisite for fatherhood – although recognised in practice that it was unlikely to be such a straightforward relationship. Defining late twenties to thirties as the best age for fatherhood was described across the sample as offering a balance between maturity and remaining 'in touch', and between life experience and energy, as the ability to engage in physical activity with their child was seen as an important aspect of fatherhood (Shirani, 2013).

> We're at the age now where you know we're not old by any means, in our thirties but we're not old … I'm still young enough when the kid's in their teens to run around the football pitch with it, so yeah it's a good time. (Joe, 31, C)

Participants were rarely critical of men who had chosen to become older fathers and they did not appear to associate any risks with this. Instead they were more likely to problematise *maternal* age, discussed in terms of a ticking body clock.

> I think she was in more of a hurry than I was to have children, I guess 'cause she's a couple of years older than me. So … for me it

wasn't a case of time running out whereas for her you start to worry. (William, 29, C)

This kind of language provides a vivid representation of the female biological clock placing temporal restrictions on the window of childbearing possibility. Talk about 'time running out' suggests that men see their partners as having a reproductive time limit, based on their chronological age. Concerns about partner's age generally began when women reached their early to mid-thirties, with most partners aged 34 and over described by the men, and often also by health professionals, as 'older' mothers, which carried negative connotations in terms of risk. Few participants felt that they had an age limit to conception themselves and overwhelmingly concerns about age were centred on partners, even when men were significantly older. This placed temporal biological restrictions decidedly in the realm of women, upholding the notion that they are responsible for reproductive decision-making.

Participants frequently discussed their understanding that the culturally acceptable age to become a parent has changed over time so the 'right time' is now later than in previous generations. Delaying parenthood until one's thirties was also seen as a middle-class trajectory, with emphasis on the moral position of choosing to become parents rather than making the transition as a result of an unplanned pregnancy.

> My wife's three years older than me and she was sort of fairly keen on it, she was a bit worried that her 'clock' was starting to tick and we didn't really want to be 'old' parents. I know parenthood's got, tends to be a lot older now, people used to have kids when they were 18, everybody did, whereas nowadays it's mainly more sort of professional people perhaps having a little bit later and also actually got through university and that sort of thing. (Vincent, 30, A)

Vincent suggests this trajectory is common amongst 'professional people', indicating that those from working-class backgrounds are seen as being on a faster track without the mediations of progressive stages and the surety of a career as time passes (Macvarish and Billings, 2010). Whilst some working-class men in the sample indicated similar linear trajectories, others articulated their own moral position based on alternative life course trajectories they could have followed.

> I and everybody I know expected it a lot sooner than what it is ... they all thought 'he's gonna be like his uncle and have kids when

he's 15/16' but I'm the good one in the family so, you know, I waited a bit longer than what they all expected ... I think I've grown up and matured quite a lot since the age of 16 because I used to go round with the wrong people and used to do stupid things, but I'm out of all that now ... I think now it's time to settle down ... I think I've lived my life a lot, at the end of the day the way I look at it is I'm not getting any younger. (Nathan, 24, C)

Although comparatively young, Nathan describes how he has delayed parenthood until he felt mature and responsible, following a period where he was able to live an individualistic life. In this way his decision to delay parenthood shows similarities to the trajectory articulated by middle-class men, although occurring at an earlier chronological age.

The next logical step

For many fathers, right time appeared to be constituted by fulfilling a sequence of age-related goals so they reached the stage where parenthood became the next logical or expected step.

> I think it sort of starts when you sort of get a job then if you move out if you meet someone, get the ring on the finger ... It's sort of ticking off little boxes and once you've sort of got a child on the way I think that's the full set for a lot of people. (Gary, 28, A)

These responses suggest that having a job, a stable relationship, owning a family home, and having time to travel and pursue interests are things which should be done before embarking on parenthood. However, these sequences are not always straightforward. For example, the couple relationship[3] must have had an appropriate duration so as not to be seen as 'rushing into things', yet must not remain childless for too long as questions about fertility were likely to be raised, which held implications for masculinity.

> Yes, very much, people are pleased and happy. In many ways it suggests that previously they were saying, 'Oh well, why not?', 'Better cover that one up', kind of thing, and 'not ask any more questions about that one' ... I suppose I'm speculating, but maybe they're thinking 'Oh well, he's not infertile then', which they might have been thinking, or 'she isn't', or something like that. (Simon, 35, A)

Simon's ten-year relationship was lengthier than most when becoming a parent and he suggests this might have led to speculation about the couple's fertility. However, a relationship of less than two years was seen as an insufficient duration, raising concerns about stability and creating perceptions of rushing. From two and a half to five years was seen as most appropriate; having some time as a couple but not waiting too long so fertility is questioned. This careful balancing highlights how variability in the duration of life course stages is restricted by these unspoken limits, indicating the extent to which the timetable of socially approved life course progression appears ingrained in the men's consciousness.

In addition to the relationship, having a stable job and appropriate housing were seen as prerequisites for embarking on parenthood. Ideally, the men would have been working in a secure job for an average of around four years, with a steady income and they would own a suitable house. Many participants bought houses or moved during the pregnancy or when trying to conceive, idealising larger houses in more suitable 'family' areas (Townsend, 2002). Throughout their responses to questions about right time, the men frequently cited the importance of being settled or secure in terms of relationship, finances, employment, and home.

The importance of having stability before embarking on parenthood was echoed throughout the interviews, including those where it was not in place. Being settled was seen as giving the baby the best start in life, and giving the men themselves the best chance of coping with a demanding task by being well-resourced. Research amongst young adults has also highlighted their views on the importance of having these factors in place before taking on the responsibility of children (Gordon et al., 2005) suggesting that plans may be made earlier in life for this sequencing of events.

One reason put forward to explain the trend towards delayed parenthood is that larger numbers of people are now investing increasing amounts of time into higher education, meaning the transition to employment and financial independence occurs later than in previous generations (Scherger, 2009). The study data would appear to provide some support for this claim; around half the men in the research had been to university, a much higher proportion than they reported for their own parents. The average age for first-time fathers in the sample who had been to university was 34, compared to 28 for non-university educated fathers. Although the ideal of the man as a sole breadwinner has apparently become less prominent in recent years and many families have dual earners, the importance of financial security before embarking on parenthood is a commonplace idea in men's accounts (Jamieson et al., 2010; Shirani et al., 2012). Thus the factors which contribute to establishing financial stability have important implications for the timing of parenthood.

> I think it's just the way men think; you think that you're the one who needs to make sure the finances are in place that you can afford to look after them, which isn't a conscious thing, it's certainly not a sexist thing, it's just the way men's minds work I think, they feel they've got to be the provider. My wife and I earn similar levels of money and both work just as hard, you just feel in yourself that it's your responsibility to make sure that's in place and so you take on a bit more of that, more of the planning and the organising, and make sure you've got the funds. (William, 30, C)

A life of one's own

The idea that parenthood and having a life of one's own were mutually exclusive was suggested by many men. The expectant fathers felt that having some time to themselves, often to go travelling and gain other significant life experiences, was something important to do before having a child, as it would be difficult to do afterwards given the restrictive and demanding nature of parenthood. This individualistic period was seen as a significant part of the life course that provided the man with valuable experiences, which would be beneficial for fatherhood. When asked to reflect on what it would have been like for them to have their children earlier, many of the men expressed a concern that they would have been missing out.

> I mean people say 'Oh, you have your children first and then your life', or 'Your life first and then your children' ... I'd rather have had ma life first and then have children, because to me it's a choice as opposed to just falling into, the kind of parent trap, you know, doing what your parents did or even doing what your parents might have told you not to do, it's to do the same as them. (Kenny, 41, A)

Kenny distinguishes his own situation from his parents' early transition to parenthood; described as a 'trap' they fell into. In contrast, the way he has organised his own life by leaving parenthood until later suggests that he is making an active choice for fatherhood rather than an unintended transition. Several of the men suggested that the fact they had *chosen* to have a child also indicated that they were mature enough to have chosen to take on the related responsibility. Consequently an unplanned pregnancy was associated with irresponsibility, contradicting notions of mature and responsible fathering.

Joining the club

Signs of socially approved timing also came from others in the same-age cohort having children; particularly friends and siblings. Going through the transition to parenthood at the same time as peers offered them a comforting sense of shared experience, which several men described as 'joining the baby club'. For Joe, other people in the same age having children acted as a signal that it was something he 'should be doing' in order to keep up with them. This highlights how standards in relation to timing and sequencing are activated through linked lives (Heinz and Krüger, 2001).

> I think you know as you get older it becomes more and more important I think because perhaps in society you see ... people who are the same age having kids and you think 'oh I should be doing that now'. (Joe, 31, C)

For other participants, younger siblings and friends having children before them indicated that they had somehow become out of place, not progressing in their lives at the same pace, and this acted as an impetus for thinking about having their own children. For many of the men who had all the prerequisite stability factors in place, friends having children provided a significant jolt to thinking about parenthood, raising the question if not now then when?

> We've now been married four years, been together six, so we've now done a lot of things that we wanted to do in our lives ... everything we've wanted we've been able to buy and now I think we both come from a family orientated background, so that was something we've always said that we would like children, and like lemmings we might as well have one. (Jason, 32, A)

Jason's response indicates that when they had done, achieved, and acquired everything they wanted there was no longer any reason to delay parenthood, so there is a sense that 'like lemmings' they decided to follow the normative pathway to parenthood and have a child because it is something they always assumed would happen (Lewis, 2006). These men appear to be following what Zinn (2004) terms a traditionalisation mode of biographical certainty, based on the reproduction of given action patterns which appear to be natural or self-evident. For such individuals, decisions about working life or family formation are not an issue of whether to go ahead but of fine-tuning; when and under what circumstances the individual events of a traditionalised sequential pattern should be realised. This suggests a strong sense of conformity amongst participants, with other timing experiences proving potentially challenging (Shirani and Henwood, 2011).

Something missing

If all the above 'pre-requisite' stages were in place, some men experienced the feeling that something was missing in their life, for most this was thought to be a child-shaped gap.

> I *had* thought, I suppose that what I thought might be lacking was just a general thirty-something lack of achievement. ... Then, I suppose, the *slow* realisation that maybe what was lacking wasn't that but was a need for something else to complete the family, to complete the family dynamic. It *may be*, having to look after someone and care for and nurture, is what we're missing. (Charles, 33, B– his emphasis)

Once they had been successful in careers and relationships, many of the men articulated the desire for a new challenge, something which would add meaning to their lives. As the reward for having children today lies almost exclusively in their emotional value (Beck and Beck-Gernsheim, 1995), parenthood was perceived as able to provide this. For these men, parenthood represented another opportunity for fulfilment, which would involve taking a back seat to prioritise the child. This could also be viewed in terms of linear progression; at a stage in their lives where they had established career, status, and relationships, having a child offered a chance to progress in another direction. Thus the new life phase is legitimated through recourse to biography as it represents a 'climb' (Alheit, 1994). For example, although William was not enjoying his job, fatherhood offered him a sense of achievement in another area of life.

> The thing that springs to mind, is just the chance to be really good at something ... to be able to look back and say 'I might not be doing a fantastic job but hey, I'm a brilliant father' and I can die a happy man knowing that I'll have done something really good in my life, (William, 29, C)

The men drew a distinction between having a child because there was something missing (implying an existing sense of stability but desire for a new challenge) and having a child because there was something wrong i.e. to fix a troubled relationship, which was viewed as inherently irresponsible.

A view from eight years later

When their first child was eight years old, Sample A participants were revisited and asked to reflect again on the timing of parenthood. Over half viewed the timing differently in light of the lived experience of fatherhood. Six men who had originally felt it was the right time continued to believe this unequivocally eight years later. Here Bruce highlights the balance between energy and life experiences, as discussed above.

> Ideal really, because looking back Jeremy came along just about when I was thirty, and I would say that's great because both Angela and I had plenty of time to do the selfish singleton thing in our twenties and did, and we're still young enough to be active and involved and to go out and kick a football round the garden or take them down to whatever various things that they want to do, be fully involved. (Bruce, father at 29)

In addition to those consistent in their perceptions of timing, four men who originally described the pregnancy as off-timed in some way retrospectively felt that they had become parents at the right time, achieving a good balance between age, relationship length, experience, and energy. Although these men were all in their early thirties when becoming fathers, and thus in line with perceptions of chronological and biological right time, three had wanted to delay further but retrospectively were glad that they had not. For example, Malcolm was working away from the family home when his wife became pregnant and for the first year of his child's life, commuting back at weekends. Although aged 32, he had wanted to delay as he did not want to be a 'part-time father', but as his partner was several years older she was reluctant to wait due to concerns about her biological clock. However, eight years later he described how he would not alter the timing as he felt he had achieved a good balance between energy and life experiences. In contrast, six men felt that they would like to have been up to five years younger at the point of having an eight-year-old child because of their perceptions that they would have more energy to be active with their children and feel less tired. These men ranged in age from 32 to 40 at the time of their first child's birth. This increasing significance of age and physical ability has implications for masculinity, and was the main reason older fatherhood was seen more negatively eight years later (Shirani, 2013). Although concerns about age were raised, all the fathers in this group justified their situation by explaining that parenthood could not have happened any earlier; they were not in the right relationship, careers needed to be established, fertility difficulties delayed conception, or they simply did not feel ready any earlier.

> I don't know 'cause part of me says I'd like to have had them younger in my life, part of me says how ready would I have been if I had? (Ray, father at 34)

Interestingly, no men suggested that they would have liked to have been older when making the transition to fatherhood[4], with younger fathers commenting that they would benefit from still being relatively young when their children left home.

> I think it's worked out well yeah. And long-term wise when they're thirty we're probably still gonna be quite robust and with it so hopefully be young grandparents but not too young. Yeah so I think it's worked out alright, so far. (Sebastian, father at 26)

Some of the men in this group illustrated an inherent problematic in timing decisions; that in waiting until the necessary prerequisites are established, which can take many years, they are older than they had hoped to be when making the transition to fatherhood, which had different implications later in life. Ray's comment highlights the difficult balance between the factors seen to constitute right time. For some these do not coincide, indicating underlying structural antagonisms which mean that there may never be a right time (Currie, 1988).

Three men talked about how they would have liked more time as a couple before the birth of their first child. This was discussed in terms of having the opportunity to do things together and form a solid relationship before the complications of children were added. However, this was generally not problematised because they expected to make up for this when the children left home; indicating a notion of having a life before and after children rather than simultaneously. These men had some of the shortest relationships in the sample prior to parenting, which, as Rick indicated, could have significant implications.

> We never really had a long courtship you know, we were kind of together for a few months then Tanya fell pregnant. Once Imogen

came along the whole parenthood thing kicked in (laughs) and we never sort of had a chance to revisit the court, the sort of spending time together really. So I think the relationship's inevitably been affected by that. We've given so much I think to the children that we probably haven't given as much to each other as perhaps we should've done. I think we're both aware of that. (Rick, father at 35)

Rick's response that he would have liked more time for the couple relationship first and would like to be younger suggests that in his circumstances the necessary prerequisites did not coincide, which complicates decisions about right time.

Conclusions

Data presented in this chapter indicate a high level of agreement between participants in relation to right time, suggesting certain assumptions about a normative life course prevail. This does not just relate to achieving the necessary prerequisites but doing this in sequence and for an appropriate duration (Rawls, 2005). Against the background of claims of increasing individualisation, many participants in this study held a long-standing assumption that they would become parents; the choice was 'when' rather than 'if'. Given the apparent pervasiveness of a normative pathway discussed in this chapter, this approach may imply that participants carefully plan their life course according to a strict schedule, but this is not necessarily the case. For example, several men found thoughts about having children only occurred to them when they were married and thus in a relationship they deemed suitable for family life. The majority of men had a notion that they would become fathers at some point in the future but this always remained in the future until they realised there was no longer any reason to wait. Participants recognised that current socially approved trajectories were different to those of previous generations; drawing distinctions from their own fathers, who had overwhelmingly made the transition to parenthood at an earlier age. Yet the men were not reflexive about how these changes had occurred and, whilst it highlighted fluctuating cultural ideals of right time, it did not appear to influence their own desire to fit in with contemporary patterns. Instead, the careful balancing, sequencing, and duration of each life course stage discussed by the men suggest an ingrained socially approved timetable of life course progression, with parenthood as a key marker of adult status. Even for those who described making an off-time transition, the concept of right time was meaningful as something they considered their own transitions against.

Whilst, for some, achieving these prerequisites appeared relatively straightforward, providing a sense of achievement and progression, particular challenges are raised for those whose lives do not unfold in this way (Shirani and Henwood, 2011). Some participants who described it as the right time discussed the way they had to negotiate these ideals; for example meeting a partner later in life and awareness of apparently ticking biological clocks could mean a transition to parenthood early in the couple relationship in an attempt to stay 'on target'. This frequently appeared to have repercussions later in life, highlighting how such institutionalisation of experience can potentially be harmful. For example, some men found that not reaching the desired stage on time was a source of considerable anxiety. Participants indicated that they felt women would be more anxious about not becoming a mother 'on target' because of the preoccupation with female biological clocks, discussed earlier. This focus on female biological clocks may therefore be one of the reasons women continue to be seen as primarily responsible for the timing of parenthood.

Much of the complexity men described about negotiating the right time for fatherhood centred on achieving economic stability through secure employment, as, despite increasingly egalitarian views and a range of employment situations, the men overwhelmingly saw themselves as holding the primary responsibility for financial provision (Shirani et al., 2012). This centrality of financial provision to good fatherhood is underlined by current UK policy and discourse, which emphasises the economic contribution of men (Lewis, 2002). The continuing salience of 'breadwinning' and long-term financial provision, and the delay in being able to establish this through longer periods in education, may be one of the reasons why men may postpone parenthood. As postponement heightens during periods of economic difficulty (Rindfuss et al., 1984), the current economic recession may prompt increasing delays.

Taking a qualitative longitudinal approach to fertility decision-making has served to deepen understanding of the issues impacting on men's timing decisions and how these are negotiated. Whilst many men clearly experience different pathways to parenthood, this does not equate to the dissolution of normative biographies as not all pathways hold equal credence. Therefore, this chapter challenges the contentions of theorists such as Beck and Beck-Gernsheim (2002), discussed in the introduction, that in postmodernity people create choice biographies. By taking a temporal approach the chapter also offers a new contribution to theoretical debates on individualisation and the life course in relation to family trajectories, illustrating how within structural constraints participants make timing decisions based on assumptions of biographical certainty and life course progression. The concept of choice in relation to reproductive decision-making is therefore called into question in the context of competing temporal pressures around reconciling prerequisites for parenthood in conjunction with the time limits imposed by the biological clock.

Acknowledgements

The author wishes to thank the Economic and Social Research Council for funding the original study and refunding it as the 'Men-as-Fathers' project, part of the Timescapes network (award numbers RO22250167 and RES 347 25 0003). The author would also like to thank Karen Henwood, Principle Investigator of the Men-as-Fathers project.

Notes

1. The original release discussed in the press has subsequently been updated to Cohort Fertility (2009) England and Wales, Office for National Statistics, Statistical Bulletin, http://www.ons.gov.uk/ons/rel/fertility-analysis/cohort-fertility--england-and-wales/2009/index.html, date accessed 7 June 2013.
2. These figures relate to situations at the time of the first child's birth but participants circumstances varied over the course of the study.
3. For a small number of men it was important to be married before having children, however most did not draw a significant distinction between marriage and cohabitation in terms of relationship stability.
4. Although some men wished they had spent more time as a couple first, they had not wanted to be any older when becoming a father. They described wishing they had met their partner earlier in life, which would have allowed more time together, rather than wanting to change the timing of parenthood in relation to their chronological age.

10
Anticipating and 'Experiencing' Birth: Men, Essentialisms, and Reproductive Realms

Tina Miller

Introduction

Shifts in relation to reproductive realms and gendered assumptions about how caring, work, and intimate relationships become organised, are increasingly discernible across western societies. These shifts signal, for some, new possibilities and ways of imagining motherhood, fatherhood, and doing family life and they emerge as a consequence of sociocultural, economic, and political transformations as well as medical developments. One response to these changes has been to reframe areas of reproduction in relation to rights and responsibilities, which in turn has prompted debates about equity, gender, caring, and parenthood in a sphere more traditionally identified as primarily and essentially (biologically) maternal. These shifts have led to more mutable ideas about fathering responsibilities and practices which increasingly encompass emotionally invested, hands-on involvement as well as continued expectations of economic provision.

Yet the legacy of more traditional reproductive and caring arrangements means that journeys into first-time fatherhood are characterised by less clear trajectories for men than those assumed for women. Using data from a UK-based qualitative longitudinal study on transition to first-time fatherhood, men's experiences of these more ambivalent spaces will be explored across this chapter (Miller, 2010, 2011). Examining the men's experiences and narrations as they move through the novel terrain of preparation in the antenatal period, the hospital birth, and then early caring, provides important insights into masculine practices of gender and men's 'choices' and how these can be articulated. The data reveal the ways in which the men engage different discourses to make sense of their emerging paternal identities – as involved and caring individuals who can at times also feel detached and uncertain – and the findings provide opportunities to theorise changes and continuities in understandings of gender, masculinities, and reproductive arenas.

Setting the context: maternal spheres and changing masculinities

In order to explore and contextualise contemporary understandings of men's paternal agency, it is first necessary to discern elements of change and continuity which have (re)configured reproductive spaces and men's location within them. This involves examining the interrelated areas of gender and power, masculinities, and men's discursive possibilities alongside their practices of agency, as they anticipate and become fathers. By focusing on the changes in 'discourses' – societal ideas and visions of how things should be – and the ways in which men draw upon these, changes and continuities in men's role in the reproductive arena become more apparent. In the UK, just as in many western societies, shifts in dominant discourses can indicate expected changes in behaviours, for example, as signalled in policy changes such as the introduction of two weeks paternity leave in 2003. For most men their role as economic worker/provider has been prioritised in 'breadwinner' discourses and remains a recognisable and accepted dimension of successful masculinities, but for men becoming fathers there is also a growing awareness of an 'involved fatherhood' discourse too (Wall and Arnold, 2007). Here an emotional, more hands-on, caring relationship is emphasised between father and child, in ways historically more exclusively associated with mothering and assumptions about women's essentialist (and natural) capacities to care. Increasingly, in men's accounts of anticipating fatherhood elements of both these discourses might be anticipated, and in practice elements – or strands – of a range of different discourses are usually drawn together when individuals are invited to give an account of their selves. And as selves are narrated, particular practices of agency – ways of being and acting in the social world – are also imagined and presented. But for the men becoming fathers in this study, practices of agency as paternal subjects are initially uncertain: how should and do men becoming fathers act?

To date, exploration of men as reproductive actors has been largely absent, or underemphasised, in historical and traditional constructions of reproductive realms. Even though recent historical analysis has challenged representations of fathers as remote figures, largely absent from emotional aspects of family life[1] it remains the case that antenatal preparation, birth, and child rearing have been much more exclusively associated with women in western societies (Dudgeon and Inhorn, 2004; Locock and Alexander, 2006; Miller, 2005). But more recent research – capturing broader social and cultural trends – suggests that changes are (slowly) occurring in how family lives, child-care practices and responsibilities, and paid work are – or could be – organised at the individual and societal level[2] (Dermott, 2008; Doucet, 2006; Featherstone, 2009; Miller, 2010, 2011). Broader structural shifts such as the feminisation of the work place open up new possibilities – for example, the masculinisation of the home sphere – and although change has been and continues to be very slow, shaped in relation to a legacy of patriarchal arrangements, increasingly there is evidence of some change (Björnberg and Kollind, 2005; Crompton et al., 2007; Hobson and Fahlén, 2009). This manifests itself through the widely recognised discourse of the 'involved father' and associated demonstrations of involvement for example through men attending antenatal preparation classes and the birth. Similarly, research has also shown men's intentions of 'being there' as an emotionally engaged, sharing father in ways that are seen to be different to their own father's generation (Deeney et al., 2012; Dermott, 2008; Miller, 2010). However, although change is discernible, the reproductive realm remains a morally circumscribed and 'policed' space in which men's place and practices of agency are ill-defined and for some, contested (Allen and Hawkins, 1999; McBride et al, 2005).

It is apparent that ideas about reproduction have been most centrally associated – conflated even – with women, appropriate displays of femininity, and assumptions of maternal 'instincts' in ways which could be argued to have excluded men. Any reframing of reproduction in order to more inclusively take account of men as paternal actors also requires consideration of the implications of change for women's lives as mothers. This is not to argue against actively promoting a more inclusive model which recognises maternal and paternal actors, but rather to acknowledge the complex gendered relations (and so power displays and 'choices') which have underpinned this area and relationships historically (Miller, 2011). Similarly, arguing for women's autonomy over their reproductive bodies has provided an enduring focus for feminists working in a range of different country and reproductive contexts (Chase and Rogers, 2001; Miller, 2005). But this does not have to be incompatible with men's involvement, or fears that they may 'appropriate the reproductive arena' (Markens et al., 2003, p. 475). Indeed recent research tends to show that it is women who mediate and facilitate their husband/partners involvement (or try to engage them) in matters related to childbirth and the reproductive sphere continues to be novel terrain for men anticipating and becoming fathers (Markens et al., 2003; Miller, 2010). To date men's historical exclusion from matters associated with maternity in the UK and other western countries has been contingent on and understood within a framework of hierarchies of masculinities where paid work operates as a highly valued measure of successful, hegemonic, masculinities and where emotionally engaged, child-caring does not (Deeney et al., 2012; Miller, 2010; Wall and Arnold, 2007). But theorisations of gender, masculinity, and gender order have been subject to continued appraisal and new theorisations and more nuanced understandings of these have emerged (Anderson, 2009; Connell, 2005; Hearn and Pringle, 2006; Kimmel et al., 2004; Seidler, 2006).

Masculinity is now much more critically understood as multiple and fluid – rather than singular – thus providing new, less traditional possibilities and opportunities for men to 'do gender' as elements of 'gender fates' are re-imagined (Anderson, 2009; Beck and Beck-Gernsheim, 1995; Connell, 2005; Deutsch, 2007; Petersen, 2003). In early scholarship which sought to theorise masculinity, fatherhood was either ignored or mentioned 'only briefly' and any notion of men as agents in reproduction was absent (Lupton and Barclay, 1997, p. 3). However, more recent theorisations of masculinities as 'inclusive', 'less hierarchical', 'caring', and 'intimate' challenge earlier hegemonic constructions and signal new ways in which men can position their selves without (as much) fear of sanction for behaviours which would previously have been read as emotionally 'weak' or 'effeminate' (Anderson, 2009; Johansson and Klinth, 2007; Seidler, 2006). At the same time these shifts are also discernible in strands of dominant discourse – for example, involved, caring fatherhood – and so provide gender-equality-inflected resources for men to draw upon when narrating emotional and more intimate aspects of their lives. But the discourses which frame men's experiences of reproduction and fatherhood do not presume or suggest biological predisposition and so in the novel terrain of antenatal preparation and hospital birth, men must negotiate a pathway which demonstrates appropriate, if uncertain, preparation. This unfamiliar context, as the data examined in later sections of this chapter will show, leads the men in this study to narrate their transition experiences by invoking elements of different discourses. These include elements of essentialist language – 'bonding' and 'instincts' – much more associated with mothers as well as emotionally embellished language, which sit in contrast to more traditional ideals of rational and autonomous masculinity. But narrating paternal identity as an 'expectant father' can be a complicated endeavour as paternal storylines in the antenatal period are nowhere nearly so clearly prescribed as those assumed of 'expectant mothers'. At the same time hegemonic worker discourses continue to provide a range of normative and, for men, acceptable storylines.

Men's practices of agency in what have been (largely) 'maternal' spaces can then be tentative as they seek a role for themselves which mediates their wife/partner's expectations and those of health and medical professionals who control the formal spaces of reproductive journeys. Research has shown that men can feel like ambivalent 'onlookers' or 'outsiders' in these spaces which are increasingly expected to accommodate these (only more recently recognised) paternal clients: and the professionals can feel similarly ambivalent about their presence (Chin et al., 2011; Draper, 2003b; Locock and Alexander, 2006). The term 'midwife' does after all mean 'with woman' – and professionally some midwives feel this should remain their singular focus. But notwithstanding this there is increasing research interest and recognition of men as 'partners in reproduction' (Lohan et al., 2011, p.1512; see also Dolan and Coe, 2011) as childbirth is anticipated in more shared and inclusive ways (Miller, 2010) and this marks a shift from men being traditionally portrayed 'as relatively unconcerned and unknowledgeable about reproductive health' (Dudgeon and Inhorn, 2004, p. 1381). Reframed ideas around reproduction have then positioned men more centrally in the reproductive arena than at any time in recent history. But the legacy of essentialist maternal assumptions, gendered hierarchies, power relations and displays, and discursive frameworks which can reinforce binary essentialist categories, render this an intricate domain to traverse as the data below illustrate.

The study

This chapter uses data from a UK study that focused on a group of men's experiences of transition to first-time fatherhood. In this study 17 men were interviewed across approximately two years in their lives in which they became fathers. The iterative research process involved interviewing the men on up to four separate occasions, followed by an end-of-study postal questionnaire to collect data on their experiences of participating in the study. This chapter is based on data collected in the first two of four interviews carried out antenatally between seven to eight months, once the pregnancy was well established and postnatally between six and 12 weeks following the birth. The semi-structured interviews covered areas around expectations, birth, fathering experiences, perceptions of self and identity, caring, work intentions, and practices. In the first interview participants were first asked to describe how they had felt when they found out they were to become a father. The subsequent interview began with a question asking participants to describe what had happened since the last interview. This approach gave the men the opportunity to narrate their accounts of anticipating and later experiencing fathering and fatherhood in the ways they chose. University ethics approval for the study required that potential participants must 'opt in' to the research and to provide written consent. In spite of using a diverse range of work places, shops, leisure, and other premises to advertise the research, the eventual sample of 17 men who participated in this (longitudinal) study were all white, employed, heterosexual, and living in dual-earner households. The mean age of the participants was 33.7 years at the time of the first interview; ages ranged from 24 years to 39 years. In the UK, the mean age of fathers in England and Wales was 32 years in 2008 (Office for National Statistics (ONS), 2008).

The interviews were mostly[3] carried out in the participant's home in the evenings or at weekends. All interviews were recorded with the participants' permission and at the end of the study transcripts of the interviews were sent to those who wanted them as a token of thanks rather than for data-checking purposes. The interviews (from across the whole project) lasted between one and three hours and were transcribed verbatim. Analysis of the data was initially individual, thematic, and temporal, involving examining how and when the men drew on different discourses to narrate their initial intentions and later experiences of the birth and early practices of caring. These individual stories were then compared and patterns identified across the whole data set. Interestingly, different elements of discourse which emphasise natural capacities, emotional investment, and biological essentialism – and which consequently have been much more associated with women as maternal subjects – are woven through the men's accounts of transition to first-time fatherhood.

The findings

Across the data from the two interviews the men can be seen to anticipate first-time fatherhood, attending (some) antenatal classes[4] and scans before witnessing the birth of their first child(ren) and then engaging in early caring activities. Engagement in caring for the new baby in the early days was facilitated by the two weeks paternity leave introduced in the

UK in 2003. Not surprisingly anticipation of the birth and fatherhood provided a dominant thread running through the men's anticipatory accounts, which were narrated as episodes of detailed personal transition and subjective experience across the postnatal data. The birth signalled a number of things; it provided a focus for the men's antenatal preparations ('but I'm definitely going to be going along to those [antenatal classes] just to find out what I can do to make her feel any better or to help her get through [the birth]'); it culminated in the arrival of a baby and so provided a physical and embodied involvement they were not able to fully share with their pregnant wife/partner ('but the emotional response of actually carrying the baby, I don't experience'); it was also symbolic and imagined as confirmatory of a new – shared – future as a family ('it's made us think about our future and be really clear about our future ... a future together'). These areas of anticipation, preparation, and birth are illustrated through the data below.

Practices of involvement and preparation

As noted earlier, the ideal of the involved father permeates representations of modern fatherhoods in many western societies, but actual practices have been slower to change (Wall and Arnold, 2007). However, ideas of involvement and practices which demonstrate preparing appropriately for the arrival of a first baby were evident in the men's antenatal accounts. The number of different sources (books, Internet, television programmes, DVDs, etc.) used to find out about pregnancy and birth, and their quantity ('a massive pile of books', 'we have just read hundreds of books', and 'our mini library of baby books'), is held up as an example of appropriate preparation and involvement. In the following extract, Ben talks about 'scans' and 'reading' as providing empowering opportunities for men's involvement:

> It is quite empowering for the bloke to read the books about that stuff you don't really know what is going on and what is going to happen so it is quite good to have an idea ... have a sense of what is just happening at all the stages. You can imagine without having had that involvement with the scans or reading about you can almost put it to the back of your mind you don't know anything about what is going on.

The potential for detachment – rather than involvement – is apparent in this extract. Across the participants' accounts there were examples of the men looking for ways in and seeking opportunities to demonstrate involvement, whilst also expressing feelings of ambivalence, unsure of how normative, masculine practices of agency and maternity and envisaged paternal identity all fitted together. In the following extract Frank talks about his immersion into an arena where all the talk 'is baby orientated':

> Everything is baby – we have two sets of friends, one's just had their second, one their first, so, everything is baby orientated, so I'm quite glad to escape every so often, and get off to watch rugby and just be a bloke for a couple of hours [laughing].

It is interesting that here Frank infers a distinction between things 'baby orientated' and his subjective sense of self as being 'a bloke' which is not associated with baby-oriented things, but is demonstrated through more recognisably masculine activities such as watching rugby. This extract helps to illustrate how different practices and strands of discourses are drawn upon – sometimes simultaneously – as the men seek to position themselves as preparing appropriately, but still engaging in behaviours which are recognisable as unambiguously hetero-normative and masculine.[5] Other aspects of (antenatal) preparation are also drawn upon by the men to demonstrate their masculine subjectivity as distinct from elements of an emerging paternal identity. For example, several of the participants used the expression 'it's not really me', to convey their sense of ambivalence or actual resistance to attempts by their wife/partner to involve them in some areas of antenatal preparation as the following extracts illustrate:

> So yeah I've tried to find out bits about it ... yeah she tried to get me on one of these dad's [internet] cafes, share your feelings with dads and things like that, but I really don't have time [laughs] or don't want to you know. I tried to get into it and tried to put something on there but I just couldn't do it [laughs]. (Stephen)

> I think she would probably really, really like it if I sort of went out and sort of read loads and loads of stuff and watched the baby channel, I know she would really like that but I just, it's not really me ... and like [wife] watches like programmes, she watches for hours but it's not really me. (Joe)

Sharing 'feelings' (emotions) with other men or watching hours of the 'baby channel' are activities that are not closely associated with traditional ideas of masculine behaviours and here these forms of preparation/involvement are avoided or rejected.

More generally reading and attending scans (an interest in the technology which accompanies birth is emphasised in some accounts) are held up as evidence of preparation and involvement by most of the participants. Reading is usually orchestrated by their wife/partner using a range of strategies ('I've had a few books forced under my nose to read' – Stephen). But having lots of 'baby books' is not the same as having read them, as Dylan makes clear 'we have just read hundreds of books ... I couldn't name any of them to you, I just flick through them'. In fact reassurance is experienced by the men through the knowledge that their wife/partner is doing all the reading alongside a belief that she will also know, instinctively, how to give birth and be a mother. But this does leave a space and a role – and a growing cultural expectation – for the men at the birth (Miller, 2013). They anticipate that as their wife/partner will be occupied giving birth, they will be able to be more active, mediating actors, being able to assert agency on behalf of their wife/partner. As Ian says, 'I think my wife will need me there' and Dylan talks of his role as 'trying to comfort ... encourage ... and distract [wife] from all the pain', Graham elaborates these points further in the following extract:

> Yes I think it's a modern convention really isn't it the father to be there at the birth and I think we always assumed that I would be. We have talked about my role and it was a key part of the class and reassuring and if necessary being a bit of an interpreter and standing up for [wife] to the professionals on her behalf and that sort of thing. I imagine it's going to be a pretty stressful experience because obviously she will go through a lot of pain and worry and [I] sort of have to go through it [laughs] precariously with her.

As noted earlier the birth looms as a pivotal event for the men, 'when I think about the birth and the baby arriving I feel quite sort of emotional about that like it is a really big thing' (Ben). The men's anticipated role at the birth will also include helping their wife/partner to adhere to the pain relief and other choices made before the birth, as the process of labour and birth unfold. In the following extract Gus describes how he will be involved in the birth:

> there are reminders pinned up around the house, ehm ... 'each contraction brings me closer to the birth of our child', these are little sound bites that I have to hit her with during the birth ... at night, she sort of says 'right, if this happens, what do you do?' ... but this is

> because I've told her 'look, you need to sort of drill it into me because otherwise I'll get there and I'll be like ... you know ... I'll be all right, I'm pretty sure I'll be all right ... it's quite sort of ... it's daunting ... because I just ... I don't know, maybe it's a pride thing, but you know, I don't want [wife] coming out [of hospital] and someone asking her 'and how was Gus during the birth' and her going 'well, to be honest, he was crap'.

The rehearsing and hesitancy which is apparent in this extract reveals Gus's concerns with managing his role as birth partner alongside a sense of uncertainty (he has not done this before and neither has his partner) and also wanting to be judged to have been successful after the event. Here Gus signals another area of change in which paternal success at the birth interestingly becomes a measure of masculine pride.

Across the antenatal data, men's preparation for birth is illuminated through various activities, which as the birth approaches can include helping to write the 'birth plan'.[6] Although this prompts mixed feeling – about whether birth plans are a good thing or serve only to promote a sense of failure – all the men are aware of their existence: 'we assumed that we had to do a birth plan just because it's mentioned actually everywhere' (Dylan). The birth plan also illuminates changing ideas about men as actors in this traditionally maternal sphere (for example, expectations they will want to cut the umbilical cord once their baby is delivered) as is apparent in the following extract from Richard:

> So in the birth plan it sort of says, I think it's the one thing we've discussed, is do I want to cut the cord and my sense is well by the end I don't think I'd care ... I think the main thing that I'll be trying to focus on, which is probably not what you should be focusing on because it should come naturally, is actually just trying to be helpful to [wife], you know trying to be the appropriately emotional and doing all that stuff, the care bit. (Richard)

In this extract, there is an implicit assumption that men's role at the birth is to be emotionally supportive and caring (the 'care bit') and that this will (hopefully) emerge 'naturally' alongside birth as a process that will be lengthy ('by the end I don't think I'd care') and is unknown. For all the men, 'it's difficult to see beyond the actual day of the birth' (James) which after weeks and months of anticipation, detachment, ambivalence, and longing for, finally arrives.

Labour and birth

The men's experiences of the birth are explored through the data from the first postnatal interviews when the babies were aged between 6 and 12 weeks. All the births were different to what had been anticipated and involved five caesareans, (three of which were performed as emergencies) and several others that involved the use of forceps and are witnessed and described as 'traumatic' and 'scary' ("and they literally dragged [baby] out on two contractions"). Most of the births are not perceived to have been 'natural' or 'easy', although one is described as being 'like a walk in the park' and 'text book'. Analysis of the early postnatal data illuminates the ways in which men's experiences of power and feelings of powerlessness operate as they occupy this historically maternal arena. The men experienced contrasting and conflicting feelings of being peripheral and central, autonomous and highly emotional actors at the hospital birth. The formal health spaces (the antenatal clinic, hospital, and labour ward) provide unique spaces in which to explore practices of masculinities alongside emergent practices of paternal agency. The men's experiences of the process of labour culminating in a birth and their own transition to first-time fatherhood challenge, and at times reinforce, hegemonic ideas of masculinities and essentialisms which are explored below.

The birth is characterised by certainty – that a baby will be born – and uncertainty in that the onset of labour and the length and detail of the birthing process cannot (normally) be known in advance. All the men have eventually either read and/or attended antenatal classes about the birth and all anticipate having a role to play. There is apprehension too as some talk of being 'squeamish' and 'not good with blood' and express concern about how they will respond as the birth proceeds. But there is an overriding sense of wanting to be supportive and protective and to keep the birth on track according to either the written birth plan or the antenatal discussions the couple have had about the process. Plans about the birth are made in hope rather than experiential knowledge as Joe says 'obviously we don't know because we had never gone through this process before'. All the births (but one[7]) took place in hospital maternity units and the foreignness of these is conveyed by Richard as he describes hearing 'another woman screaming in the same way [as wife] a strange mooing scream that women do, but I'd obviously never heard before'. Gus too described his response to the alien environment of the maternity suite as he tries 'to make it look homely, but it just wasn't happening and I felt "oh, I don't know what to do"'.

It is not only the men who had not done this before but their wife/partner too (Miller, 2005, 2007). However, assumptions expressed in the antenatal interviews about their wives/partners instinctive maternal capacities, alongside encouragement to help 'plan' the details of the birth process, provide the men with a (false) sense of reassurance that the process is natural and so somehow (relatively) uncomplicated. Experiences of witnessing the birth and seeing their wife/partner in unexpectedly uncontrollable pain as the birth proceeds are for most shocking, as Sean says 'it was just really horrible, not the birth, but seeing [wife] in so much pain. That in itself was just a horrible, horrible, experience'. Feelings of helplessness were clear across much of the data as illustrated in the extracts below:

> But it wasn't that comfortable to see it, it was horrible, knowing that there's nothing ... you normally, if she's in pain, I would comfort her, stroke her hair, hold her, but she didn't obviously want me to touch her. (Gus)

> They were doing things to her which [were] really hurting her and I found it really hard just to see her go through all this pain and know it wasn't going to stop. (Stephen)

The men experience feelings of powerlessness as they realise they are unable to protect their wife/partner from the pain of childbirth as labour unfolds and they can only look on and witness, rather than ameliorate, the pain. In the antenatal accounts all the men had recognised the normative expectation that they would be at the birth and they had mostly anticipated being active partners in the process. But as the actual process of birth is encountered there is a sense of confusion over how central or helpful they can be and this is reflected upon in the postnatal data. As Sean says 'but I felt, no spare part's wrong, but in real terms I didn't actually do anything'. In the following extract Richard too alludes to a sense of his own confusion about his role at the birth:

> It was just a very weird experience to witness this thing and sort of clearly to understand that it's quite important for me to actually just to be there. There's nothing for me to really do. (Richard)

The ambivalence around paternal actors in maternally etched spaces is evident in these extracts as the 'importance' of being at the birth is acknowledged on the one hand, alongside feelings of being peripheral when it turns out that there's nothing to really do, on the other.

During the antenatal interviews most of the men have ideas about how they envisage their role at the birth unfolding. These ideas are informed by the various preparation sources and activities they have engaged with ('we went to an active birth class') and mostly involve seeing themselves as adopting an active 'mediating' or 'interpreting' role between their wife/partner and the professionals managing the birth. During labour and birth this envisaged role, which implies positioning themselves as in control, rational, and autonomous is tested, invoked in rare instances (e.g. demanding to be moved to a larger maternity unit in a different hospital) and, as labour and birth proceed, is in most cases abandoned ('I was just a mess a blubbering mess'). Concealing unexpected or alarming aspects of the birth was one practice some of the participants alluded to ('I was pretending that it was alright, even though I thought everything was going horribly wrong'). Retrospectively 'traumatic' and 'horrible' experiences of birth and feelings of being powerless and/or things being out of control led some to question professional expertise and information sharing in the antenatal period:

> And the only question I would have is the baby was obviously big and there was, and [wife] obviously must be quite small, so why was it not thought of sooner that there might have been a problem? (Sean)

Reflecting on the professionals who have managed the births elicits mostly positive, grateful responses ('basically the midwife was really good', 'I kind of stood there not really knowing exactly what to do and I think the midwife was absolutely fantastic', 'I'm glad there's loads of doctors there, 'No I didn't speak to the doctors whatsoever', etc.). The men's reflections on their own performances elicit a more varied range of descriptions ('she says I was useful', 'well I sort of tried not to get in the way', 'they gave me a pillow to hold because I was so nervous', 'I'm like "Oh, I don't know what to do"') and some feel they could have done better. Not keeping to the birth plan or pre-birth discussions, especially in relation to pain relief, leads some to express a sense of personal failure as they reflect on their inability to control events as the uncertain process of birth had unfolded. Uncertainty and failure are not characteristics readily associated with dominant ideals of masculinity yet are woven through many of the men's accounts including Ian's below who feels that he 'did fall down':

> He was [delivered by] forceps because after we got an epidural she had tried not as many positions as we would have liked, and that is where

> I did fall down a little bit. Because we had practised more positions and I think I should have persuaded her that she should have moved around a bit more ... And again that's another thing we put down in the birth plan that ideally his umbilical cord should stop pulsing before it was cut, that all went out the window. (Ian)

Clearly writing birth plans in the antenatal period induced a sense of preparing appropriately and being ready for the event of birth. But the uncertain process of birth means that things may not to go 'to plan' ('We did write a birth plan, won't bother next time'). In the early postnatal data there is a discernible sense of having been unprepared for labour and birth which is not 'text book' even though classes have been attended and books read. As Richard observes, 'the other thing the preparation classes don't say ... they don't say you will be standing there and you'll feel really useless'.

Following labour which unfolds in uncertain and varied ways and results in different types of (assisted) birth, live babies are eventually delivered. The birth is a momentous and pivotal moment in which emotions and new responsibilities converge as the men become fathers. The clarity for some of the men of how their life had shifted in a single fragment of time is conveyed in the following extracts as new responsibilities and instant emotional connections are described:

> And there was quite a mental switch from worrying about [wife] and how she was doing to you know more than I thought there would be, suddenly there was an obvious switch to yes worry about [wife] but also worry about the baby as well and how's he doing and that was an interesting moment that I hadn't really anticipated. (Graham)

> I was crying my eyes out, floods of tears ... It is, it's I can't even, oh, I wasn't thinking about I was crying, it was just like "I don't even know what I'm doing", it was amazing, yea, it was utterly amazing, and then they wrapped him up and they plonked him down, ... and I was just holding him, and as soon as ... he was crying his eyes out, and then they put him into my arms, and he stopped crying instantly ... that was nice ... and then he just lay there, eyes wide open and stared straight at me for, I don't know, five minutes. (Gus)

> And [the baby] came out and I was sobbing with tears of joy, nobody could just console me at all you know the best and worst day of my life it was amazing absolutely amazing. I got first hold of him while they stitched [wife] back up. (Stephen)

Whilst holding and meeting their first-born child is an emotional event for all the fathers, it evokes a range of responses. These range from claiming an immediate, instinctive connection, or 'bond' (in ways more often assumed of women) to their new baby, to more tentative emotional accounts as seen in the following extract where William describes first seeing his baby daughter:

> You know massive affection but not a massive bond. I wouldn't say there was like our eyes met and it was just the most amazing thing. I wouldn't say that, you know but clearly affection, lots of affection but concern for [wife] as well.

The tentative tone in which William describes his feelings could be argued to be more acceptable as an expression of early paternal identity than would be the case for a new mother. Societal and cultural expectations do not associate men with essentialist ideas of immediate attachment/bonding to their child and so more ambivalent ways of describing new parental relationships are acceptable, which is not the case for women (Miller, 2007). But, interestingly, descriptions of attachment and bonding are found across the men's early postnatal accounts – 'I think the biggest surprise is the bond between me and baby son', 'I thought I wouldn't necessarily know what to do ... but it's kind of instinctive' – as their individual journeys into first-time fathering begin.

Conclusions

This chapter set out to explore men's experiences of emerging paternal identities and associated behaviours in a sphere of the social world most essentially (and historically) delineated as maternal and feminine: reproduction. The findings from the UK-based qualitative longitudinal data presented here provide evidence of both some change and yet some elements of continuities as the complex interplay of assumptions around gender, biology, and appropriate and acceptable reproductive behaviours unfurls. The data point to these men's recognition of a greater emphasis on particular paternal practices (their preparation for the birth through reading, attending antenatal classes, contributing to birth plans and decisions concerning types of pain relief and details of delivery) which lead them to envisage having a particular supportive, mediating role at the birth. In part this is further supported by the introduction of paternity leave in the UK in 2003 and a greater emphasis in policy debates and the media on men's involvement in fatherhood. But births which follow uncertain trajectories (as almost all births do) can position the men as passive witnesses rather than the active mediators for which they had prepared. An implicit but mostly unspoken element of their envisaged role leading up to and at the birth, involves recognisably masculine and patriarchal ideals, being in control and providing protection. But when they find that they are unable to protect their wife/partner from the extremes of childbirth pain, or ameliorate its affects, or adhere to the birth plan, their own vulnerability comes into focus. Yet whilst vulnerability runs counter to hegemonic ideals of successful, strong, and singular notions of masculinity, in this context it illuminates a range of emotional responses which confirm ideas of masculinities as fluid, caring, and changing.

The findings confirm that the men embark on journeys into first-time fatherhood drawing upon elements of involved fatherhood discourse and ideas of gender equity especially in relation to anticipating caring for a child, but in which normative ideas of rational, autonomous, hegemonic masculinity feature as a dominant strand. However, the process of witnessing the birth of their first child and becoming a father evokes for most profoundly emotional responses which are expressed in the language of 'instincts' and 'bonding', more traditionally associated with maternal rather than paternal subjectivities. It is apparent then that in the UK, reproduction and those who have historically populated and managed these spaces are undergoing change. The analysis of paternal experiences revealed in this chapter challenges and confirms, normative, assumptions and practices which operate across reproductive realms and highlight a need to think more carefully about how paternal subjects are included into this arena.

Notes

1. For example, see Bailey (2011).
2. The classed, 'raced', and gendered dimensions of these arrangements are of course important to note (see Gillies, 2009).
3. A minority of participants chose to be interviewed at either their workplace or my university office.
4. Men in the UK are able to attend/accompany their wife/partner to antenatal classes and scans provided through the National Health Service (NHS). In addition some couples make private provision to attend a series of preparation classes or one-day courses (e.g. on active birth) provided by organisations such as the National Childbirth Trust (NCT) (see http://www.nct.org.uk/courses).
5. Of course pregnancy is new for women too when they contemplate first-time motherhood, but their socialisation and all the messages which circumscribe

ideals of femininity, maternity, and reproductive bodies in western, pronatalist societies make this a more anticipated space for women – which does not mean that it is not also experienced as confusing and perplexing (Miller, 2005, 2007).
6. 'A birth plan is a record of what you would like to happen during your labour and after the birth' (see National Health Service, 2013).
7. One birth was a planned home birth.

Part III
The (Global) Reproductive Marketplace

11
Putting 'Daddy' in the Cart: Ordering Sperm Online

Lisa Jean Moore and Marianna Grady

> NEW Graduate Donor Alert!! Donor 2775! CMV Positive Donor 2775 is a mature and good-looking man. He is a devoted father who very much wants his daughters to experience, as he has, the richness of our world. Adventurous at heart, he has travelled to and lived in many parts of the world including a tour in the Peace Corps. He is an open-minded and enlightened individual. He is now an ESOL [English to Speakers of Other Languages] teacher and very invested in helping children succeed.

Like a birth announcement or a product launch, this happily punctuated description, found on the Fairfax Cryobank Facebook feed,[1] conveys the excitement of a company that is introducing brand new inventory. A click away, the YouTube channel for Cryos International offers 'what you need to know' videos about the intricacies of home versus intrauterine insemination and fertility timing. Shifting to a different computer screen, the California Cryobank's Facebook page displays users' photographs of adorable cryobabies and provides a venue for questions about donors, such as 'Anyone else use donor 11397?' Social media's commercial thrust has entered the reproductive marketplace, and these examples attest to the fact that purchasing human semen is not as unusual or uncommon as it once was. As businesses, sperm banks must contend with competition – competition for the best and brightest donors, and more importantly, competition for market share, otherwise known as customers, users, or recipients. The growth of social media outlets has spurred the development of new competitive strategies available to businesses, a development that has influenced the practice of several cryobanks. As a result, the use of these websites has begun to reorient social norms surrounding the procurement of semen. These shifts have created new frontiers for parents and have modified the ways in which reproduction, masculinity, and kinship are understood in the early twenty-first century.

In 2007, Lisa Jean Moore's *Sperm Counts: Overcome by Man's Most Precious Fluid* argued through content analysis of printed donor catalogues that masculine hierarchies and hegemonic masculinity are enacted through advertisements and advertising strategies produced by sperm banks. Key findings showed how sperm banks reproduce masculine ideals through their presentation of phenotypic, biological, and social characteristics. These groupings categorise men in very specific ways that highlight:

> [S]ocial and physical power. Height, weight, body build, and favourite sports provide indicators of a donor's health and ability to be physically dominant. Categories of occupation, grade point average, and years of college provide indicators of social survivability and social dominance ... the categories tend to reify the power differences among men (Moore, 2007, p. 109).

Since the publication of *Sperm Counts*, the sperm banking industry has expanded and enhanced its own promotion both by emulating shopping websites and by utilising social networking websites. This essay revisits the arguments presented in *Sperm Counts* through interpreting the ways that the semen banking enterprise works to stay relevant on the Web.

Using online sperm bank materials, this chapter explores the digital representation of sperm for the next generation of consumers. We investigate how semen is stylised and narrated through online platforms in Web 2.0[2] and analyse how this information creates different cultural meanings for sperm, masculinity, and childrearing. Questions we explore include: How has Web 2.0 changed the sperm banking industry? How have sperm banks responded to the growing use of social media? In what ways have sperm banks made their websites more interactive? In what ways do Web-based modes of information, such as user-generated content and social media, affect the understanding and consumption of sperm for reproduction? Most significantly, we argue below that new forms of biomedicalised intimacy emerge between users, while simultaneously dematerialising and therefore erasing the actual man from the process of kinship.

Cyberethnography of reproductive commercial spaces

Our study is primarily an exploratory, qualitative, and interpretive analysis of Web-based sperm bank websites that engage in social networking. Our data were collected over a six-month period in 2012 and interpreted based on a modified grounded theory methodology.[3] By triangulating data sources about five US sperm banks and the means they use to represent their product on several websites, we were able to establish various points of comparison to explore multiple concepts about sperm in different social medial environments. This empirically based methodology combines grounded theory techniques with content and discourse analysis as a way to develop theoretically rich explanations and interpretations of semen, sperm banking, and advertising strategies. We were able to track both the frequency of certain terms and representations, as well as the interpretative meaning and significance of these expressions. Similar to other qualitative research, content analysis can be both exploratory and descriptive, enabling partial insight into why significant relationships or trends occur. The aim is not toward standardisation of facts into scientific units, but rather to appreciate and explore the range of variation found within a given phenomenon. Outliers (representations that do not fall neatly into the most common themes and concepts) are useful because they enable analysts to capture the range of variation and dimensions of the concepts.

Our methodology was also informed by Kuntsman's (2004) articulation of 'cyberethnography'. Conducting a cyberethnography engages with the Internet as a 'physical' location that may be explored and interpreted. Our field for data collection consisted of Web-based spaces that allow users to virtually interact with sperm bank operators, who in turn give users access to representational donors, since many are unlikely to be available in 'real-time' in these online spaces. Obviously our study is not exhaustive of all sperm banks and users. However, we have attempted to collect data from sperm banks representative of a range of variation. Furthermore, since we did not interview individual users of these websites, we cannot make claims about users' experiences of the websites. Despite these limitations, our interpretation does reveal the diversity of techniques that sperm banks use to attract users, and it addresses the unintended ways that users engage with sperm bank platforms to investigate their product.

For our study, we analysed the company websites of five sperm banks and the several social media-related websites that these banks, described below, used to reach consumers. For the company websites, we analysed the arrangement of information and the ways in which the sperm banks chose to display their product. The inclusion criteria for social media websites were defined by the capabilities of commenting on posts, 'liking' specific posts and comments, reposting posts and comments, or responding to posts and comments. While each social media website does not offer total access to each of these features, we found that Facebook, Twitter, YouTube, and various blogging websites have some, if not most, of these features. In analysing these criteria, we were able to examine how each of the five sperm banks used such features to attract shoppers.

The biomedicalisation of sperm: from consumerism to kinship

The US sperm banking industry is at a unique crossroads in the early twenty-first century. As the use and diversity of fertility services grow, sperm banks are able to offer their customers more services than ever before. Likewise, the development of Web 2.0 has provided sperm banks with distinctive pathways to new and current consumers. Furthermore, this confluence of biotechnological innovation and Web-based social media marketing is not exclusive to the sperm banking industry; these shifts affect all biomedicalised industries in the early twenty-first century. Science studies scholars explore the political, economic, and social effects both inherent within and produced by biomedicalisation. Biomedicalisation is a term that describes:

> the increasingly complex, multisited, multidirectional processes of medicalisation that today are being both extended and reconstituted through the emergent social forms and practices of a highly and increasingly technoscientific biomedicine. We signal with the 'bio' in biomedicalisation the transformations of both the human and nonhuman made possible ... in new and complex, usually technoscientifically enmeshed ways (Clarke et al., 2003, p. 162).

Deciphering 'technoscientific identities' greatly impels our research, more specifically through the ways in which it relates to the processes of (re)formulating identities through the constantly shifting biomedicalised sperm banking industry. The creation of technoscientific identities produces, in turn, 'transformations of both the human and nonhuman' (ibid.), constructing identities for possible families through the imagined disembodied sperm donors and the yearned-for potential babies.

In relation to donor identities, Moore (2007) explores the conjuring of sperm donors through the presentation of phenotypic, biological, and social characteristics. The importance of such characteristics is ranked as they are communicated to potential clients, a gesture that mirrors contemporary understandings and values of fatherhood and masculinity. Additionally, Schmidt and Moore's (1998, p. 25) analysis of sperm bank marketing techniques suggests that banks work to effectively create cyborgified 'technosemen' that dematerialises, distils, and disembodies masculinity. However, the 'technological manipulation of semen' does not end with sperm becoming physically modified. The Internet provides consumers with access to information about technosemen, the processes through which technosemen is manufactured, and the new improvements in technosemen that sperm banks provide (Clarke et al., 2010, p. 177). In order to keep up with current demand for spermatic biomaterial, the sperm banks must equally be knowledgeable of the advancements described on the Internet to stay competitive in the online marketplace.

Mamo (2010, p. 178) alludes to the importance of the Internet in shopping for sperm, noting that buying sperm has predominantly become an 'online trade'. The Web-based commodification of sperm serves neoliberal consumer tendencies towards 'a consumer society marked by ideals of ownership, presumed individual choice, and consumption as means to fulfil one's desires, identities, and life goals' (pp. 189–190). When choosing a sperm donor, consumers are also portraying themselves and their prospective families to their respective families and in their social surroundings. Hertz (2002, p. 2) notes that there is an apparent difference in how mothers describe identity-release donors and anonymous donors to their children; anonymous donors are described through the phenotypic, biological, and social characteristics that the mothers glean from their donor profiles, and identity-release donors are described through 'more concrete, personal knowledge'. Sperm banks uphold the sense that these atomised donors can be transformed into a father-like image, which can be seen throughout the Web-based donor descriptions available for prospective families (p. 8).

Both anonymous and identity-release donors are described to children to establish the child's sense of self, which Mamo (2005, p. 251) also describes and notes as 'a potent narrative in US culture'. But in what ways are these typified characteristics of donors really constructing a father-like figure? Pieces of donor information are being conflated with a man, which is ironic because, ultimately, sperm is not just a product, but rather is biomaterial from an actual person (Hertz, 2002). Do these donor descriptions actually provide a sense of corporeality? In her exploration of the commodification of male and female sex cells, Almeling (2011) addresses some of the consequences of creating technosemen in terms of the identities of sperm donors. In her research, Almeling finds that male donors feel 'objectif[ied], alienat[ed], and dehumaniz[ed]' as their lives become represented by sex cells (p. 170). The (lack of) identity for sperm donors illustrates the ways in which the biomedicalised sperm banking industry commodifies health and biomaterials in detrimental ways. Chiefly, the sperm banks' company websites give potential parents the ability to search for possible donors with unprecedented speed and detail. It is also impersonal, since, unlike many cases of commercial egg donation, recipients never meet the sperm donor prior to pregnancy (and only sometimes after the child's eighteenth birthday) and the ease of clicking accelerates access to a profusion of online profiles. Like any online shopper, a user can quickly discard a donor when a different one catches her eye. Nonetheless, this access leads to much thought and negotiation among parents and single women respectively. Mamo (2005, pp. 237–238) investigates the ways in which lesbian couples 'negotiate and construct' their desired child by focusing on how the listed sperm donor characteristics aid in the illustration of possible 'affinity-ties'. She defines affinity-ties as families imagining futures with children based on donated biological and social relations.

In her findings, Mamo discerns that this form of biomedicalised kinship is mutually dependent on the users, the information and services provided by the sperm banks, and the selection of a sperm donor. Consumers consider many important details when formulating these affinity-ties, such as how their health histories may correspond with the donors, the donor's physique and education, 'shared histories, physicality, and social subjectivities with donors' (Mamo, 2005, p. 237), the availability for the child to know his or her own genesis through identity-release donors (p. 252), and other factors. All of these details are available in the search options on the sperm banks' websites, through which users can select the features that they want their child to possess or have access to. Importantly, Mamo notes that while this technoscientific production of sperm may dictate hegemonic understandings of reproduction, it is ultimately the lesbian couples who 'interpret, respond [to], and modify' this information to their advantage (p. 240). Sperm banks may organise the websites (and thus command which phenotypic, biological, and social characteristics are understood as significant), but it is in interacting with these choices that users produce multifaceted meanings for both the biomaterial and for the sperm banking industry as a whole.

The rise of Web 2.0 has created interactive online spaces that facilitate connections between families who are related via sperm donors. Hertz and Mattes (2011, p. 1152) highlight the important role that the Internet is playing in formulating relationships. Their research reveals that Web 2.0 directs and informs the donor-created families, allowing them to both reach out to other donor-created families and negotiate the level at which they interact with these other families. Importantly, these donor-created families use social media websites such as Facebook for these interactions – to share photographs, to message one another, and to comment on each other's profiles. In our research, we found that the predominant online outlets that sperm banks utilise for outreach are the sperm banks' Facebook pages and the forums available on their company websites. Sperm banks provide users with these outlets to record their journey of creating a family and to connect with other users going through the same processes.

Significantly, sperm bank-sponsored outlets were created after other donor-sibling websites (such as Donor Sibling Registry and certain Facebook and Yahoo groups) were already established; it is therefore important to analyse their specific effects. Hertz and Mattes (2011) note that, prior to the advent of the Web, sperm banks would have never imagined their users connecting with each other in the ways that they do today via the Internet. While websites offered by the sperm banks give users access to create kinship relationships with other donor-related families (p. 1152), they also show a strong sense of capitalising on consumer wants. For example, Moore (2007) describes that in tracking consumer's desires (demand), sperm banks have height requirements of donors (supply). One of the strategies many commercial websites employ has been to keep users on the website for as much time as possible. The sperm bank websites that we analysed appear to be using similar tactics, offering their users yet another way to co-opt technologies in order to modify and generate meaning; families utilised the Internet for their advantage in any medium, whether the venue was generated by the sperm bank or an outside source. This evokes the changing and continuously shifting accesses to and understandings of social and familial relationships that have been influenced by the rise of the Internet (Hertz and Mattes, 2011, p. 1141).

Branding the sperm bank online

The five banks that we investigated were The Cryobank of California, Xytex Cryo International Sperm Bank, California Cryobank, Fairfax Cryobank, and Cryos International. We selected these sperm banks primarily as a convenience sample with an eye toward the breadth of the field of sperm banks. These particular sperm banks were among the most common to appear in a variety of search terms on Google. We also attempted to select sperm banks based on geographical location (although all ship internationally), as well as their use of social marketing, defined by the sophistication of their websites and their utilisation of Facebook, Twitter, YouTube, blogging, and other new media strategies. We considered the ways in which these five sperm banks use social media as a means of establishing a 'brand personality,' and how this branding offers a competitive edge in the pursuit of consumers (also known as recipients or users) beyond the mere quality of the spermatic inventory.

Blackston (1993, p. 113) argues that brand personality and brand image seek to reflect and emulate consumer values, and that this likeness helps consumers to create a relationship with the brand based on the brand's attitude. These 'brand relationships' are not necessarily based on how the brand actually perceives the consumer, but rather is based on how the consumer recognises the brand's perception of the consumer. This complex engagement with consumers exemplifies the nuances that exist within brand relationships, an engagement that extends beyond a simple catchphrase. The sperm banks that we analysed have intricate brand relationships with their users, and these relationships are sustained through the marketing techniques employed on their company websites and the various social media platforms that these sperm banks utilise. These five sperm banks use the Internet to reach their customers primarily because it is expected in today's marketplace. Multifaceted brand relationships depend on brands to be 'proactive' and 'supportive' (pp. 121–122). Examples of these concepts were observed throughout our research. Of the sperm banks that use all or most of the social media outlets that we analysed, they almost always framed their pervasive social media presence as an effort to address their users' needs.

Sperm banks 'package' both themselves and their sperm donors via their websites, making an effort to personify the semen and stylise the sperm banks. Generally, each of the sperm banks suggests a brand image that is polished, highly inviting, yet quality-conscious. While presenting themselves as down-to-earth, they reassure their users that they are receiving top-quality products from a friendly and relatable company. The main page of all five sperm bank company websites contains similar images of babies, while four of the five (80 per cent) sperm banks include images of donor-created families, and three of the five (60 per cent) include a male or multiple males who represent the sperm banks' pool of donors. These presentations are important because they embody the company's brand image. For example, The Sperm Bank of California highlights how they are 'Proudly serving lesbians & singles' in creating families, and they thus display multiple images of these sorts of families on their main webpage. Likewise, Xytex Cryobank states on their website's main page that 'Xytex focuses on a diverse donor catalogue and we're constantly adding new donors'. The company denotes this through the display of their various donor-selecting products and images of the donors on their main webpage. Actions such as these generate reactions within users, impelling them to respond to these brands in significant ways.

Table 11.1[4,5] describes the variations in Web presence by each of the five sperm banks. The findings show that each sperm bank has some form of a Web presence. The sperm banks that create exposure in each area of social marketing create a multi-sited and striking presence that enhances their dominance. Likewise, those that do not develop and re-develop a presence on each of the various platforms are difficult to find on the Internet. Three out of five (60 per cent) of the sperm banks offer websites that allow their customers to shop for sperm online. Additionally, each of the sperm banks has a blog and a Facebook fan page, while three of the five (60 per cent) sperm banks have a Twitter page and a YouTube page. Interestingly, the sperm banks use each website to promote the other Web-based outlets that they utilise.

These findings articulate the various ways that Web 2.0 platforms command the current marketplace for commercial goods. Facebook, Twitter, YouTube, and proprietary forums are time-consuming services to create and maintain, but each sperm bank appears to be in a race to establish their dominance on the Web by utilising a full range of Web-based modes. Most commonly, sperm banks lead with their 'ace in the hole' and employ social media to boast about the babies created by their products. Baby tweets, photographs, and status updates direct users to advertisements about the products and services offered by the sperm bank. Clickable feature stories are informational and educational, and news feeds aggregate stories about reproduction and fertility. Each sperm bank employs its own combination of social media websites,

Table 11.1 Variations in web presence by each of the five sperm banks studied

	Website	Able to create a profile/ account	Favourites/wish list/shopping cart/ purchase history	Voice verification/ baby pictures of donor	Online forum	Facebook	Twitter	YouTube	Company blog
California Cryobank	www.cryobank.com	Yes	Favourites: Yes. Shopping Cart: Yes	Yes/Yes	Yes	2,198 Likes/followers Page Created: 11 June 2008	953 followers	25 subscribers/7,023 views Latest activity: 14 August 2012 Date Joined: 12 March 2010	Last updated: 22 November 2011
The Sperm Bank of California	www.thespermbankofca.org	No	No	No/Yes	No	23 Likes/followers	No	No	Last updated: 20 September 2011
Xytex Cryo International Sperm Bank	www.xytex.com	Yes	Favourites: Yes Purchase History: Yes	Yes/Yes	No	819 Likes/followers	43 followers	No	Last updated: 4 August 2011
Cryos International	ny.cryointernational.com	No	No	No/Yes	No	223 Likes/followers Page Created: 20 April 2010	No	29 subscribers/45,257 views Latest Activity: 3 March 2011 Date Joined: 20 February 2010	Last updated: 1 April 2011
Fairfax Cryobank	www.fairfaxcryobank.com	Yes	Favourites: Yes Shopping Cart: Yes	Yes/Yes	Yes	671 Likes/followers Page Created: 10 May 2011	690 followers	0 subscribers/113 views Latest Activity: 23 April 2012 Date Joined: 23 April 2012	Last updated: 12 September 2012

which suggests an awareness of the kind of customer service they would like to provide in their brand relationship with consumers.

Regarding the company websites of the sperm banks, three out of five (60 per cent) of the sperm banks offer advanced consumer-driven websites through which consumers 'create a profile,' subsequently allowing a user to save searches, monitor purchase histories, and 'put daddy in the cart.' In addition, users can purchase what we call additional *sperm donor previews* (including pictures and audio files of the sperm donors) that are only a click of the finger (and a credit card entry) away. These features are unique in that their functionality depends on being Webbased; refining a search or placing an order via computer is made easy for consumers, and viewing donor profiles is effortless when users can save their searches and have multiple tabs open in their Web browser. Amplifying the already established affinity-ties of biomedicalised kinship, these sperm donor previews and the imagined 'daddy in the cart' become increasingly 'real' through the proliferation of data about him. 'Daddy' has been reduced to bits of code, sonic waves, and biological cells – all handily added to the cart for a later check-out.

Two of the five (40 per cent) sperm banks provide users with a forum in which they can post. These forums allow users to interact with other users of the same sperm bank, as well as with the moderator designated by the sperm bank. Both forums are identically formatted, in that they provide users with access to information about the sperm bank, frequently asked questions, and company announcements. More importantly, users are able to engage with the more personalised, user-specific subsections of the forums. These forums include subsections related to single mothers, same sex couples, couples with fertility problems, and information about 'selecting a donor'. In addition, satisfied customers share their personal stories about using the sperm bank and their journey to pregnancy and familyhood. Interestingly, these forums allow users to connect with one another and increase opportunities for biomedicalised kinships beyond the sperm bank (Hertz and Mattes, 2011; Mamo, 2005). Users ask each other which sperm donor they are using or which ones they have used to see if there are shared donors. This sort of dialogue is similar to the types of interpersonal connections made on websites such as The Donor Sibling Registry, except that these forums allow users to connect before a child is even produced. In fact, the user bypasses any need for a Donor Sibling Registry, as it is an anticipated biomedicalised connection that is being narrated. The immediacy of providing this forum-based information for consumers is exemplified by the fact that Fairfax Cryobank now offers a mobile phone application to access their forum. Although originally intended for the forums, posts are also present on the Facebook pages, demonstrating the ways in which users modify these Web-based spaces to fit their needs.

Importantly, a common theme with the social media outlets, the company websites, and the forums is that they allow the sperm banks and the users to establish a 'buzz' for past, present, and future donor profiles. The sperm banks utilise these three venues to present 'new product launch coming attractions' for soon-to-be-available donors, describing their features and release dates. Additionally, sperm banks will let users know when a previously unavailable donor has been restocked. To promote past, present, and future donors, sperm banks post details on their Facebook pages and on the forums, even offering discounts for specific donors on these websites. In addition, Twitter congratulates expecting parents, which surreptitiously boosts confidence in the sperm banks' products. For example, on 6 June 2012 California Cryobank exclaimed, 'Another Scorpio CCB baby is on the way. That's not morning sickness; it's an overflow of passion from that confident, little winner!' Interestingly, we found that much of the buzz around donors comes from Facebook and forum posts submitted by sperm bank users. The implications of this user-generated buzz are described below.

Fronting for the donor

Men are everywhere and nowhere on sperm bank websites. Their imagined bodies are described and evoked, their reproductive material is enhanced and marketed, and their offspring are featured and celebrated. But the men themselves are mute. Who is doing all the 'talking'? Rather, consumers and sperm banks become agents or virtual barkers for men who donate sperm. Through posts, comments, and tweets, male bodies are described using socially desirable attributes. For example, as Fairfax Cryobank tweets, 'NEW Donor 4538 = Total cutie and outdoorsman! He is handsome with sparkling beautiful light blue eyes contrasted' Donor 4538 is now reduced to not only a vial but a virtual presence to be retweeted and texted.

Sperm banks assiduously work to manage their user-friendly, (re)productive websites through carefully crafted branding, but they are not the only users generating content on their sites. Impression management, as detailed by Goffman (1959), is generally understood to be the subconscious or conscious work we do to create a presentation of the self to others that achieves the appropriate reaction. Our lives are a constant engagement in impression management at our jobs, in our communities, and in our intimate relationships – all working so that we can present ourselves as we would like to be read. Sperm banks use impression management strategies combined with their brand identity to establish a relationship with users and to manage the user's relationship with particular donors. Users are encouraged when they know that a particular bank's spermatic inventory is ample and works efficiently to create cute babies. A familiar platform, such as an online shopping template, manages the impression of the user by signifying a common experience; 'I know how to do this. I look for the variables that I like. I can read the user reviews and place what I want in my cart.' Any sense of stigma or fear of the unfamiliar is assuaged by the use of familiar online shopping routines combined with the carefully crafted impressions of donors.

There is a word-of-mouth community in sperm bank social media platforms that mirrors user reviews, similar to those found on shopping websites. However, instead of reading and writing reviews of clothing and other consumer goods, users are focusing on spermatic biomaterial. The reviews are primarily user-driven, and the sperm banks' social media and forum moderators chime in only occasionally to correct or direct a user in a particular direction. For example, it is common on the California Cryobank, Fairfax Cryobank, and Xytex Cryobank Facebook pages for users to ask other users general questions about donors. (For instance, on 1 June 2012 a user posted on Xytex Cryobank's Facebook page, 'Hi there we are proud parents after using donor bfm 5054. Anyone else used him?') Posts such as this also appear on the forums. Dialogues about prospective donors are commonplace on forums and Facebook. Interestingly, donor numbers are suggestive of product numbers, so even the mention of 'Donor #4562' invokes a similar sense of consumerism. Users also post pictures of their newborns with heartfelt sentiments of thanks to the sperm bank and their donor (usually indicated by the vial number). It appears that these Facebook and forum posts are unintentionally providing review-like information, which in effect generates a greater desire for specific donors. These unintended consequences of social networking exemplify the complexities within online sperm marketing and the Web-based sperm market: Users effectively become advertisers for the sperm bank through their online posts. They are vouching for the donor's reputation, affirming their own reproductive choice, and promoting the sperm bank in one fell swoop.

Thus, a homosocial female community emerges, providing information and advice for one another about the pending purchase in a forum disquietingly similar to that of the community of helpful fellow shoppers at Zappos.com, a popular US online shoe outlet. These virtual relationships with other women indicate that there is some type of a vibrant and developing community in pursuit of the normative and (relatively) socially valued status of being pregnant (Mamo, 2005, p. 256). For example, women who post about infertility issues on the forums are received with well wishes from other users and are provided tips for better results. Significantly, the variety of consumer's interaction with the sperm banks, other consumers, and the imagined donor collaborate to create a mediated community in which single women and couples are interactive and the donor is static.

Facebook and forum users are able to choose their level of participation (Hertz and Mattes, 2011), and, in fact, the number of views on the forum posts indicates that more users lurk than post. In our careful sifting through data, there did not seem to be any sperm donors (or men for that matter) posting on any of the websites. However, the men could be some of the 'lurkers', checking in on their relative rank among the users.

YouTube channels are a popular source of information for consumers due to the speed at which the videos disseminate information. The videos present on the YouTube pages centre around providing information, but they inadvertently represent the sperm banks' capitalistic enterprise. An example of this is the production values of the videos; some sperm banks, such as California Cryobank, have highly produced and stylised videos, while sperm banks such as Cryos International and Fairfax Cryobank have under-produced videos that appear to have been made with a handheld camera.

Furthermore, the YouTube videos direct users to understand the information through the sperm banks' modes of representation. Thus Schmidt and Moore's (1998) understanding of technosemen is retooled via the multimedia format of donor-selection guides and the detailed informational videos describing the semen-cleaning processes. In addition, sperm banks weave a narrow and normative narrative around their product, more centred on the shopping experience than the sociocultural implications of the product's connection to fatherhood. The sperm banks focus on the 'fun' aspects of choosing a sperm donor – what he looks like, what his hobbies are, how these factors may affect your imagined baby, and the ways in which you can imagine a fully realised donor daddy (Mamo, 2005, p. 253). They do not highlight the consequences of using their product, such as the possibility of biomedicalised kinship formations (Hertz, 2002, p. 6; Hertz and Mattes, 2011, p. 1136). These biomedicalised kinship formations, impelled by social media platforms, are transforming the notion of fatherhood and creating greater slippage between fatherhood, a social role, and paternity, a biological contribution.

As Almeling (2011, p. 167) highlights, the sperm banks make their donors out to be a 'daddy' insofar as the donor does not impede on the family purchasing the sperm. In atomising and disembodying the sperm from the male, sperm banks are creating and supporting a narrative concerning contemporary fatherhood and masculinity that places donors even farther from the flesh (and the industry). By providing the raw material, sperm banks obviate the need for the presence of men, further dematerialising the male body, as they distil it and then generate content all around it. Actual men are disassembled and then reassembled by the users who create narratives about them, conversing over forums and posting status updates. Virtual realities stand in for actual men, and sperm vials are traded over Internet connections. This dematerialisation shifts cultural understandings of masculinity and fatherhood through distancing of men from an embodied experience of reproduction; commerce, social networking, and assisted reproductive technologies all coalesce to transform the ways that sperm, not men, is seen as necessary for reproduction. In doing this, men become mere proxies and commodities, not active participants.

To reiterate, our chief question when conducting our cyberethnograpy has been this: How do the Web and its social networking components change the nature of sperm banking as they reframe ideals about kinship, gender, and the role of social media in the reproductive marketplace? We argue that in using Web 2.0 platforms, the five sperm banks are able to capitalise on speed and ease to provide a direct-to-consumer biomedicalised product. In doing this, sperm banks seek to control their image and their product by using tools unique to the Internet, effectively producing company and product branding. However, the consumer-based reviews and dialogues that are available on the sperm banks' forums and on their Facebook pages challenge this goal. There is an expansive narrative of the donor sperm that creates a collaborative understanding of new forms of kinship and fatherhood. This escalation of social media and marketing both atomises the male body and diminishes the participation of the actual man.

Implications

As argued, sperm banks have afforded users a greater level of control in their reproductive choices through the availability of commerce-driven websites and the utilisation of social media platforms. From posting on forums to acknowledging pregnancies through Facebook and Twitter, users are able to generate more content about donors than ever before. Furthermore, there are multiple websites where users can gain access to information about pregnancies, fertility, donors, and siblings. Kinship networks can be created in anticipation of a family prior to a pregnancy even being attempted. Significantly, while the imagined and customised 'daddy' might be put in the cart after careful consideration of users' reviews, the actual man has become further removed from the reproductive marketplace. They are silent in the online chatter in these homosocial spaces for women, where sperm is discussed and traded without the actual inclusion of the man himself. The discursive production of male essence, his sperm, is representative of all that man has to offer, his aptitudes and his form. Social marketing opens the potential for superfluity of description of donor attributes.

As feminist researchers, we were struck by the ways that the sperm banks' Web presence replicates the hyper-consumptive, neoliberal market-based agendas of late capitalism. Users can literally spend hours investigating the attributes of donors, visualising possible offspring, and making wish lists of patently transparent economic incentives of bodily and social attributes. Will there be a new data aggregator that says, 'if you like Donor 2241 for his blue eyes and multi-lingual talents, then we suggest you try Donor 5432'? Furthermore, there are clearly eugenic implications of these marketplaces since social class and financial constraints limit users' access, while social preferences and medical 'standards' limit men's universal participation.

However, just as assisted reproductive technologies create opportunities to move away from exclusively heteronormative reproduction, social networking (especially user-generated) opens new opportunities for creating kinship networks, locally, globally, and across generations. On a final note, it is important to be clear that we are not arguing that those who use sperm banks, and consequently Web-based commerce sites, are in some way different from others who reproduce. More traditional means of human reproduction are also riddled with conscious and subconscious instrumental decision-making concerning desirable variables for offspring. We are suggesting that semen banking in general, and the companion websites in particular, make for a more transparent practice of distilling the wants and desires of potential families, especially in regard to the male contribution to fertility as sperm banking removes and potentially disentangles the embodied and fleshy men from reproductive marketplaces. We can buy anything on the Web, from doctors to cocktails, shoes to sperm. Persuaded by 'strangers' feedback', we are accustomed to putting purchases in the cart. Sperm banks and their users curate a stable of sperm and, by extension, elucidate our changing desires for types of imagined men and possible kinship networks. Ironically, the proliferation of Web-generated data about these families and men both sensationalises and objectifies. Putting daddy in the cart and rating your purchase through user feedback both reduces the role of men and increases the specificity and precision by which we judge the 'male contribution'.

Acknowledgements

The authors would like to thank Monica Casper, Andrea Grady, Kristen Karlberg, Mary Kosut, William Levine, Laura Mamo, and Lara Rodriguez for their careful reading and Francis Myles for preparing Table 11.1.

Notes

1. Fairfax Cryobank (2012) *NEW Graduate Donor Alert!! Donor 2775!*, http://www.facebook.com/permalink.php?id=108795209208019&story_fbid=29242476751172, date accessed 4 July 2012.
2. 'Web 2.0' describes the development of the Internet since roughly 2004, when user-centred design, information sharing, and the interoperability of online services grew prominent.
3. Grounded theory is a deductive process whereby analysts incorporate as much data as possible so the formative theories can be used as deductive tools. Through the writing and rewriting of analytic memos, grounded theory aims to incorporate the range of human experiences (see Strauss and Corbin, 1994).
4. The information within the chart was updated on 23 September 2012.
5. *Website for Company*: provides information about the company and its products.
   Creating an Account/Profile: gives signed-in users more personalised shopping privileges, including saving favourite donors, saving specialised searches, and posting access in forums.
   Favourites/Wish List/Shopping Cart: allows users to save their favourite sperm donors on the website. The Shopping Cart is where users move their preferred donor. Each term is actually used on the websites, mimicking those listed on general shopping websites.
   Voice Verification/Baby Pictures of Sperm Donors: allows shoppers to listen to the donor answer questions, giving them insight into the 'personality' of the donor. Baby pictures allow shoppers to visualise their possible children.
   Online Forum: allows shoppers to further research their possible donor and connect with other sperm bank users.
   Facebook Page: 'Liking' allows users to follow updates and discussions on the Facebook page.

Twitter Page: allows sperm banks to directly connect with users via tweets about newborns, babies who are 'on the way,' and news stories about reproductive health. Banks retweet posts from users.

YouTube Page: provides videos explaining products, services provided, and success stories. Some YouTube pages take a more scientific approach, while others are more infotainment-centered.

Company Blog: creates a journalistic narrative for users to read and follow; provides a variety of information, including fertility FAQs and tips, sperm donor information, and product information.

12
Reciprocity in the Donation of Reproductive Oöcytes

Margaret Boulos, Ian Kerridge, and Catherine Waldby

Introduction

Decisions that individuals make about giving in their everyday lives are socially situated in that they are constrained by the social and legal norms of their times. In contemporary Australian society, human body parts are circulated between individuals and institutions for therapeutic and research purposes. Tissue donation, broadly referring to a range of body parts, can include myriad 'things' such as blood, bone marrow, gametes or embryos, DNA, cancerous tissue, and organs procured before or after death.

Gametes are reproductive cells such as sperm and eggs, and both can be transferred between people like blood but do not pose risks of potential infection for the recipient. Rather, the recipient, donor, and children born from the practice must negotiate social norms about relatedness and identity (Kirkman, 2003; Orobitg and Salazar, 2005; Pollack, 2003; Shaw, 2007, 2008a; Turkmendag, 2012). Almeling (2009) writes that the first publicised case of assisted conception with donated sperm occurred in 1909, while the first baby to be born from donated eggs occurred in Australia in 1984. In many parts of the globe, the practice of egg donation has expanded (Yee et al., 2007). However, in Australia and New Zealand it remains a marginal practice accounting for five per cent of all fertility treatment cycles (Wang et al., 2011). In both countries, the practice occurs on the basis of altruistic donation as the exchange of any human tissue for money is illegal (Shaw, 2007). The donor is able to receive reimbursement for expenses such as travel and childcare costs but is not covered for loss of earnings.

This fact challenges a mainstream narrative of postmodernity regarding the increasing commodification of those areas of life in which natural social relations continue to exist (Hoeyer, 2007). Postmodernity is an era of neoliberalisation, a paradigm which contends that individual freedom is facilitated by the market rather than the state (Stilwell, 2006). Neoliberalism was the basis of government reforms in the UK, US, and Australia from the 1980s and continues to some degree today. It involves the deregulation of markets and the dismantling or severe economic rationalisation of state-provided services in areas of health, education, and welfare (Waldby and Mitchell, 2006). Commodification involves creating a market price for objects. Commodification often evokes consternation that the dignity of a person or people is lost because commodifying the human body involves its objectification (Hoeyer, 2007).

Human tissue donation remains altruistic in Australia, in part, as a result of the influence of British social thinker, Richard Titmuss (1997 [1970]) who argued that *giving* blood needed to be maintained for the best interests of both society and the individual. While neoliberal economic thought emphasised negative liberties (Berlin, 1969), Titmuss (1997 [1970]) argued that opportunities for positive liberty, such as an individual's 'right to give', was fast disappearing and should be conserved.[1] However, Titmuss maintained that this could only occur within anonymous contexts. This chapter examines one form of modern reproductive technology, the donation of eggs, to explore instances of gift giving and questions of anonymity, altruism, and gender in contemporary Australian society.

Egg donation in Australia and New Zealand

Women's reproductive biology differs from men's in that women are more likely to experience rapidly declining fertility from their late twenties while environmental factors and lifestyle can affect sperm (Oliva et al., 2001). If a woman presents to a fertility clinic and her egg 'quality' is deemed to be poor, the clinician may suggest using donor eggs. Egg donation usually involves the transfer of eggs from a younger woman whose fertility has been proven through previous pregnancy. The donated eggs will then be mixed with sperm and any subsequent embryos will be implanted in the recipient's uterus.[2]

Whether gamete donation occurs in identified or anonymous contexts is different throughout the world. Shaw (2008a, p. 14) reports that in New Zealand a culture of openness between donors, recipients, and children has been fostered 'since the late 1980s'. For example, the donors in her study noted that they were able to establish some sort of relationship with their recipients. In Australia, fertility clinics do not act as fertility 'brokers' and thus have instituted an informal process of donor recruitment between individuals. Fertility clinics actively encourage their patients to find their own donors through their social networks or by advertising. Community publications such as *Sydney's Child* are popular resources where advertisements for egg donors can be placed.[3] If the donor and recipient are not known to each other outside the context of egg donation, a number of discussions may occur before both parties agree to go forward with the process. Once their agreement is made, the recipient and donor will both attend the fertility clinic so that the donor may undergo a series of physical and psychological tests, as well as mandatory counselling sessions prior to providing informed consent.[4]

Altruism and anonymity: giving to the universal stranger

In his germinal work Titmuss (1997 [1970]) argued for the continued voluntary and altruistic organisation of blood provision in the UK on the basis of economic efficiency and medical imperatives. Comparing Britain's blood provision services with the US, Titmuss argued that the safe and efficient provision of blood could only be sustained through altruistic donation, drawing on data that showed vendors in the US were usually socially and economically deprived, selling their blood as a means of income. He also drew on studies showing that this 'skid-row' population was more likely to have serum hepatitis than others in the general population (p. 129).

Titmuss's views about blood donation were not just technical but also ethical. He defended the 'right to give' on the basis that it was an important positive liberty. Titmuss's defence of donation is complex. On the one hand, donors are motivated because they can perceive a universality of need. On the other hand, if donors actually needed blood, they can only *hope* they will be the beneficiary of the kindness they have shown to others and they cannot *expect* it. Anonymity, Titmuss argued, would preserve the fair distribution of blood because donors were giving to an undifferentiated recipient. Need was based on medical criteria not social divisions such as class, ethnicity, or gender (p. 306). The altruistic donation of blood allowed donors to actively embrace the universal stranger and this could have important implications for social cohesiveness generally. It allowed donors to find their inner 'moral compass' of altruistic behaviour towards countless others whose only commonality is their basic human frailty (ibid.).

There are also sociocultural norms that inhibit the examination of the selection of recipients by donors. Bauman (1990, p. 90) has argued:

> [I]n the case of the gift ... the needs and the rights of others are the main – perhaps the only – motive for action ... The goods are given away, the services are extended merely because the other person needs them and, being the person it is, has the rights for the needs to be respected.

Echoing Titmuss's (1997 [1970]) celebration of the 'free' gift, Bauman's (1990) perspective suggests that recipients should be seen as an undifferentiated mass of needy people. Protocols of anonymity reinforce the idea that the inherent need of recipients should not be questioned or evaluated and should be enough to motivate the donor.

Titmuss' insistence that blood provision in industrialised societies could be a gift was made in light of Marcel Mauss' essay *The Gift* (2001 [1954]), first published in 1923.[5] Mauss imagined reciprocity as an essential part of gift giving. In the Melanesian societies that Mauss studied, gifts were exchanged in face-to-face interactions between inhabitants of the Trobriand Islands at designated times and were not so much an act of spontaneous generosity but a form of self-interest and aggrandisement. While gift giving created relationships through the obligation to reciprocate, it also created and sustained power relationships. Gifts, therefore, as Frow (1997, p. 109) has argued, can be used to create indebtedness and act more as loans because they are always returned, oftentimes with interest. Empirical studies of contemporary tissue donation practices correspond to this latter conceptualisation. As Shaw (2012, p. 298) observed, reciprocity is rarely enacted precisely in equivalent terms but 'failure to return a gift, or say thank you for that matter, symbolizes a refusal to cement the social and moral bond' (see also Shaw 2008b).

Thus in contrast to Mauss (2001 [1954]), who considered that obligations to give, receive, and reciprocate were the overt or latent motives behind the exchange of gifts, Titmuss (1997 [1970]) radically insisted that the free gift, given in the spirit of generosity and spontaneity, could prevail. The free gift was not a social fiction but rather had to be instituted under the condition of anonymity. Even though there are significant differences between the approaches taken by Mauss and Titmuss, both seem to agree that gift giving between people known to each other will occur as cyclical reciprocity.[6]

Women and altruism

Feminist approaches to social relations generally, and those of egg donation and surrogate motherhood specifically, are critical of constructions of altruism within gift relations. Normative ideas of femininity tend to emphasise altruism and self-sacrifice (Shaw, 2007). In light of our earlier point that individual decisions are difficult to disentangle from their social contexts, it is necessary to unpack the connections made between altruism and gender. An important question is, do women have a 'right to give' or are they under obligation to do so?

The connection between gendered social norms and the practices of gamete provision is difficult to ignore. Pollack (2003) and Almeling (2009) have demonstrated that despite the practice of egg provision being conducted as frank commercial exchanges, narratives of altruism continue to be important to the practice in the US. In her study of egg donors in Massachusetts, Pollack (2003) argued that recruitment agencies and fertility clinics construct the recipient woman as normatively feminine because she desires a family whereas the egg provider is 'selling' motherhood and this is problematic. Pollack contends that the 'problem of the unfeminine motivations of the donors is solved by the construction of a narrative of altruism' (2003, p. 255). In her study of semen and egg donation, Almeling (2009) asserted that potential donors were screened during the recruitment process at some clinics and women who expressed self-interested motivations such as financial reward or inappropriate feelings towards the child were often excluded from participation. In the context of sperm donation in the same recruitment agencies, it was deemed suitable for men to express interest in the financial reward they received (ibid.). These studies indicate that if asked about their motivation to provide eggs, women may feel pressure to suppress their 'true' feelings and appear altruistic.

Much research (Blyth, 2004; Haylett, 2012; Kirkman, 2003; Nahman, 2008; Warren and Blood, 2003) has sought to understand motivations behind the provision of eggs for reproductive purposes. In Shaw's (2008a, p. 18) examination of what she terms generally as 'bodily gifting practices'[7] including surrogate motherhood (both gestational and traditional) in New Zealand, participants reported their desires to become egg donors or surrogates for a variety of reasons 'including empathy for other women who want to have children ... being generous and wanting to help someone else ... and familial love, obligation or responsibility'). Kalfoglou and Gittelsohn (2000, p. 799) reported that payment was the primary motivating factor for participants in their

US study. Thus by interviewing egg donors or sellers outside the context of fertility clinics, research indicates that women do undertake this practice for a variety of reasons which can be characterised as altruistic or instrumental.

Identified egg donation: complicating donor's narratives about egg donation

The preceding discussion of gender, altruism, and egg donation for reproductive purposes sheds light on the complex experience of the practice, particularly the ways in which the virtue of altruism may be deployed. Yet isolating the decision to become an egg donor, while intellectually helpful, tends to conceptualise gift giving in a rather abstract sense. For instance, categorising the reasons to donate as *either* altruistic or instrumental insufficiently captures the fact that the process involves a lot of minor actions over a period of time that could take up to six months. How do donors maintain their commitment? These approaches also tell us little about the *relationship* between the donor and recipient. There are, of course, good reasons for this: with the exception of participants in Shaw's (2007, 2008a) studies into egg donation in New Zealand, scholars such as Pollack (2003), Almeling (2009), and Nahman (2008) rely on data from donors giving within anonymous contexts. Decisions to donate eggs for reproductive purposes are complex. We suggest that paying attention to the ways in which gift giving is *enacted* can provide a more holistic understanding of egg donation for reproductive purposes. We argue that in contexts of identified egg donation, selecting the recipient *is* a place where value judgements about 'good' parents can be made. Moreover, this does not usually lead to the establishment of social relations. This suggests that we cannot assume giving human tissue in identified contexts will inevitably lead to cycles of reciprocity. Rather, we must understand the specificities of giving eggs *within* contexts of identified donation to build nuanced knowledge of tissue gift giving practices in postmodernity.

Study context

This chapter draws upon the results of a qualitative study examining issues around egg provision for Somatic Cell Nuclear Transfer (SCNT) research.[8] Ethics approval was obtained from Sydney South West Area Health Service and the University of Sydney. Participants were recruited through the project's industry partner, a non-profit fertility clinic attached to a major Sydney teaching hospital. As researchers from the University of Sydney, we did not have access to the clinic's patient database; instead, with the assistance of clinic staff, we mailed letters to potential participants introducing the study, including a 'contact for consent' form. Individuals mailed written notification to indicate their interest, after which interview times were scheduled by telephone.

In total, we interviewed 43 women in the greater Sydney region of New South Wales (NSW) between August 2009 and August 2010. While there were three cohorts in this study (women undergoing their own fertility treatment, reproductive egg donors, and the 'healthy-donor' [women aged between 18 and 30 are considered biomedically to be at their peak fertility]), only data from the reproductive egg donor cohort are reported here. This cohort includes five women all of whom donated eggs between 2007 and 2010. The database from which they were drawn had a total of 35 donors, two of whom were excluded from participating because they lived in remote areas. The women who participated in this study were aged between 18 and 50 and all had completed secondary education with most obtaining further qualifications. Semi-structured interviews with reproductive eggs donors lasted 90 minutes and usually took place in the participant's home.

All interviews were tape-recorded and transcribed with consent and pseudonyms were given to preserve the anonymity of participants. Using NVivo software, a rigorous theoretical coding process was applied to the data (Flick, 2006). Open codes were broadly applied during the initial stages where transcripts were read and reread in an effort to comprehend the perspectives of participants (Ezzy, 2002). Comparisons were made between our own codes and interpretations with existing literature. Themes which emerged related to cohort experience and social responsibility. The qualitative research methods used generated an enormous amount of data. However, given that this sample is very small, even by qualitative social science standards, the development of causal relationships between variables cannot be based on these results. Despite this limitation, the data collected resonate with other reports of egg donation in Australia (Kirkman, 2003; Warren and Blood, 2003) and has the potential to contribute to further research into egg donation for reproductive purposes.

Of the sample, three donors (Jenny, Paola, and Donna) gave eggs to women they knew outside the context of donation: Jenny to a couple in her extended social network and who were not considered to be 'great friends' with sporadic contact over the year; Paola donated to a work colleague in her late 40s who had started a relationship with a new partner. Donna donated eggs to a close friend. Two donors, Raja and Agnes had thought about donation prior to instigating the search for a recipient: Raja found an advertisement in the local newspaper and Agnes responded to a number of advertisements about egg donation before selecting her recipient. Not all donation attempts were successful: of the sample, pregnancy ensued for the recipients of Raja's, Agnes' and Donna's eggs. Both Raja and Jenny undertook the process of extraction twice while the others did so once.

The decision to provide eggs: finding 'good parents'

For egg donors in this study, the first step was the identification of a suitable recipient. Donors indicated that they exercised considerable care in selecting an appropriately worthy recipient. Vetting potential parents and making value judgements about their suitability, in some ways, resembled a less regulated version of the assessment involved in adoption (Family & Community Services, 2011). Participants in our study tended to represent their recipient as 'deserving' because it would redress a sense of imbalance in the recipient's life. In some cases, they also implied that recipients who had not 'struggled' (to conceive) or did not display the required desperation may be less deserving of their help. In this respect, donors represented their chosen recipients in terms of their need – the more needy the recipient and the greater the obstacles that they have faced, the more appropriate the gift.

> I did want to meet them. I want to, sort of, know that my eggs are going at least to someone who's worthy of them, in a sense. (Agnes)

> I remember just feeling very intensely that if I could help them, I would like to, because it's something that she wanted so much, and I just thought myself how awful it would be, having had two children, to feel that I couldn't have fulfilled being a parent, because it's important to me and it was obviously very important to her as well. (Jenny)

> Um, I think because they were so great to kids anyway. And I know they'd been trying [to have a child] – they were so good to my first, my Benjamin, my son; and I thought, 'God! You'd make great parents.' And they're just so caring, and they give to everybody else – I was like, 'Oh well, it's time that you got something back.' And they've been trying as well. (Donna)

Donors have their own qualifications about whom they consider to be a 'good parent'. Importantly, as the excerpt from Agnes suggests, recipients who placed conditions on the donor they sought *disqualified* themselves by not appearing to be unconditionally grateful.

Agnes: And a lot of the ads – it was quite funny. They were very specific.
Interviewer (MB): In what way?
Agnes: 'You have to be a non-smoker. You have to have fit' [sic] – like, I know they recommend to have your own children first before donating eggs, but a lot of them were very specific in regards to that. 'You have to be Asian, or you have to be over – no, you have to be under 25', and ... I'm like, whoa! 'But completed your own family!' I'm just, like, 'Alright. You're a bit too fussy!' Then there were some other ads where they'd had their first child but wanted a second one, and I was, like, No, I don't want to go for them, because they've at least got one. I want to go for someone who hasn't.

In the excerpt below, Paola represented the woman recipient, Angela, as a 'good parent' precisely because she already had children. Paola used the evidence of Angela's previous mothering experience as a basis for her claim that Angela will make a 'good parent'.

> Apparently ... there's a couple of local magazines where you can actually put an ad in and say, 'I'm willing to donate my eggs', but I'd rather do it with someone I know ... I've known Angela for a while, I've met her husband, I've met all her children ... and I know for a fact that Angela would be a good parent ... She's quite strict with her children. If she'd been my mum, I might have finished school, you know? Very strict. So any kid having that kind of parent would be a good child. (Paola)

Other elements that figure in decisions about the right recipients are more socially constituted traits such as financial status, relationship stability, and personality. These attributes represent Australian norms about 'good' parents that reflect some of the diversity of contemporary society. For example, as the excerpt from Raja's interview shows, sexual orientation is not a basis to exclude people, but donors in our study ultimately remain committed to an idealised version of relationships as long-term and child-focused.

> I wouldn't just donate and not know ... I sort of need to know the couple's financially secure and that their marriage, or relationship – I don't care if they're gay or whatever – that their relationship is strong. They're really two important factors for me. (Raja)

These responses show that for our participants, donating eggs in identified contexts enables them to select a specific recipient among the undifferentiated mass of infertile individuals or couples. These findings resonate with a study of a Spanish fertility clinic reported by Orobitg and Salazar (2005, pp. 46–47). As Spanish law prohibits egg donation between known individuals including friends, participants were only allowed to donate eggs in anonymous circumstances.[9] However, in the absence of known recipients, donors reported imagining their recipient as a woman in their life whom they knew was struggling with infertility. These responses also indicate that donors in our study felt some responsibility to the child they were helping to create. As we discuss below, this responsibility does not involve an ongoing relationship.

Sustaining motivation

There is a considerable time lag between making the decision to donate and completing the cycle as the donor undertakes a physically arduous process that involves many smaller decision points. Murray and Golombok (2000) found that three-quarters of potential UK egg donors changed their mind about donating after receiving information on the procedures involved. Egg extraction involves the self-administration of drugs including injections and nasal sprays over a four- to six-week period. Initially drugs are used to stop ovarian stimulation (Weeks 1–2) before the injection of gonadotropins which are used to simulate 'the development of several egg-containing follicles' (Pearson, 2006, p. 608). The final stage of medication involves hormones to mature the eggs. During this time, donors will visit the fertility clinic to have multiple blood tests and pelvic ultrasounds. At the end of the cycle, the donor is admitted as an outpatient and undergoes surgery where she is intravenously sedated and given a local anaesthetic. The eggs are extracted through a large glass needle inserted into the ovary through the vaginal wall. The eggs are then mixed with sperm *in vitro* in order to create embryos that are subsequently implanted into the recipient woman's uterus. There can be side effects to the medication and women can experience symptoms including 'depression, short-term memory problems, insomnia, bleeding, hyperovulation stress syndrome [and] weight gain' (Ikemoto, 2009, p. 770). How did our donors, some of whom gave their eggs to women whom they did not know outside of the donation context, address these issues?

For donors in our study, the relationship between the donor and recipient (usually the woman) was an important aspect of the process. As discussed above, egg donors make moral judgements about their recipients and this selection process can engender a specific relationship where each woman may be emotionally and physically vulnerable to the other. We suggest that egg donation in identified contexts is characterised by a distinctive intimacy which, after Plummer (2003, p. 13), we understand as: 'a complex sphere of 'inmost' relationships with self and others [that are] not usually minor or incidental (although they may be transitory), and they usually touch the personal world very deeply'. This sense of intimacy is encapsulated by Agnes's description of her recipient:

> And it was just bizarre. She was like a 43 year old version of me. Like ... it was very interesting; the same personality, temperament ... we're both nerds; we sit at home, reading a million books, mm, very nice. (Agnes)

While not all of the donors in our study described the same sense of personal identification with their recipient, all of them had formal and informal relationships with each other. This arose out of initial meetings and private negotiations, as well as mandatory counselling sessions at the recipients' fertility clinic. Clinical regulations contribute to this peculiar intimacy by subjecting each party to formalised scrutiny. For example, counsellors must explicitly discuss how each party perceives the disclosure of donor-assisted conception. Importantly, individual clinics rarely act as moral gatekeepers by defining who is a 'good' parent or worthy recipient.

Irrespective of how this intimacy is achieved, donors were strongly focused on the needs and desires of the recipients, with whom they felt considerable empathy. We suggest that the donor constructs her relationship with the recipient as an important motivating force. Knowing the donor facilitates the construction of an affective framework through which the extensive medical procedures can be experienced and tolerated. The medical procedures outlined earlier are a means to an end, not ends in themselves. Helping a woman to become a mother is the end. Raja asserted that knowing the recipient was an important factor in psychologically sustaining her until the goal of the recipient's pregnancy was reached.

> I think what made it easier was that the couple was really nice. That made it a lot easier. If it had been just an anonymous donation, and having that similar experience, I don't think I would have gone through with that [two cycles of ovarian stimulation]. I think I would have stopped. But it's just because the couple were really nice. (Raja)

This point assists in moving beyond categorising decisions to donate in one way. Data from our cohort indicate that practices of egg donation for reproductive purposes can encompass *both* altruistic and instrumental motivations. Orobitg and Salazar (2005) also argued that labelling egg donation as either altruistic or instrumental inadequately captures the diversity of experiences for the donor. Although they are described as 'donors', women who provide eggs for reproductive purposes in Spain receive 'compensation' in cash. Yet some participants reported that although they were initially attracted to the financial reward, during the process they understood themselves as helping another woman. Still, other participants who were initially motivated by altruism found that the compensation was a fair recognition of their effort which they found put them at much 'inconvenience and discomfort' (p. 44).

Kinship, genetics and motherhood

In this final section, we want to discuss egg donation for reproductive purposes as a way of establishing social relations. This relates directly to what Titmuss (1997 [1970]) wished to avoid in a context of blood donation: reciprocity and the obligation to return the gift. We suggest that by looking at this specific instance of gift giving, we can connect complex practices of contemporary reproductive technology with more conventional views of the family. Importantly, our participants did not express a strong interest in having ongoing relationships with the child they were hoping to create. While donors were concerned with finding 'good parents' they did not express the belief that they had any reason or responsibility to ensure that their recipients could or would guarantee this. Nor, did they consider the contribution of the egg and the donor's genetic material to be the basis of a socially significant relationship with the child:

Interviewer: Did you discuss that beforehand ... What kind of contact there would be?
Raja: Yeah. The lady said that she would like contact and to be friends afterwards and everything, and I said, 'OK, look, I don't want to be a second mum or anything, but if you give me photos or an update every year or whatever, I'd be happy with that, just to see that everything's OK.'
I explained to him [donor's partner] I don't consider from the moment I was doing it that these are my children, they've got my DNA but that's as far as it goes. [Later in the interview]

These attitudes correspond with findings from studies by Kirkman (2003) and Nahman (2008) that revealed that participants distinguished between the offspring of donated eggs and their 'own' present or future children. In doing so, donors severed the connections between kinship and genetics upon which Euro-North American societies are based (Orobitg and Salazar, 2005). This was evident in our own study, as donors emphasised gestation and rearing as the *constitutive* elements of motherhood:

And some people don't really understand it's just an egg; but a lot of people think 'well, that's almost like that's your child as well?' and it's not. Because I've seen it grow in somebody else, and it's all their child – so for some people it's a bit ... 'oh, that's a bit weird!' (Donna)

I never see it as 'my child' or 'half my child', I guess it's all just perspective. It's in her. She's going to be pregnant, she's going to give birth to her. (Agnes)

Social scientists have often celebrated that Assisted Reproductive Technology (ART) has enabled rather radical kinship formations, particularly in relation to individuals whose sexuality or marital status does not conform to traditional notions of parenthood (Franklin, 2007; Shaw, 2007). Normative scripts about the nuclear family (one mother and father) often remain in legal contexts which can be slow to catch up to technological change. Yet this orthodox narrative is also embodied by participants in our study, who constructed the contribution of their eggs as minimal and virtually insignificant, despite the fact that without them, no pregnancy could ensue. These elisions are also reported in relation to recipients of donated gametes as many parents do not disclose the fact to their children (Grace et al., 2008; Skoog Svanberg et al., 2003).

Discussion: altruism in identification of egg donation

To conclude, we return to Titmuss (1997 [1970]) whose anonymous blood provision service would institute a nationalised flow of 'free gifts' – a small sphere of gift giving in an increasingly market-driven world. Already we have demonstrated that there are many ambiguities to the model of altruistic egg donation and gaps between the social expectations of the practice and individual experiences (see Shaw 2007, 2008a). Understanding donor motivation is a complex issue, particularly in relation to ideologies of gifts giving, altruism, and women.

The findings from this small study suggest that despite the anxieties and misgivings of Titmuss or Mauss, altruism does not necessarily need to be based on anonymity. In the context of egg provision for reproductive purposes, there are both structural and personal reasons for this. Within contemporary legal structures, gamete donation involves the creation of a separate and sentient human being, subject to rules of guardianship and property. In this context, legally recognising an egg donor is a complex, often fraught, affair because egg donation challenges *legal* definitions of maternity in a way that semen donation does not. This, as Jackson (2006) has argued, is because within English common law (from which Australian law is derived), the attribution of maternity, prior to the use of ART, has been an uncontested fact while the attribution of paternity is technically a legal fiction; this is based on historic modes of reproduction during which eggs remained *in vivo* while semen was always externalised.

Just as donors emphasised gestation as constitutive of motherhood, so too does NSW law, offering the following definition in *The Status of Children Act (1996 No 76)*:

> 3) If a woman (whether married or unmarried) becomes pregnant by means of a fertilisation procedure using an ovum obtained from another woman, that other woman is presumed not to be the mother of any child born as a result of the pregnancy.

It is interesting to note the consequences of motherhood in the context of gestational surrogate motherhood, where the recipient woman contributes her genetic material (the egg) but a third party gestates the foetus. Under this definition, the woman who undertakes gestational surrogate motherhood is the mother until the recipient woman adopts the child.

In line with such legal structures, our data suggest that the donation of eggs was not always intended to produce ongoing social relationships because of the threat such relationships might pose to norms about families. As Pollack (2003) and Almeling (2009) found, *if* the egg provider suggested a long-term relationship, her motives may be considered deviant and thus the repression of these desires signals an adherence to social norms. This is critical in understanding gift giving in specific contexts and exploring the *gendering* of such practices. To some extent, this echoes Haimes's (1993) claim that the symbolic representation of gamete donation tends to adhere to traditional gender norms. Policymakers in the UK, whom Haimes interviewed, tended to view semen donors as potentially threatening to the nuclear family because 'semen donation is presented [in relation to] inappropriate sexuality, such as masturbation, adultery and illegitimacy [while] egg donation on the other hand, is presented as a new and complex procedure [and] firmly located in a clinical setting and is essentially asexual' (p. 91). Thus the relations created by the 'gift' are context-specific and while the act is *sometimes* used to produce cycles of reciprocity, at other times the gift is given as an isolated gesture.

An important dimension to understanding the relationships which emerged in this study of identified egg donation is the construction of the recipient as a person with 'urgent' needs. In their narratives, donors in our study largely constructed the recipient as deficient in her capacity to be fertile, and it was the donor's role to assist in correcting the significant imbalance identified in the recipient's life and returning it to equilibrium. In effect, the recipient was expected to be fulfilled by the act of the donor and they needed to do little apart from accepting the gift they had been offered. This suggests that once the recipient was designated as worthy because of the 'severity' of her needs, the norms of formal reciprocity may be suspended.

For Mauss and Titmuss, gift giving in identified contexts invariably led to obligations of reciprocity but in the data presented here, donors did and could not impose norms of reciprocity. The narratives of donors in our study show that they emphasised the affective framework in which the donor and recipient met rather than the objects which were given. Donors gave away not only their eggs but considerable amounts of time and effort, becoming involved in interviews, counselling, tests, and medical procedures. Yet they did so in the service of someone who lacked eggs, not in an abstract sense. Hence questions of reciprocity are not simply based on whether the donor and recipient are known to each other but how the recipient is constructed in the first place.

Finally, it would be remiss not to situate these narratives within broader debates about the distribution of health in postmodernity. Egg donors do not imagine a claim on 'their' eggs once they are given, expressing the view that they expect no reciprocity for their actions and for some scholars this may be explained by arguments about pressure for women, however subtle or overt, to be generous and self-sacrificing. However, we would also suggest that the data indicate that acting within altruistic contexts can also involve the implementation of other instrumental concerns, namely the selection of recipients. Donors in our cohort did not understand parenthood to be a universal expectation or 'right' to be granted simply because one wishes it. This is a significant diversion from Titmuss's blood recipient whose health was to be considered a universal right, provided by the welfare state. Donors in this study believed that fertility patients are not a homogenous group but that some are more deserving than others. This process of hierarchically organising recipients can be considered a reflection of neoliberal reforms which have delegitimised the provision of universal health care. Governments, particularly in the Anglo-sphere, are no longer required to provide services to all citizens because this may encourage idleness and rorting, precisely the 'undeserving' charity case. In the context of reproductive egg donation, there is scope for some forms of non-traditional families. However, donors ultimately favoured couples whose parenting abilities were primarily connected to their struggles with infertility and their ability to provide a high level of material comfort.

Conclusion

Our aim in this chapter has been to explore the practice of egg donation within contexts of identified exchange as it occurs in Australia generally and NSW specifically. Egg donation is a very recent practice relying on sophisticated technological advancements. By separating the role of the mother into two women and facilitating infertile women to have children, egg donation can be situated within postmodern narratives about the flexibility of families, the malleability of bodies, and the increase of individual 'choice'. However, if we look beyond these debates, we see that the precise nature of this 'choice' and the attendant relationships are contested. Should utilising these facilities be made on the basis of altruism as donors and recipients, or does the free market meet our needs as transactors? Titmuss (1997 [1970]) argued that by maintaining altruistic and anonymous blood donation, people could exercise their 'right to give'. In the midst of demographic transformation in 1970s Britain, Titmuss carried an important message to encourage people to embrace the 'universal stranger'. However, drawing on narratives of our cohort, we have shown that donors are careful in their selection of recipients, evaluating them on the basis of personal and social characteristics that are the embodiment of 'good parents'. While Titmuss (1997 [1970]) and Mauss (2001 [1954]) did diverge in their attitudes towards gift giving, they shared scepticism that in settings where individuals are known to each other, debts of reciprocity are difficult to ignore or reject. However, we have shown that relationships between donors and recipients did not necessarily lead to expectations of reciprocity; the recipient's need, while not undifferentiated as the universal stranger, was urgent enough to release them of the bonds of obligation. Postmodernity offers specific technological bases to subvert social conventions but practices of egg donation in Australia reveal that while ideas of motherhood are changing, some norms about families as nuclear and bounded prevail. Thus we have found that a contextual approach to giving tissues is necessary.

Acknowledgements

This research was funded by an Australian Research Council Linkage Project Grant 'Human Oöcytes for Stem Cell Research: Donation and Regulation in Australia' (LP0882054). Margaret Boulos wishes to thank Dr Franklin Obeng-Odoom for his insightful commentary and analysis.

Notes

1. Berlin distinguished between two dimensions of liberty: negative and positive freedom. Negative freedom is the freedom from interference or coercion to engage in activities and aptly characterises libertarian views of liberty. In contrast, positive freedom is a fuller expression of autonomy, to be 'the instrument' of oneself (1969, p. 131).
2. The recipient woman may be using sperm from her partner or sperm from a bank or another donor.
3. *Sydney's Child* is a magazine published 11 times per year and is distributed freely. Other Australian cities have their own versions. The magazine contains parenting-related articles by professional writers and information on child-related issues and events. The classifieds section contains advertisements for health and educational professionals as potential recipients advertising for egg donors. From an irregular scan of the magazine,

advertisers seeking egg donors are mainly heterosexual women of various ethnic backgrounds.
4. This information was obtained from staff at the clinic where we conducted this research.
5. *Essai sur le don* was published in French in 1923. The first English translation was published in 1954.
6. Titmuss's (1997 [1970]) approach to gifts and tissue provision has also been criticised for failing to adequately capture the complex relationships that have emerged in contemporary globalised societies. Waldby and Mitchell (2006, p. 22) argued that blood can now be fragmented to such a degree that whole blood transfusion is rare and that one extraction will result in a multiplicity of recipients.
7. Shaw (2011) uses this term to refer to a range of contexts where human tissue or substances are transferred between two or more people including breast milk, surrogate motherhood, and ovarian eggs.
8. The data are drawn from a three-year Australian Research Council (ARC)-funded project (Human Oöcytes for Stem Cell Research: Donation and Regulation in Australia, 2008–2011). This study investigated the social and bioethical issues related to new legislation introduced into the Federal Parliament in 2007 which permitted the technique of therapeutic cloning called Somatic Cell Nuclear Transfer (SCNT).
9. Spanish law prohibits knowledge of the donor's identity to the recipient adults or the conceived child (Baetens et al., 2000).

13
Expressed Breast Milk as Commodity: Disembodied Motherhood and Involved Fatherhood

Victoria Team and Kath Ryan

Introduction

Exclusive breastfeeding for at least six months is the recommended method of infant feeding (World Health Organization [WHO] and United Nations Children's Fund [UNICEF], 2003). Breastfeeding is being promoted as the social norm in the US, UK, and Australasia in an attempt to replace formula feeding (American Academy of Pediatrics, 2012; Australian Health Ministers' Conference, 2009; The Royal College of Paediatrics & Child Health (RCPCH), 2011), which became popular from the 1940s (Stevens et al., 2009). Breastfeeding women often feed expressed breast milk (EBM) to their own infants although EBM feeding is different from breastfeeding. Some women share, donate, and/or sell excess EBM.

In this chapter, we discuss contemporary EBM commodification and related issues identified from in-depth interviews with breastfeeding women in the UK. We emphasise that, in postmodernity, male and female reproductive roles in relation to infant feeding have changed as a result of the commodification and commercialisation of breast milk. In the UK, EBM feeding is thought to offer breastfeeding women more 'choices' in relation to the involvement of the father in caregiving. In contrast, we argue that paternal involvement in feeding is problematic in its alignment with neoliberal agendas which encourage women to 'choose' to outsource infant feeding to someone else. As we demonstrate, multiple anxieties and challenges are related to mother–infant separation and involvement of a third person. Our discussion of women's embodied experiences of breast milk expression, EBM feeding, and paternal involvement in infant feeding and related motives contributes to a deeper understanding of contemporary reproductive roles.

Human milk markets

In contemporary western cultures, the sale of human milk is one of the 'taboo trades', similar to the sale of virginity, sex, oocytes, and surrogacy; women are the 'primary suppliers' for these markets (Krawiec, 2010, p. 1742). Historically, human milk markets have always existed, reflecting socio-economic and racial inequalities and indicating that breast milk was commodified a long time ago (Fentiman, 2009). Wet-nursing was practised in ancient Greece, Rome, and Roman Egypt, and wealthy families employed as many wet-nurses as they could afford (Abou Aly, 1996). African slave women were forced to feed White children before their own (Rody, 1995). Across the centuries, wet-nursing was practised in France, Britain, and the US (Fildes, 1988). Since the mid-twentieth century, there have been many attempts to reframe the meaning of breast milk on its own from commodity to marketable product. For example, in England, in the late twentieth century, breast milk was defined as a 'gift' (Raphael, 1955) and was accepted only from middle-class women, 'who were immune of suspicion of financial motive' (Law, 2010, p. 24). Relatively recently, North American women have been encouraged by Prolacta Bioscience (a for-profit human milk processing company) to donate EBM for African babies infected with HIV. However, only 25 per cent of donated breast milk was sent to Africa, while 75 per cent was re-formulated by the company and sold to neonatal units nationally (Fentiman, 2009).

Many human milk banks operate on a not-for-profit basis. However, the sale prices of banked human milk to hospitals are very high at USD$100 per litre in Norway (Grøvslien and Grønn, 2009) and USD $3–5/ounce in the US (Palmer, 2009; Woo and Spatz, 2007). Prices established by milk banks are usually attributed to the costs related to donor screening and donated breast milk collection, processing, testing, storage, and delivery (Grøvslien and Grønn, 2009; Hartmann, 2011; Tully, 2000). Thus EBM undergoes a standardisation process and higher EBM prices introduced by human milk banks supposedly correspond to improved EBM quality. Institutional, particularly corporeal, involvement in facilitating and transferring reproductive gifts suggests the commercialisation and commodification of women's reproductive bodies, body parts, tissues, and fluids (Shaw, 2008a). Breast milk commodification overemphasises EBM as a product, to the detriment of breastfeeding as a process, thereby equating it with formula milk – both 'products served in the same container but coming from different sources' (Van Esterik, 1996, p. 273).

In the industrialised world, EBM black markets, or the unregulated sale of human milk for profit, continue to grow (particularly via the Internet) (Akre et al., 2011; Dawson David, 2011; Vogel, 2011; Woo and Spatz, 2007). In most developed countries, human milk sale is not legally regulated (Dawson David, 2011; Hartmann et al., 2007). For example, in Australia, it is not regulated because of inconsistencies in allocation to either organ and tissue transplant or therapeutic goods (Hartmann et al., 2007). There have been calls for enforceable legislation of human milk sale for ethical, epidemiological, and public health reasons. Ethical concerns relate to protecting women from market-related coercion linked to the dangers of commodification of women's bodies, body parts, and body fluids (Krawiec, 2010).

Non-normative use of EBM also suggests its commodification. EBM has been occasionally used a gastronomic product and delicacy (for example, in ice cream) (Fallon, 2011), in medicinal (Hallgren et al., 2008) and home remedy (Giles, 2004) products, as an adult industry product (fetish) (Hartmann, 2011), and even as an entertainment product in drama, where milk expression and selling human milk-based foods are part of the scenario (Van Esterik, 2009). Contemporary consumers of EBM are primarily the mothers of sick and/or premature infants who wish to feed their infants breast milk, in spite of having lactation problems themselves. Although, some men purchase EBM-filled bottles and bags of frozen EBM for use as a fetish to increase sexual arousal (Hartmann, 2011). The pornography industry has even developed a separate branch – 'lacto-porn' – in which lactating women's bodies are featured (Giles, 2005). There are also reports of breastfeeding prostitution (Giles, 2004).

Methodology

This chapter is based on qualitative research that was conducted collaboratively in 2005–2006 between Bournemouth University and the Health Experiences Research Group at the University of Oxford with the aim of eliciting women's experiences of breastfeeding for the Healthtalkonline website (www.healthtalkonline.org). Ethics approval was obtained from the UK Multi Region Ethics Committee. Participant recruitment was facilitated by an expert advisory panel and national voluntary breastfeeding support groups. Recruitment criteria were (1) breastfeeding at the time of recruitment or/and (2) breastfeeding within the previous two years.

Forty-nine women participants were purposively selected to increase maximum variation in breastfeeding experiences and socio-demographic backgrounds. The selected participants varied in age from 19 to 40 years (with an average age of 32 years), and were from a wide range of socio-economic backgrounds. Some women worked in paid employment, others were either unemployed or on maternity leave. Their educational levels ranged from no qualifications through to postgraduate degrees. The participants' ethnicities included White and Black British, Bangladeshi British, Colombian, French, Indian, Pakistani, Sri Lankan, Thai, and Ugandan. Women were of various religious backgrounds, including Christian, Jewish, Muslim, Hindu, and Buddhist. Women's marital status included single, married, and co-habiting. Most women were breastfeeding their first child while some had also breast- or formula-fed previous children. Breastfeeding experiences ranged from positive (easy, long-term feeding) to problematic (ill mothers, ill infants, and multiple infants).

Face-to-face in-depth semi-structured interviews were conducted by the second author in the women's homes in all but one case. They lasted approximately 90 minutes and were video- and audio-recorded with consent. The participants were asked to tell the story of their breastfeeding experiences and then prompted to reflect on the issues that they raised, such as early breastfeeding experiences in the hospital, coming home with a new baby, breastfeeding and working, breastfeeding and maternal chronic illness, breastfeeding in public, and breast milk expression. All interviews were transcribed and coded thematically, using the constant comparative method (Pope et al., 2000). Some themes, written up as topic summaries and illustrated with video clips, appear on the Healthtalkonline website, the primary outcome of the project. In this chapter, we undertake a secondary analysis (Heaton, 2004) of the data and focus on the implications of EBM commodification and commercialisation through parental sharing of infant feeding.

Findings

Expressed breast milk as product

As noted, EBM has become increasingly commodified, commercialised, and constructed as a consumer product that has many values (Ryan et al., 2013). In addition to its great biomedical value, both nutritional- and immune-related, EBM as a product has several other values, including physical, social, financial, and economic. Financial and economic values construct EBM as 'commodity' in the direct classic meaning of this concept as a product to be exchanged for money (Marx, 1993 [1867]). Physical values as well as social and situational values form the basis of the construction of EBM as 'commodity' in its contemporary meaning as a social relationship associated with commodity rather than commodity as a thing (Appadurai, 1996).

EBM measurability, storage quality and attached social values
In our study, breastfeeding women talked about the physical values of EBM in its measurability and storage qualities. Measurability as a characteristic of EBM was valued by women because it allowed them to: (1) measure their infant's intake, that is, the amount of milk consumed, which is difficult to measure with breastfeeding; (2) measure the amount of breast milk they produced; and (3) measure their EBM stock. Women compared the available amount of EBM with the required amount for the time of their absence, and then calculated the amount they needed to produce. Generally, breastfeeding women are frequently concerned about their infant's intake and their ability to produce sufficient quantities of breast milk (Avishai, 2007; Dykes, 2002; Johnson et al., 2009; Labiner-Wolfe et al., 2008). The social value of EBM related to measurability in that it increased women's self-confidence, a perception often exploited by breast pump manufacturers and marketers (Buckley, 2009).

Many breastfeeding women mentioned the convenience and security of having an emergency supply of EBM. Freezing allowed for longer storage and increased shelf-life of the product, to use food processing terminology, as one woman said 'expressing for the freezer.'

> I had stashed an enormous amount in the freezer, just heaps of [it]. We've got one of these huge freezers with drawers, and a whole drawer was just full of milk, which was such a relief. (Rachel, 32 years)

One consequence of the social value of frozen EBM is that the role of a breastfeeding woman has shifted from producer to consumer: the more that she produces, the more she will be able to use. In the context of political economy, breastfeeding women are frequently described as 'producers' or 'workers', breast milk as 'product', and breastfeeding as a process of labour or production (Dykes, 2006; Ebert, 1996; Tabet, 1996). Although women are socialised 'to view themselves as nurturers', they are also socialised to be 'producers', 'consumers', and 'workers' (Retsinas, 1987, p.131). Davis-Floyd (1992) noted that the shift in cultural perceptions of breastfeeding from reproductive production to reproductive consumption parallels shifts related to pregnancy, labour, and maternity care. Women are offered 'the pleasures of reproduction construed as consumption' (Taylor, 2000, p. 397). Some of the women we interviewed did not use their frozen EBM and later discarded it.

In this way, overproduction and overconsumption of EBM mirrored the tendency for general overconsumption in the contemporary developed world (Taylor, 2000).

EBM currency and maternal return to work

The financial value of EBM as product lies mostly in its relationship to women's employment. In our study, career advancement motivated some women's return to work. However, for most women, financial constraints were the main reason for returning to work.

> Unfortunately these days, having a mortgage and even providing the roof over [laughs] your head is a joint salary. We both have to work, and I've had a year off ... I am the major bread winner. My husband is a freelance worker, which means he goes from contract to contract, so there isn't always the guaranteed income. My job and my salary provides for us as a family, so I know I have to return, but I have nightmares about it. (Rose, 38 years)

The amount of breast milk left in the freezer by working women was translated into their earning potential. In our study, the measurability of EBM allowed women to plan ahead and manage their work intensity, either full time or part-time, based upon the quantity of milk they managed to save in the freezer.

> I pumped for a bit and by then I was already starting to fill the freezer for going back to work anyway ... I started expressing for the freezer because I knew from my experience with my son how valuable it was to have a really large stash in the freezer. And so I knew there was heaps and heaps of milk ... and I went off to work full-time. (Rachel, 32 years)

> I went back to work when the baby was five and a half months old, [eh] part-time but [em] it meant that there was something [sufficient amount of EBM] at nursery that he was able to have as well. (Ann, 32 years)

Ongoing re-stocking allowed women to ensure that there was sufficient EBM left at home, however, when milk expression was unsatisfactory or difficult women gradually stopped breastfeeding and kept working.

> I never got on with the expressing. I don't enjoy doing it and it takes me a lot longer to express than to feed her. It takes me about half an hour just to get one or two ounces off. I only have half an hour for

a lunch break. I think you can ask for time to express at work, but being part time, I don't like to do that because ... I know the stresses on the [workplace] as well. So I'm trying to keep work happy and keep the baby happy as well. (Lisa, 23 years)

According to Roe et al. (1999), in the beginning, the amount of EBM left in the freezer influences the intensity of maternal work. Over time, it is the intensity of work that affects the duration of breastfeeding. Our data reflected this pattern. Other researchers have also found maternal return to work to be a barrier to breastfeeding, and employment to be associated with premature breastfeeding cessation (Ahluwalia et al., 2005; Haughton et al., 2010; Labiner-Wolfe et al., 2008; Ogbuanu et al., 2011). The duration of paid parental leave might also significantly affect the duration of breastfeeding (Roe et al., 1999), reducing 'the opportunity cost', that is, family financial loss attributed to maternal preference to stay at home and to breastfeed for longer (Akre, 2009, p. 12). Financial constraints are usually the motivators for most breastfeeding women's earlier than desired return to work (Hofferth and Curtin, 2006; Rojjanasrirat, 2004). Across many countries poor breastfeeding initiation rates and shorter duration have been linked with women's low socio-economic status (Amir and Donath, 2008; Flacking et al., 2007; Griffiths et al., 2007; Li et al., 2005; Wright et al., 2006). Data from the 2010 UK Infant Feeding Survey revealed that 90 per cent of women weaned their infants prematurely (McAndrew et al., 2012).

Physical, social, and financial values of EBM allowed women to continue infant feeding with breast milk in absentia – to provide a bodily substance to their infants in the absence of their maternal body. The term 'disembodied' motherhood (Bartlett, 2005, p. 104; Blum, 1999, p. 60) was applied to women who were physically separated from their infants as a consequence of pursuing their studies and professional careers. They replaced breast feeding with EBM or/and formula feeding. Breast milk expression and EBM feeding were described as processes that allowed women freedom, flexibility, and a sense of 'good' mothering (Johnson et al., 2013) because women believed that EBM was a valuable nutritional substance similar to breast milk.

EBM feeding and paternal involvement

Paternal involvement in infant feeding: profits and losses

The societal trend of 'involved' fatherhood has developed concurrently with the trend of 'disembodied' motherhood. In contemporary western cultures, an 'involved' father (Goodman, 2005) is a construct of positive parenthood (Bailey and Pain, 2001; Greene and Biddlecom, 2000; Schmidt, 2008). Men are expected to be involved in reproductive decision-making, to attend antenatal classes (see Miller, this volume), to be present in delivery rooms, to participate in breastfeeding decision making, and to be involved in infant care (Coleman et al., 2004). Following this trend, paternal involvement in infant feeding has become culturally acceptable in the industrialised world and, in some cases, it shapes feeding decisions (Lee, 2011).

Many contemporary antenatal classes and parenting books promote EBM feeding as a father–infant bonding activity. For example, Florell and Wilson (2010, p. 79) implore 'the father of a nursing infant [to] ask the mother to pump milk so that he can periodically feed the infant with a bottle'. Some researchers have described infant feeding by a father as a positive experience for both parents (Lee, 2008; Schmidt, 2008).

In our study, some women included the baby's father or other relatives in feeding because they planned to return to work, as discussed earlier, or wanted personal time for social activities. Women who decided to re-commence either work or studies found paternal involvement beneficial for their career development, but not always enjoyable.

> I, of course, expressed, and my husband was the one who was feeding him. I think at that moment he felt much more involved and becoming closer to our baby. So the fact is that I had to express, and he was supporting me as well, not only being a Mum but keeping up with my career ... Ah, it was, pretty easy and pretty difficult. It was pretty easy because I'd only work five hours a week ... On the other hand, for me it was very difficult to leave them ... It was difficult because any time you have to leave your baby, even if it is with his Dad, my God, it's like you're just tearing your heart apart. (Maria Alejandra, 35 years)

Paternal involvement in infant feeding was beneficial when it gave women free time. However, feelings of guilt for utilising this time for themselves were almost always reported in women's experiences.

> If I wanted a break or if I had to go somewhere or I would sometimes go out and stay with my family in order to get some catch up sleep ... I went back to my beautician's, to get my waxing done, whatever, just

little things to give me the bit of a break away from the breastfeeding. (Lily, 31 years)

You can express and still go out and do things. I mean in the evenings, okay he was reluctant to take a bottle, but you can still do it. You want to be able to go out and have a good drink sometimes, and to be able to have a night out. You feel a bit guilty almost even; I used to have a glass of wine in the evening. I think people say that's okay, but you still feel a little bit [guilty], and you sort of want your body back. (Katie, 30 years)

Some women believed that breastfeeding restricted the father's involvement in care and made an effort to overcome this.

I know he did feel isolated because we were securing such a bond and it was our first born child together ... I think he kind of did feel fairly isolated. I got him to do nappy changing and bathing. I think I got him to engage more in baths and things like that so, he sort of had his role to do and then I was just there sort of with the milk on tap. (Rebecca, 34 years)

A few women were prepared to replace breastfeeding with EBM feeding, or to even cease breastfeeding altogether because of the father's desire to participate in infant feeding or to give him experience of feeding, reduce his isolation, and promote father–infant bonding.

I came to terms with the fact that my husband needed to bond with our baby and needed that relationship with her and, actually I thought ... about stopping breastfeeding. (Sarah, 26 years)

Women's experiences of father's involvement in EBM feeding varied from enjoyable to frustrating; frequently women reported mixed experiences. Some women noted that EBM feeding might not be a pleasurable experience for everyone. They shared feelings of guilt, concern, and frustration.

I wanted it to be only my husband that did feed a [EBM filled] bottle ..., but my son didn't really want it, ... and my husband got quite frustrated ... so that wasn't a very good experience for him. (Katie, 30 years)

My husband was sitting feeding him a bottle of this expressed milk, and I felt terrible. I felt so guilty, oh there's my baby drinking out of a bottle, even though it was my milk, just gave me the like oh, you know ... I should be feeding my baby, and I was worried that he [baby] would get confused. (Melissa, 34 years)

Women who had positive experiences of paternal involvement in breastfeeding were pleased to see their husbands bonding with their babies through EBM feeding.

> When [my son] was six weeks I did start expressing to a bottle so, my husband can feed him, because it's nice [for my husband] to feed him. He really likes it; he really does enjoy it ... I just wanted [my husband] to get that connection, 'cause it's a very close time when you're feeding your baby and I didn't want him to miss out ... share in that little look that they give you. (Claire, 29 years)

As noted, most women conveyed mixed feelings about paternal involvement, reporting positive and negative aspects at the same time, such as having free time, but also having uncomfortably full breasts because of missed feeds and, in the worst case, developing mastitis.

> She takes milk from a bottle very well ... and she gets left with anybody [laughs], usually it's her dad – I do like to sort of get out occasionally. But at the same time he enjoys giving her milk. There've been times, Mothers' Day, for example, where my husband wanted me to have a lie in, and he gave her a bottle of expressed milk, and that was fine. It wasn't such a benefit because I woke up feeling rather full [laughs], and the first thing I needed to do was feed the baby. (Melinda, 27 years)

> Well just in case I ever wanted to go out for an evening or my husband wanted to feed him, 'cause I wanted eventually to be able to get my husband to feed him in the evening ... I don't think I should have done that 'cause that was probably what sparked the mastitis. Then I stopped the expressing completely. (Katie, 30 years)

The practice of EBM feeding, purely for the experience of feeding and to promote bonding, contributed to maternal guilt and self-blame because EBM feeding was perceived by some participants as being 'not what the baby wants.'

> I tried everything, I must have bought so many different makes of teat thinking that maybe one would be more palatable than another,

none of them were. I remember horrendous times when I left her with my husband and expressed milk in a bottle thinking, if he gave it to her when she was really hungry, she would take it; and she just wouldn't ... It's amazing that even at that tender age she showed those traits. (Emma, 35 years)

For some women, EBM feeding to promote father–infant bonding undermined breastfeeding as a process:

I said to him 'if you want me to bottle feed then I'll bottle feed for the sake that, you know, you can bond' 'cause he wants to bond with his daughter, he really does. (Tanya, 19 years)

In the literature, milk expression, EBM feeding by fathers, and gender roles are controversial (Rowland, 1985). Chodorow (1999, p. 217), who supports the notion of shared parental responsibilities, has pointed out that equal engagement in parenting requires 'the whole parenting of warmth, contact, and reliable care, and not the specific feeding relationship itself'. Moreover, in the industrialised world, the traditional divisions of child care responsibilities are changing (Vincent and Ball, 2006). In 2008, 91 per cent of surveyed UK fathers took some time off from work when their child was born, including their two-week ordinary paternity leave alone or in combination with other leave (Chanfreau et al., 2011). In 2011, the UK government introduced additional paternity leave up to six months, entitling fathers to additional paternity pay of £135.45 per week or 90 per cent of their gross average weekly earnings (whichever is lower), if their partner returns to work before the end of maternity leave (Government UK, 2011).[1]

Historically, paternal involvement in infant care was dichotomised, particularly in feminist debates. Liberation feminists and equality feminists who supported women's equal participation in paid employment (Deutsch, 1999; Ehrensaft, 1990; Okin, 1991) have viewed shared parental involvement in infant feeding as beneficial because of the potential to increase women's 'choices', improve their career prospects, and reduce the burden of care. In contrast, equality feminists who studied gender power relationships (Boswell-Penc, 2006; Kahn, 1989) have viewed paternal involvement in infant feeding as controlling and oppressing. Tomori (2009) has suggested that these debates do not contribute to the understanding of postmodern issues related to paternal involvement in infant care, which are focused merely on supporting women with breastfeeding. Positive aspects of paternal involvement in infant feeding described in the literature include breastfeeding support and promotion (Sherriff et al., 2009) and EBM feeding when a woman returns to work (Tomori, 2009).

Paternal involvement in infant feeding as a market-expanding technique

Providing deeper insight into contemporary EBM feeding practice, we argue that the promotion of paternal involvement in infant feeding under the guise of father–infant bonding is a market-expanding strategy in line with neoliberalisation; one that separates women from breastfeeding babies and eventually leads to the introduction of infant formula. Gender diversification of breastfeeding used by profit-motivated breast pump and related accessories manufacturers and retailers reinforces breast milk as a commodity. Breast pumps are advertised not only for medical reasons but also for psycho-social reasons. They are presented as gifts, included in infant care packages, and distributed during antenatal education classes and breastfeeding promotion programmes (Cohen et al., 2002) creating an impression that a breast pump is a 'must' for every woman who has or even plans to have a baby (Avishai, 2007). In our study, women pointed out that breast pumps and related accessories were promoted in antenatal classes:

> I went to one [antenatal class] which was specifically for breastfeeding, which was probably about two weeks before I was due with my first baby, which was fantastic. It gave hints and tips ... and the benefits of having a breast pump, which I had not really considered up until the antenatal class. (Emily, 30 years)

Some women said that they did not have any medical indications for breast milk expression or plans to be separated from their baby due to work or other reasons. However, a breast pump and related accessories were purchased by their husbands together with other maternity care products.

> My husband actually did buy the sterilising tank and all this and all that [breast pump and related accessories], and I said 'that's fine no problem, if I have to, I'll express and I'll use it,' but it's just in the box how it was bought. (Ruth, 25 years)

To increase the desirability of breast pumps and related accessories EBM feeding practice is promoted as offering women a 'choice' and men 'experience'. We contend that this neoliberal marketing strategy, which undermines breastfeeding and makes more work for women, is similar to established infant formula marketing messages in which formula feeding is positioned as offering 'equality of rights for men and women', 'freedom to pursue careers', and 'liberation from domesticity' (Boswell-Penc, 2006, p.118). To illustrate, we noted one caption on an infant formula advertisement that read 'For you and for them' beneath a photograph of a father bottle feeding his baby during the night while his partner slumbered beside them – a slogan that is very hard for women to argue with. Although EBM feeding might indicate men's involvement in infant care, it does not reflect the degree of their involvement. In reality, 'few fathers actually do take the whole responsibility of infant care and most artificial [and EBM] feeding is still done by mothers' (Palmer, 2009, p. 86).

Similarly, Hausman (2008, p. e3) has observed that women's choice of the method of infant feeding 'has not liberated women from the burdens of maternity, although many women have benefited from entry into waged labour made possible under current constraints by replacement feeding'. The American Academy of Pediatrics (2012) stated that breastfeeding is a public health issue and not a personal lifestyle 'choice'. We contend that this message is in danger of being drowned out by the commotion of manufacturers and advertisers analogous to that of smoking cessation messages that were muted by advertisers in the early stages of tobacco control (Smith, 2004).

Conclusion

EBM is firmly established as a commodity in line with other female body parts, tissues, and fluids. In this chapter, we have discussed the social and economic values of EBM that have led to its commodification and contributed to changes in gender roles in relation to parenting. Among other qualities of EBM as a product, measurability and storage quality were valued by women. These EBM qualities allowed women to continue breast milk feeding while they resumed work or studies or took a break from breastfeeding. Having a stock of EBM in the freezer increased their confidence that they would not run out of EBM. Increases in women's confidence and paternal bonding are frequently highlighted by breast pump manufacturers and sellers and promoted as benefits of breast milk expression and EBM feeding, suggesting further commercialisation of EBM (Avishai, 2007; Buckley, 2009).

This chapter contributes to a reframing of reproduction within sociology by providing additional evidence for the global trend for commodification of reproduction and parenthood in postmodernity. Contemporary understandings of gender roles in western societies are changing. In late modernity, gender roles are described as being in a state of transition during which society questions gender role beliefs and integrates gender roles in a less stereotypical way (Weiten et al., 2010). In line with this and in conjunction with our UK data, we discern that the new cultural practice of EBM feeding can have the positive effect of promoting father–infant bonding and offering women the ability to participate more fully in social life, including the return to paid employment.

Nevertheless, male and female reproductive roles cannot be viewed in isolation from western social institutions which influence and shape these actions (Browner and Sargent, 2011). The established practice of EBM feeding by the father when a breastfeeding woman decides to return to paid employment or have 'time out' is a cultural response to the attempts of social institutions to control male and female reproductive bodies in relation to employment. Short paid maternity leave, return to work policies, and lack of childcare facilities are some of the elements of this institutional control over breastfeeding that require women to express breast milk and involve other people in infant feeding. These neoliberal policies promote 'disembodied motherhood' (Bartlett, 2005, p. 104; Blum, 1999, p. 60) and 'involved fatherhood' (Goodman, 2005, p. 190). These policies have also reframed parental reproductive roles in relation to infant feeding. These constructions contrast sharply with earlier gender norms in relation to infant feeding that defined a woman who prioritised breastfeeding over her career as a 'good mother' and a breadwinning father as 'absent' (Schmidt, 2008, p. 61).

The elements of patriarchy are still common in postmodern societies and their institutions, where paternity is placed at the centre of social relationships (Rothman, 2008). Although there might be family benefits related to paternal involvement in infant care, we argue against father–infant bonding being promoted at the expense of the mother–infant relationship and the replacement of 'nursing' with 'the delivery of a nutritional substance' (Hausman, 2003, p.104), both of which can potentially contribute to mother–infant separation and eventual breastfeeding cessation. Likewise, maternal, infant, and societal benefits of breastfeeding should be taken into account in feminist debates on shared parental involvement in infant feeding (Galtry, 1997).

Acknowledgements

The work was funded by Bournemouth University, UK, and conducted in collaboration with the Health Experiences Research Group, University of Oxford. We would like to acknowledge the women who participated in this research and the advisory panel who helped with recruitment.

Notes

1. There is no statistical information available on the proportion of fathers who have utilised the recently introduced paternity leave.

14
What Does Not Kill You Makes You Stronger: Young Women's Online Conversations about Quitting the Pill

Elizabeth Arveda Kissling

The birth control pill is often credited with midwifing the women's movement, but that's not really accurate; it was just in the right place at the right time. The ability both to prevent and plan pregnancy allowed women and girls to take advantage of new opportunities in education, work, and public life that feminist activism was beginning to make possible (May, 2010). The pill is currently used by 12 million women in the US, and more than 60 million women worldwide (Jones, 2011; Philipson et al., 2011), and there is little doubt that access to consistently reliable contraception increased educational and occupational opportunities for women, as well as support for feminism in western industrialised nations.

The history of the pill and its social impact, including its emergence as the first lifestyle drug and its political history as a tool of population control, is well documented (Briggs, 2010; Eldridge, 2010; Ghazit, 1999; Goldberg, 2010; May, 2010; Tone, 2001). Available in the US since 1958 in evolving strengths and formulations, it is the only drug known simply as 'the pill'. It is one of the most intensely studied drugs in history (Boston Women's Health Book Collective, 2011), and it is believed to be among the safest – safer than aspirin, as an editorial in the *American Journal of Public Health* noted 20 years ago (Grimes, 1993).

Despite those numbers and more than 50 years of use, the pill continues to generate controversy in the US. A new generation of conservative politicians opposes access not only to legal abortion, but also to birth control. Five US Republican Party 2012 Presidential candidates signed a so-called 'Personhood Pledge' that would redefine legal personhood as beginning at fertilisation, despite well-established scientific and medical practice defining conception and pregnancy as beginning at implantation (Toobin, 2012). The eventual Republican nominee, Mitt Romney, vowed with numerous members of his party to ban public funding for Planned Parenthood and other family planning clinics. Additional public debate about contraception erupted in the US in early 2012 when the United States Conference of Catholic Bishops (2012) objected to national health insurance reform because it would entitle employees of Catholic hospitals, schools, and service agencies to coverage of contraception, just like others who have health insurance provided by their employers.

Pill use appears to be declining in some places, such as Canada (Lunau, 2009), and in the last five years, vast numbers of young women are increasingly posting their personal testimonies of quitting the pill online. Women are quitting for a variety of reasons: to restore feelings of psychological and emotional health, to regain lost libido, to relieve cardiovascular symptoms and disorders, or to ease anxiety about these or other health issues. They are finding one another and sharing their stories of pill misery, sometimes finding valuable support from one another that is seldom available anywhere else as they go through what can only be characterised as withdrawal symptoms as they quit the pill. They are writing on health care and beauty forums, personal blogs, magazine and newspaper sites, and any place the topic of contraception and hormones arises. This chapter will examine several of these online conversations, looking especially at themes of postfeminism and the political contradictions confronted by North American women in the twenty-first century as they seek control of their fertility, their health, and their lives. Although online conversations are accessible globally, these websites are primarily US-based, and from context, most of the respondents appear also to be US-based, although a few self-identify as Canadian or British. The responses have not been classified according to origin, for obvious reasons.

What's so bad about the pill?

While new iterations of the pill are purported to be safer, a new generation of young women are questioning the risks and safety of the pill, as reports of cardiovascular problems, blood clots, depression and related mood disorders, and declining libidos increase, as do lawsuits against manufacturers. At the time of writing, 11,900 lawsuits involving about 14,000 plaintiffs against Bayer (manufacturers of Yaz and Yasmin, which contain the fourth-generation synthetic progestin, drospirenone) are pending (Feeley and Cronin Fisk, 2012). More suits were filed as more women learned of the US Food and Drug Administration (FDA) announcement that combined oral contraceptives containing drospirenone are linked with venous thromboembolisms, known commonly as blood clots (U.S. Food and Drug Administration (FDA), 2012a, 2012b). Bayer began to settle some of these suits in the summer of 2012, and publicised that they were doing so, in hopes of keeping their stock from falling. News of settlements led even more women to suspect Yaz/Yasmin as a cause of their blood clots, and file still more suits (Seedol.com, 2012).

The 2011 FDA review of drospirenone-containing oral contraceptives ultimately concluded in early 2012 that the pills are associated with a higher risk of blood clots than other combined oral contraceptives (FDA, 2011). Manufacturers are now required to add information about these studies to package labels (FDA, 2012b). An FDA panel debated pulling drosprienone-containing contraceptives from the market entirely, but the panel determined by a four-vote margin 'that the drugs' benefit outweighed the risks' (Grigg-Spall, 2012a). Three members of the panel had financial ties to Bayer, and a fourth had financial ties to a company that manufactures a generic version (Grigg-Spall, 2012a). The agency apparently denied the request from the Project on Government Oversight to convene a new committee.

These controversies occur in a vastly different context than the early days of the pill. Thanks largely to their feminist foremothers, most young middle-class women in the US have greater independence and sexual autonomy than earlier generations as they seek alternative birth control, and they are able to draw upon the expertise of 40 years of an active women's health movement. The birth control pill ignited the women's health movement, in response to the Nelson Pill Hearings in 1970. Investigative Congressional hearings about the safety of pharmaceutical products were convened by US Senator Gaylord Nelson, partly inspired by Barbara Seaman's 1969 book, *The Doctors' Case Against the Pill*. Medical researchers famously testified that 'Estrogen is to cancer what fertilizer is to wheat' (quoted in Kissling, 2010). Feminist activists, including Alice Wolfson, who eventually co-founded the National Women's Health Network with Seaman, interrupted the hearings when the Congressmen refused to respond to their queries:

> Why weren't we told about side effects?
> Why aren't any women testifying?
> What happened to the women in the Puerto Rico study?

Why are you using women as guinea pigs?
Why are you letting the drug companies murder us for their profit and convenience? (Kissling, 2010)

One of the concrete outcomes of the Nelson Pill Hearings was FDA-mandated Patient Package Inserts (PPIs), the printed information about risks, ingredients, and side effects included in pill packets, first required for oral contraceptives and then for all prescription drugs. Today young US women have direct-to-consumer pharmaceutical advertising to contend with; the pill is marketed to them for such non-contraceptive purposes as weight loss, clear skin, and menstrual suppression. Although pill advocates often claim the pill, at 50 years old, is among the safest drugs in history, new formulations are introduced frequently, as manufacturers seek new patents, leading to new profits.

Young women also are enmeshed in a media culture saturated with postfeminism, framing their every decision as a personal, *consumer* 'choice', devoid of broader political implications. As I have argued elsewhere (Kissling, 2013), current advertising campaigns for birth control pills are keen examples of this trend, framing this prescription drug as just another consumer product, desirable for its side effect of menstrual suppression as much for its main effect of pregnancy prevention. For example, ads for Seasonique emphasise women as individuals responsible for their own decisions ('Who says?' is repeated eight times in one television ad). The ads depict women acting alone, not consulting doctors, partners, or peers about birth control choices (Kissling, 2013). The earliest public discussions of the pill were partly shaped by the nascent women's health movement and influenced by critical voices such as Seaman, and the Boston Women's Health Collective, creators of *Our Bodies, Ourselves*, the bestselling women's health reference book. Today, the dominant voices in public discussions about hormonal birth control in the US include anti-choice politicians, and pharmaceutical companies. Postfeminist themes of choice, individuality, and consumerism recur in many of these messages.

The term 'postfeminism' has been contentious since its emergence into common parlance in the early 1980s. Although the prefix *post-* suggests postfeminism refers to a time after feminism, I follow the more narrow usage of critical media scholars such as Angela McRobbie (2004) and Rosalind Gill (2007, 2008). McRobbie explains that postfeminism 'refer[s] to an active process by which feminist gains of the 1970s and 80s come to be undermined' while appearing to engage feminism, especially through tropes of freedom and choice (2004, p. 255). In other words, postfeminism is not 'after feminism' but an appropriation of feminism. Postfeminism is further characterised by the treatment of femininity as a bodily property, a shift from sexual objectification of women to representing women as sexual subjects with desires of their own, an increasingly sexualised mainstream media and culture, an emphasis on individualism and choice, the need for constant self-monitoring and surveillance, a focus on consumption and commodities, and a reassertion of the importance of sexual difference (Gill, 2007).

Whereas feminism proposes a complex political identity and strategies for social change, postfeminism assumes feminism is no longer necessary and re-rationalises it according to neoliberal values that replace classic liberal values of human rights, equality, and liberty with primacy of the contract and the marketplace (Treanor, 2005), and the individual. Postfeminist texts thus repudiate feminism and propose women's achievements as well as their failures as products of individual effort rather than collective action or structural impediments.

Notes about methodology

I stumbled serendipitously on three online conversations at *No More Dirty Looks, xoJane*, and *Dr. Sugar*, in my role as editor of *re:Cycling*, the blog of the Society for Menstrual Cycle Research. Fascinated, I started looking and soon found numerous online conversations about quitting the pill. There are more at eHealth forums, *Aphrodite Women's Health, Our Bodies Our Blog*, the discussion forums for *BUST* magazine, and many other sites, most from the US. The same concerns expressed about the pill recur: depression, moodiness, lost sex drive, acne, weight gain, and worries about cardiovascular health. For those who have already quit the pill, additional concerns include the unavoidable breakouts, return of ovulation and menstruation, and the feeling of being on an emotional roller coaster.

I examined these three discussions in detail because they were recent (the *Aphrodite Women's Health* thread on quitting the pill contains several thousand messages and has gone on for several years), and they are on popular, widely read, women-oriented US sites. As noted below, while there are a few positive comments about successful, unproblematic pill use interspersed into these discussions, satisfied users seldom feel the need to reach out or to protest or complain.

Researchers who regularly study online communication do not agree about whether such online communication is considered public or private (Eysenbach and Till, 2001; Hookway, 2008). However, these three sites are all clearly public sites, and readers post comments fully aware that their words will be seen and read by others; in many cases, they were actively seeking readers, as they asked questions and sought advice. With today's Internet surveillance technology, it is difficult to argue that any online spaces are truly private. In addition, commenter anonymity is preserved, as these users are known to me only by their chosen screen names, and only when those are available. Furthermore, it is extremely unlikely the commenters could be harmed by having their words quoted in this chapter.

This study is not a systematic content analysis of online discussions of oral contraceptives, but an exploration of emergent themes that arise when women discuss their dissatisfaction with the pill[1] with one another. Examination of online discussions such as these permits access to a far larger number of pill-using women than most individual researchers (outside of large medical practices) could reach. Examining commentary posted voluntarily by pill users, in their own words and unmediated by interview questions from researchers or medical professionals, provides the opportunity for *their* concerns to predominate.

While in some cases I counted how many messages in each group focused on certain themes, such as emotional reactions or loss of libido, it is difficult to quantify the responses. Fitsugar.com permits anonymous posting,[2] while the other two sites do not, and in all three sites, some individual users posted more than once. My interests are not in the numerical data, but the messages – the fact of them, and the recurrence of them. I have examined these discussions using Hall's (1967) model of cultural analysis in three interconnected phases: (1) close, textual analysis of cultural material; (2) consideration of the effects of the cultural material on the society; and (3) placement of the material in its specific social and cultural contexts to produce an interpretation of cultural meaning and significance.

From this method, questions emerged: What does it mean that hundreds of women are quitting the pill out of misery, after it was promised to them as the safest and most effective birth control available? Why are they so unsatisfied with the answers they receive from their health care providers that they prefer to seek the advice of strangers on the Internet? I have scrutinised their words with a feminist critical lens, seeking themes that reoccur and that can move us closer to answering those questions.

Findings

Women quitting the pill

Threads of postfeminism and neoliberalism are woven through the online conversations among young women quitting the pill. For example, a November 2011 post on the eco-friendly beauty blog, *No More Dirty Looks*, titled 'What's your take on the pill and what happens when you go off it? A girl in Paris needs help!' received 84 responses, all but ten within the first week. The Paris reader sought advice about how to cope with side effects of quitting the pill, especially the inevitable disfiguring acne. The 84 responses included 21 comments from readers who had experienced breakouts upon quitting the pill. Some respondents mentioned having begun the pill initially more for its skin-clearing properties than for its contraceptive effects – not terribly surprising from readers of a beauty blog. Nineteen contributors also mentioned emotional or psychological effects, and offered support.

But the discussion was framed from the start by one of the blog's hosts, in postfeminist, neoliberal terms of individualism:

> There's no judgment – implicit or explicit – on anyone who is on or has been on birth control pills. Some people love them, some people have to take them for medical reasons, some people abhor them. Here, we want to talk candidly about what happens when you go off them. Because, whoa. That can be hectic. (Siobhan)

Whether commenters were dutifully following this instruction or simply failed to make any political connection is unclear, but postfeminist neoliberal individualism emerged in several comments. For example, near the end of a 505-word comment detailing numerous serious health problems stemming from pill use, Jen wrote, 'This has really made me realise I need to be more proactive on what I am putting into my body and educate myself better'. Instead of attributing her difficulties to an inadequate health care system or corrupt pharmaceutical industry or other systemic cause, Jen concludes that both her difficulties and the solution are her own responsibilities.

Another contributor, Paige Worthy, wrote,

> This has actually moved me to start doing some research on my own birth control pill (Kariva, the generic of Mircette) and find what it's been doing to my body. If I find out the hormones in this pill

are somehow related to the bizarro [sic] depression I've been going through for the past ... really long time ... [ellipsis in original] and I could actually fix a LOT of problems by going off it? Man, I'll be kind of annoyed. And empowered.

Michelle, who recommended that women chart their cycles, wrote, 'I hope that more women who are questioning the pill will think about their options and take time to really get to know their bodies & do what is best :)'.

The weight of 84 comments suggests that the problems with birth control pills are shared, located in the pills and a health care system that distributes them readily and easily, for many different reasons, perhaps not always with thorough examination and evaluation first. Current research indicates that nearly 60 per cent of US pill users take it for non-contraceptive reasons, such as for cramps or other menstrual pain, menstrual regulation, acne, endometriosis, as well as for prevention of unintended pregnancy. Fourteen per cent of US pill users (more than 1.5 million women) take birth control pills *solely* for non-contraceptive reasons (Jones, 2011). Jaime, the fifth commenter, wrote, 'It is making us all sick and it is not healthy to be on the pill. I wish I knew what I know now about our "health" system and that it is really about keeping us sick and loaded up with drugs, that's where the money is'. However, Jaime is the exception among the women of *No More Dirty Looks*; the rest all saw solutions to problems with the pill in terms of individual responsibility and empowerment. Even though there was a sense of community among the readers, there was no sense of any need for collective action. Many of the solutions involved increased consumerism, as they recommended particular skin-care products to one another – herbal, earth-friendly products, to be sure – or to seek the counsel of a naturopath. One of the last commenters, Anna, even blamed feminism for her disappointment with the pill: 'In my experience, the pill wasn't this liberating token of feminism that it is marketed to be' – which is a bit ironic, since she said that she began taking the pill at the urging of her male casual sex partners.

Another online discussion a few months later also featured postfeminist themes, and even stronger concerns about side effects of the pill. In January, 2012, the online magazine *xoJane* regular column 'It happened to me' featured a story by Sarah LaDue titled 'My birth control gave me a pulmonary embolism'. The 26-year-old author wrote of her surprise to find herself in an emergency room with leg pain and coughing, and diagnosed with a deep vein thrombosis that ran from ankle to hip and spread to her lungs. The physicians treating LaDue informed her that it was not as strange as she thought: '[t]hey shook their heads and informed me that they regularly encountered otherwise healthy young women with blood clots, almost all caused by birth control' (LaDue, 2012).

Until the 2011 FDA review, clinical studies on blood clot risk and hormonal contraception concluded that such side effects were rare, although acknowledged that the risks vary with the amount of estrogen and type of progesterone in the product: risks are higher with so-called fourth-generation synthetic progesterone (such as drospirenone, found in Yaz, the brand used by LaDue) than the older forms (Raymond et al., 2012; Rott, 2012). However, it is seldom acknowledged that the risks in these studies are compared either to other hormonal contraceptives or to pregnancy – *not* to using effective non-hormonal contraceptives (Grigg-Spall, 2012a).

Within three days, LaDue's post had garnered 100 comments, many from other women who had quit the pill for similar reasons. The total included six women who had also experienced pulmonary embolisms, five who had similar experiences of deep vein thrombosis, one who had had a stroke related to hormonal birth control, three others who reported knowing someone who had had a hormonal birth control-related stroke, and 12 reports of quitting the pill because it 'made me an emotional wreck' or 'totally made me crazy' or 'fucked me up ... crazy bad emotional symptoms' [ellipsis in original].

Depression, identified as a significant side effect of birth control pills in the early 1960s, is the most frequently cited reason for quitting the pill (Kulkarni, 2007). Though it is a well-known and frequently studied side effect, depression is not considered a contraindication for a pill prescription (Böttcher et al., 2012), in part because the many studies are inconsistent in how depression, depressive symptoms, and mood changes are defined and measured across studies.

The commenters at *xoJane* seemed to be even more of a close community than those at *No More Dirty Looks*, with frequent nested sub-conversations and expressions of support for one another in this thread. Many contributors prefaced their own remarks with personal comments for LaDue like, 'So glad you are fine!' or 'I'm so glad to hear you made it out ok!' and similar statements. Much like *No More Dirty Looks* readers, all of their support and proffered help for one another with pill problems was framed in terms of individual responsibility and empowerment:

> I hate how birth control pills are kind of thought of as benign these days ... they definitely aren't. They made my blood pressure go way

up (it's normally kinda low) and I will never take them again. Thanks for writing the article and raising awareness on this issue! (Maria) [ellipsis in original]

Thanks for this. Such a good reminder to take care of myself. I think a lot of people often feel like it's somehow weak or unnecessary to go see a doctor, but it's things like this that remind me to check in. (Lindsey Keefner)

But this article makes me happy because maybe women will continue to educate themselves. (Katie Garrity)

I am completely on board for getting the word out about blood clots and birth control. My gynecologist was incredibly irresponsible with my treatment, tried to prescribe me yet another hormonal birth control, and now isn't returning my calls. So gals, please make sure to educate yourself and ask as many questions as you can think of. (Alfenia)

Even though some of the comments show an awareness of systemic issues, the women discussed solutions only in terms of individual action. Commenter Whatwhatque did present a feminist analysis but quickly changed direction and mocked herself as an 'overdramatic' conspiracy theorist:

I am starting to think that the the [sic] issues with hormonal BC are swept under the rug because they make MEN's lives so much easier (no condoms? are you on the pill? LET'S BONE!) while making so many women's lives worse! Conspiracy! I'm being overdramatic but it really does seem like as awesome as the pill is for some people, more and more I hear about the health risks and side effects, the more it makes me wonder why there aren't more doctors working to improve them/come up with alternatives. This is scary! I'm already off estrogen BC because [sic] of my migraines (increased chance of stroke!) and progestin BC makes me a crazy, zitty wreck. Maybe one day I'll have insurance and get an IUD (although I hear horror stories about that at as well!)

One commenter agreed with Whatwhatque, then she and a third commenter quickly turned the discussion to sexually transmitted diseases and personal responsibility, and men's and women's differing attitudes toward sex – another characteristic of postfeminist sensibility:

I kind of agree with you on this. What REALLY bothers me is that so many men now pretty much expect women to be on hormonal birth control and they truly have NO IDEA about possible side effects and complications. All they know is that it makes condom free sex possible. I understand that condoms reduce sensitivity, but wearing one won't kill you like DVT and PE will. And further, I think they generally have no idea about how women's bodies work. It's really frustrating and sexist. (qkim)

The worst part about guys not thinking about wearing condoms is that they're not thinking about STDs, either. I wonder if anyone has put a chart together to compare the rate of STDs and the rate of women on the pill. I have a sad feeling that they'd be similar looking lines. (Alex)

Seriously – do men not worry about this shit? Like I've had guys that were all ready to go at it without a condom before even asking whether or not I was on the pill, if I've been tested, etc. I just feel disrespected when that happens. (qkim)

Same! It shocks me every time, but I swear, most guys I sleep with don't seem to be concerned about it. (Jess)

In more than 100 comments, there was no mention of contacting pharmaceutical companies, organising a petition or protest, or any other kind of political or collective action. No one suggested reporting side effects to the FDA, although information about how to do so is included on the package inserts. Commenter Savannah seemed typical, in her view of only herself, not a political movement, not a health care team, not even her own physician, as responsible for taking action:

One of the problems with warnings about side effects is that both the patients and more worryingly, physicians don't seem to be bothered by them and physicians can give a false sense of security surrounding the drugs they are prescribing. I've been on medication and the only way I know what the side effects are is to actually read the prescription bottle.

Another online discussion that began March 16, 2010, at fitsugar.com (part of the PopSugar empire of women's fashion and health websites based in the US) as a Q & A with 'Dr. Sugar' about 'What will going off the pill do to my body?' has continued for more than two years, without any additional input from Dr. Sugar. A user known as danakscully64 stepped in and offered reassuring advice to numerous women with questions about side effects of the pill and of quitting, ovulation, and related issues. As of this writing (July, 2012), commenters were still posing questions. Danakscully64 does not appear to be affiliated with the site, and identifies herself as another dissatisfied birth control pill user early in the thread. She also makes clear she is not a medical professional, just experienced and well read – and I would add, generous and compassionate, to continue to answer questions, often posted by anxious teenagers, on a thread long abandoned by the original host. After the first year and a half, the discussion focus shifted from concerns about quitting the pill to more general questions about the pill, conception, and contraception.

Like the women at *No More Dirty Looks* and *xoJane*, the commenters at fitsugar.com are giving up on the pill due to frustration with side effects. The most frequently cited reason in the discussion was emotional effects, with 34 of the 256 comments describing mood changes. Some women reported feeling overly emotional, like Kast1: 'I'm an emotional MESS. And I mean a mess. Everything makes me sad, I feel like I subconsciously search for things to be upset about', the anonymous commenter who 'was way beyond moody', or Kelseynk, who said that on the pill, 'Everything makes me sad, I cry at the drop of a hat'. Others describe a flattening of affect, such as DBRN, who wrote, 'now I feel like emotions are dulled. I'm not on an emotional rollercoaster, but I don't feel joy anymore. I feel sort of paralyzed emotionally'. Danakscully64 said that she 'didn't realize how terrible I felt when taking the pill until I got off'. The commenters at fitsugar.com also cited breakouts, declining libido, and weight gain as frequent reasons for quitting the pill. There were also many comments about symptoms experienced *as a result* of quitting the pill, especially irregular cycles. Women wrote asking if it is normal to have no bleeding for months upon quitting, and if it is normal to have vaginal bleeding that continues for days, or even weeks, on end.

It should also be noted five commenters reported successful pill use, such as OneStopMom, who used the pill for 13 years before stopping for a planned pregnancy, and resuming pill use less than a year after giving birth: '[M]y daughter is now herself a teenager. For me the pill has been a non-issue'. An anonymous commenter reported being on the pill for nearly 30 years with no problems, stopping now to see where she is 'in the menopause phase of life'. These commenters seemed to want to reassure others that the pill can be used safely and successfully by offering themselves as counterexamples.

A few contributors also remarked on the volume of negative comments about the pill and suggested the 'Yelp effect' was in play – a reference to the popular global web company that allows everyone to be a restaurant critic. Many people believe that users are more motivated to log in and post negative reviews than positive reviews, although a recent Harvard Business School study has suggested that once reviews accumulate, 'the variance goes down', according to researcher Michael Luca (quoted in Blanding, 2011).

Postfeminism, subjectivity, and the pill

In calling out the postfeminism and emphasis on individual action in these conversations, I am not blaming young women for a lack of political initiative or feminist ambition, nor positioning them as cultural dupes, victimised by pharmaceutical companies or the medical-industrial complex. Speaking out, even on the Internet, can be a form of action. Our neoliberal era is very different than the days when a US senator would demand investigations about questionable pharmaceutical products and practices. Among many other social changes, US drug manufacturers hold more influence over both legislators and consumers, now spending nearly twice as much on promotion as they do on research and development (Gagnon and Lexchin, 2008). An examination of the lobbying influence of the pharmaceutical industry is beyond the scope of this chapter, but it must be noted that it is a powerful influence on US public policy. The industry also influences the practice of medicine: Gagnon and Lexchin (2008) used proprietary databases to supplement the information available from IMS Health, the pharmaceutical market intelligence firm that the US General Accounting Office relies on, to determine that the pharmaceutical industry spends approximately USD$61,000 per physician per year promoting drugs to doctors.

It is unsurprising that traditional protest activities such as petitions, filing complaints with manufacturers, or even reporting adverse effects to the FDA seem futile to these women. Even after it was shown that Bayer withheld data about Yasmin and Yaz, the FDA subcommittee voted not to remove the drug from the market (Grigg-Spall, 2012b). The privatisation and deregulation that characterise this neoliberal climate began in the Reagan years in the US (and the Thatcher years in the UK), as the state withdrew support for social and human services. In this contemporary neoliberal framework, all are believed to be on an equal footing of individuality, seen through the lenses of rationality and market values. As Brown (2005, p. 42) has argued, '[N]eoliberalism normatively constructs and interpellates individuals as rational, calculating creatures whose moral autonomy is measured by their capacity for 'self-care' – the ability to provide for their own needs and service their own ambitions'.

Postfeminism resonates with these neoliberal values (Kissling, 2013). Rosalind Gill (2008) has further suggested that the female subject constructed in postfeminist media culture is the ideal neoliberal subject.

The postfeminist sensibility of the twenty-first century is characterised by a greater preoccupation with women's bodies, near the point of obsession. The feminine body is a project forever in development; women and girls are always working on constructing an idealised femininity as well as an imagined perfection of health (most often manifested as an aesthetic or moral ideal). The discourses of body projects are those of 'freedom' and 'choice', often framed in terms of 'feeling good about oneself'. Gill (2008, p. 441) explains, '[i]n this modernised neoliberal version of femininity, it is imperative that all one's practices (however painful or harmful they may be) be presented as freely chosen – perhaps even as pampering or indulgence. This seems to me to be doubly pernicious'. Hormonal birth control is one means of constructing the idealised feminine body that does not bleed and is fertile only at will. The rejection of the pill is a protest, but in the current political and cultural context, it is a consumer boycott, not a political or feminist action. It becomes another example of 'the way in which power and ideology operate through the construction of subjects, not through top-down imposition but through negotiation, mediation, resistance and articulation' (Gill, 2008, p. 439). This negotiation is ongoing; young women want to control their fertility, but on their own terms, trapped in a culture and context in which individual control is illusory.

Birth control pills are promoted more heavily than ever, 'handed out like candy', as Holly Grigg-Spall (2013), the author of *Sweetening the Pill*, says. The pill is nearly always the first-line birth control, although a new study has shown that IUDs are 20 times more effective (MacMillan, 2012; Winner et al., 2012). Women reject this prescription in self-defense. They protest, but postfeminist protest is a depoliticised, individual effort. They have been told continually that everyone responds to pill differently (and encouraged to just try another brand, another formula), so their rejection of the pill feels even more individual. In seeking solutions online, perhaps they are also seeking community; may they also find collective strength.

Acknowledgements

The author wishes to thank Patty Chantrill, Holly Grigg-Spall, Meredith Nash, and anonymous reviewers for valuable feedback on previous drafts of this chapter.

Notes

1. Many of the same concerns apply to other hormonal contraceptives, such as the patch, the ring, and injectable contraceptives, but for the sake of simplicity and consistency, I have limited my focus to the pill.
2. In the strictest sense, all three sites permit anonymous posting, as users are not required to use real names. But fitsugar.com is the only site with unsigned posts; *xoJane* and *No More Dirty Looks* require users to register with an email address before comments will be posted on their sites.

Conclusion: Where Do We Go From Here?

Meredith Nash

Reframing Reproduction approaches reproduction as a complex location and inherently interdisciplinary topic – it is a book about the changing meanings and experiences of postmodern reproduction for women and men in the contemporary West. Embedded in the title of this volume is a vision for a 'reframing' of reproduction that involves opening up the study of reproduction in sociology, in particular, to further critical interdisciplinary interrogation and methodological innovation. This vision also flags the need for more substantive engagement with feminism(s) and, in particular, its applied political dimensions. After all, an examination of reproduction is, for me and many of the contributors in this book, an act of feminist commitment. By gathering the voices and visions of a broad range of scholars and placing them in conversation with one another, this volume extends the discursive boundaries of 'reproduction' in order to account for the significant social changes and shifts in gender relations in the West that have accompanied neoliberalisation and its effects. Methodologically, I hope that a significant contribution of this work is its effort to showcase a variety of methods that illuminate our understanding of reproductive experiences, technologies, and practices.

Yet critical questions remain and will determine where we go from here. What does a 'reframing' of reproduction *mean* when it is so deeply entangled in the ambivalences and uncertainties of postmodernity? How will the increasing commodification and globalisation of reproductive 'choices' affect women and men and current configurations of stratified reproduction? What new obstacles will this bring? What controversies will it ignite? This book makes it apparent that there are no straightforward answers. However, what remains uncontested is that reproduction will always be under sociological scrutiny and the contributions of other fields will keep it open to conceptual re-imagination.

In the remaining paragraphs, I shall highlight just a few domains in which the 'reframing' of reproduction has potential to develop.

One domain can be found in the terms 'empowerment' and 'choice' and the evolution of their meanings for women and men. As Annandale (2009) has suggested, we must be wary of inclinations to suggest that people in the West in the twenty-first century are now wholly 'empowered' or 'liberated' by the increased range of consumer 'choices' available to them in the global reproductive marketplace. She flagged this as a vexed, highly visible issue evident across many representations and discourses of reproductive health, especially for women. This book speaks directly to these ambivalences. A number of chapters have problematised 'empowerment' for women and men in the reproductive realm in view of neoliberalisation, commodification, and globalisation, and especially in relation to technologies such as prenatal screening (Chapter 3), birth control (Chapters 4 and 14), assisted conception (Chapters 2 and 11), and gamete donation (Chapter 12). Although reproductive technologies have made new spaces for parenthood, they also entail multiple consequences that affect individuals and social groups differently and within power relations, an idea forged within much of feminist theorising on reproduction. The chapters in this book remind us of the necessity of expanding the terrain upon which 'empowerment' is at stake including how various aspects of identity like gender, race/ethnicity, sexuality, (dis)ability, and social class intervene in reproduction.

Also at stake in future developments is how reproduction is politicised within sociology and how gender is (re)constituted in postmodernity. Although feminism has been present in the development of reproduction as a subfield of sociology, 'feminism' tends to make fewer appearances (in name, at least) in sociological writing on reproduction. Feminism tends to lurk in the background. As Skeggs (2008, p. 681) has noted, there are still sociologists who, in spite of having an awareness of 'gender', manage to 'protect themselves from any of feminism's difficult questions'. This volume brings this issue to the fore. Profoundly, what this book has shown is that the same questions that feminists raised in the 1970s about reproductive 'choice' and medicalisation loom large over several chapters (e.g. Chapters 3, 4, and 5): What is the meaning of 'choice' and how is 'choice' enacted? How does technology affect reproductive 'choices'? What are the spaces for resistance? What are the strategies for politicisation? The analyses in this book illuminate that how 'choices' are made and by whom remain at the forefront of contemporary gendered experiences of reproduction. In light of this, pregnancy is a good case in point – pregnancy is an experience in which women's bodies are considered to be simultaneously indistinguishable from and separate to unborn bodies. Yet, increasingly, women are seen to be responsible for the well-being of unborn 'babies' and women's reproductive 'choices' are constrained as a result (Chapters 4 and 6). A key point that remains crucial is that women are not simply 'empowered' or 'oppressed' by this. Rather, women's 'choices' are multiply constituted and they are dependent on the discourses that are available to them at a particular time.

Feminist arguments about gendered power relations and experiential knowledge similarly linger as key nodes in which sociological scholarship and feminist activism meet. Chapters 6, 7, 8, and 13 make it apparent that experiences such as pregnancy, birth, and breastfeeding produce embodied experiential knowledge that is shaped through complex social processes. Although women's embodied subjectivities are influenced by gendered biomedical discourses, their embodiments are not produced exclusively through medical power relations. While poststructuralist gender theories have been critical in allowing for 'sex' and 'gender' to be examined as unfixed categories, it becomes apparent that (women's) biological bodies still have political weight (Chapters 6 and 8). Although, binaries are problematic, dispensing with them altogether does not overcome the fact that essentialism is still operative in the formation of reproductive identities for women (Annandale, 2009). In this sense, the future of reproduction in sociology demands what Malson and Swann (2003, p. 199) refer to as a 'double focusing' of feminist and sociological theories in the interrogation of the category 'woman' and the political economy of gender. The 'personal is political' should not be a casualty of postmodern deconstruction.

This book has similarly flagged that we must ask ourselves why men and masculinities continue to be marginalised in the sociology of reproduction and how we can change this in the future. Men are central actors in reproduction – as men are more completely drawn into sociological discussions of reproduction and reproductive health, new frameworks are necessary to organise and explain the reproductive roles that men play and the reproductive issues they experience (e.g. Chapters 9 and 10). Allowing men to define reproduction for themselves will undoubtedly open up aspects of their seemingly 'invisible' reproductive lives. Revisions of hegemonic masculinity are likely to continue to inform future work on men and reproduction and, as this book has shown, addressing the complexity of masculine subjectivities is now timely. Current debates around gay men and family formation, for example, would be a good place to further develop this analysis (e.g. Chapter 2). In what ways do the reproductive practices of gay men both accommodate and destabilise hegemonic masculinity? How do new reproductive technologies structure the 'choices' available to gay men? How are their reproductive 'choices' regulated and stratified? What is the role of consumption in producing gay family forms?

Finally, as this book has been focussed primarily on the West, invariably, there are topics and locations that were not considered and that clearly warrant debate and discussion. I am hopeful that this collection will motivate others to raise their own questions and to consider other reproductive issues and experiences using critical perspectives beyond what has been discussed here. Without a doubt, reproduction will remain a fertile topic in sociology for many years to come.

Bibliography

Abate, M. A. (2010) '"Plastic makes perfect": My beautiful mommy, cosmetic surgery, and the medicalization of motherhood', *Women's Studies: An Interdisciplinary Journal*, 39(7), pp. 715–746.

Abel, E. K. & Browner, C. H. (1998) 'Selective compliance with biomedical authority and the uses of experiential knowledge', in M. Lock & P. Kaufert (eds), *Pragmatic Women and Body Politics*, Cambridge: Cambridge University Press, pp. 310–326.

Abma, J. C. & Martinez, G. G. (2006) 'Childlessness among older women in the United States: Trends and profiles', *Journal of Marriage and Family*, 68(4), pp. 1045–1056.

Abou Aly, A. (1996) 'The wet nurse: A study in ancient medicine and Greek papyri', *Vesalius*, 2(2), pp. 86–97.

Adam, A. (2001) '"Big girls" blouses: Learning to live with polyester', in A. Guy, E. Green & M. Banim (eds), *Through the Wardrobe: Women's Relationships with Their Clothes*, Oxford: Berg, pp. 39–52.

Adkins, L. (2002) *Revisions: Gender and Sexuality in Late Modernity*, London: Open University Press.

Agigian, A. (2004) *Baby Steps: How Lesbian Insemination is Changing the World*, Middletown: Wesleyan University Press.

Ahluwalia, I. B., Morrow, B. & Hsia, J. (2005) 'Why do women stop breastfeeding? Findings from the pregnancy risk assessment and monitoring system', *Pediatrics*, 116(6), pp. 1408–1412.

Akre, J. (2009) 'From grand design to change on the ground: Going to scale with a global feeding strategy', in F. Dykes & V. Hall Moran (eds), *Infant and Young Child Feeding. Challenges to Implementing a Global Strategy*, West Sussex: Wiley Blackwell, pp. 1–31.

Akre, J., Gribble, K. & Minchin, M. (2011) 'Milk sharing: From private practice to public pursuit', *International Breastfeeding Journal*, 6(1), http://www.biomedcentral.com/content/pdf/1746-4358-6-8.pdf, date accessed 20 July 2013.

Alcoff, L. (1988) 'Cultural feminism versus post-structuralism: The identity crisis in feminist theory', *Signs*, 13(3), pp. 405–436.

Alcoff, L. & Potter, E. (eds) (1993) *Feminist Epistemologies*, New York: Routledge.

Alheit, P. (1994) 'Everyday time and life time. On the problems of healing contradictory experiences of time', *Time and Society*, 3(3), pp. 305–319.

Allen, G. (2011) *Early Intervention: Smart Investment, Massive Savings: The Second Independent Report to Her Majesty's Government*, London: Cabinet Office.

Allen, S. M. & Hawkins, A. J. (1999) 'Maternal gate-keeping: Mothers' beliefs and behaviours that inhibit greater father involvement in family work', *Journal of Marriage and the Family*, 61(1), pp. 199–212.

Almeling, R. (2007) 'Selling genes, selling gender: Egg agencies, sperm banks, and the medical market in genetic material', *American Sociological Review*, 72(3), pp. 319–340.

Almeling, R. (2009) 'Gender and the value of bodily goods: Commodification in egg and sperm donation', *Law and Contemporary Problems*, 72(3), pp. 37–58.

Almeling, R. (2011) *Sex Cells: The Medical Market for Eggs and Sperm*, Berkeley: University of California Press.

American Academy of Pediatrics (2012) 'Breastfeeding and the use of human milk', *Pediatrics*, 129(3), pp. e827–841.

American Association of Clinical Endocrinologists (AACE) (2006) 'Women who have experienced temporary amenorrhea at the time of spinal cord injury may still achieve pregnancy', http://www.newswise.com/articles/women-who-have-experienced-temporary-amenorrhea-at-time-of-spinal-cord-injury-may-still-achieve-pregnancy, date accessed 31 January 2012.

American College of Obstetricians and Gynecologists (ACOG) (2007) 'ACOG practice bulletin no. 77: Screening for foetal chromosomal abnormalities', *The American College of Obstetricians and Gynecologists*, http://www.ncbi.nlm.nih.gov/pubmed/17197615, date accessed 20 November 2012.

American Society for Reproductive Medicine (2008) 'Definitions of infertility and recurrent pregnancy loss', *Fertility and Sterility*, 90(suppl. 3), p. S60.

Amir, L. H. & Donath, S. M. (2008) 'Socioeconomic status and rates of breastfeeding in Australia: Evidence from three recent National Health Surveys', *Medical Journal of Australia*, 189(5), pp. 254–256.

Anderson, E. (2009) *Inclusive Masculinity: The Changing Nature of Masculinities*, New York: Routledge.

Anderson, G. (1999) 'Nondirectiveness in prenatal genetics: Patients read between the lines', *Nursing Ethics*, 6(2), pp. 126–136.

Anderson, M. (2005) 'Thinking about women: A quarter century's view', *Gender and Society*, 19(4), pp. 437–455.

Annandale, E. (2009) *Women's Health and Social Change*, London: Routledge.

Annandale, E. & Clark, J. (1996) 'What is gender? Feminist theory and the sociology of human reproduction', *Sociology of Health & Illness*, 18(1), pp. 17–44.

Appadurai, A. (1996) *Modernity at Large: Cultural Dimensions of Globalization*, Minneapolis: University of Minnesota Press.

Arditti, R., Duelli Klein, R. & Minden, S. (1984) *Test-Tube Women: What Future for Motherhood?*, London: Pandora Press.

Arms, S. (1975) *Immaculate Deception*, New York: Bantam Books.

Asch, A. (2000) 'Why I haven't changed my mind about prenatal diagnosis: Reflections and refinements', in E. Parens & A. Asch (eds), *Prenatal Testing and Disability Rights*, Washington, DC: Georgetown University Press, pp. 234–260.

Atkins, L. (2010) 'When's the best time to have a baby?', *The Guardian*, 27 May, http://www.guardian.co.uk/lifeandstyle/2010/may/26/best-time-to-have-a-baby, date accessed 27 May 2013.

Australian Associated Press (2009) 'Abortion bill passes with bipartisan support', *Courier Mail*, 3 September, http://www.couriermail.com.au/news/breaking-news/abortion-bill-passes-with-bipartisan-support/story-e6freonf-1225769231225, date accessed 30 September 2012.

Australian Broadcasting Corporation (ABC) (2009) 'Queensland rising', *Q&A*, 30 July, http://www.abc.net.au/tv/qanda/txt/s2634705.htm, date accessed 30 September 2012.

Australian Broadcasting Corporation (ABC) (2010) 'Abortion on trial in Queensland', *Background Briefing*, 1 November, http://www.abc.net.au/radionational/

programs/backgroundbriefing/abortion-on-trial-in-queensland/2982710, date accessed 30 September 2012.
Australian Broadcasting Corporation (ABC) (2013) 'Abortion pill RU486 and three cancer drugs added to the Pharmaceutical Benefits Scheme', *ABC News*, 30 June, http://www.abc.net.au/news/2013-06-30/abortion-pill-ru486-and-three-cancer-drugs-added-to-the-pharmac/4790158, date accessed 8 August 2013.
Australian Bureau of Statistics (ABS) (2005) 'Births, Australia, 2005', http://www.abs.gov.au/AUSSTATS/abs@.nsf/Lookup/3301.0Main+Features12005, date accessed 23 June 2013.
Australian Bureau of Statistics (ABS) (2011) 'Births, Australia, 2011', http://www.abs.gov.au/AUSSTATS/abs@.nsf/mf/3301.0, date accessed 23 June 2013.
Australian Health Ministers' Conference (2009) 'The Australian National Breastfeeding Strategy 2010–2015', http://www.health.gov.au/internet/main/publishing.nsf/Content/49F80E887F1E2257CA2576A10077F73F/$File/Breastfeeding_strat1015.pdf, date accessed 3 June 2013.
Australian Institute of Health and Welfare (AIHW) (2011) 'Australia's Mothers and Babies 2009', http://www.aihw.gov.au/publication-detail/?id=10737420870, date accessed 25 September 2012.
Avishai, O. (2007) 'Managing the lactating body: The breast-feeding project and privileged motherhood', *Qualitative Sociology*, 30(2), pp. 135–152.
Ayalah, D. & Weinstock, I. (1979) *Breasts: Women Speak About Breasts and Their Lives*, London: Hutchinson.
Baetens, P., Devroey, P., Camus, M., Van Steirteghem, A. C. & Ponjaert-Kristoffersen, I. (2000) 'Counselling couples and donors for oöcyte donation: The decision to use either known or anonymous oöcytes', *Human Reproduction*, 15(2), pp. 476–484.
Bailey, A. (2011) 'Reconceiving surrogacy: Toward a reproductive justice account of Indian surrogacy', *Hypatia*, 26(4), pp. 715–741.
Bailey, C. & Pain, R. (2001) 'Geographies of infant feeding and access to primary health-care', *Health & Social Care in the Community*, 9(5), pp. 309–317.
Bailey, J. (2011) 'Masculinity and fatherhood in England c. 1760–1830', in J. Arnold & S. Brady (eds), *What is Masculinity? Historical Dynamics from Antiquity to the Contemporary World*, Basingstoke: Palgrave Macmillan, pp. 167–188.
Baird, B. (1996) '"The incompetent, barbarous old lady round the corner": The image of the backyard abortionist in pro-abortion politics', *Hecate*, 22(1), pp. 7–26.
Baird, B. (1998) 'The self-aborting woman', *Australian Feminist Studies*, 13(28), pp. 323–337.
Baird, B. (2008) 'Child politics, feminist analyses', *Australian Feminist Studies*, 23(57), pp. 291–305.
Baker, J. (2008) 'The ideology of choice. Overstating progress and hiding injustice in the lives of young women: Findings from a study in North Queensland, Australia', *Women's Studies International Forum*, 31(1), pp. 53–64.
Barthes, R. (1972) *Camera Lucida: Reflections on Photography*, New York: Hill & Wang.
Bartlett, A. (2005) *Breastwork: Rethinking Breastfeeding*, Sydney: UNSW Press.
Bauman, Z. (1990) *Thinking Sociologically*, Oxford: Basil Blackwell.
Beauchamp, T. L. & Childress, J. F. (2009) *Principles of Biomedical Ethics*, New York: Oxford University Press.

Beaulieu, A. F. & Lippman, A. (1995) '"Everything you need to know": How women's magazines structure prenatal diagnosis for women over 35', *Women & Health*, 23(3), pp. 59–74.
Beck, U & Beck-Gernsheim, E. (1995) *The Normal Chaos of Love*, Cambridge: Polity Press.
Beck, U. & Beck-Gernsheim, E. (2002) *Individualization*, London: Sage.
Becker, G. (2000) *The Elusive Embryo: How Women and Men Approach New Reproductive Technologies*, Berkeley: University of California Press.
Becker, H., Stuifbergen, A. & Tinkle, M. (1997) 'Reproductive health care experiences of women with physical disabilities: A qualitative study', *Archives of Physical Medicine and Rehabilitation*, 78(12 Suppl. 5), pp. S26–S33.
Beckman, L. J. & Harvey, S. M. (2005) 'Current reproductive technologies: Increased access and choice?', *Journal of Social Issues*, 61(1), pp. 1–20.
Bekker, H. L. Thornton, J. G., Airey, C., Connelly, J., Hewison, J., Robinson, M. B., Lilleyman, J., MacIntosh, M., Maule, A. J., Michie, S. & Pearman, A. D. (1999) 'Informed decision making: An annotated bibliography and systematic review', *Health Technology Assessment*, 3(1), pp. 1–168.
Bell, A. V. (2010) 'Beyond (financial) accessibility: Inequalities within the medicalisation of infertility', *Sociology of Health & Illness*, 32(4), pp. 631–646.
Bell, K., McNaughton, D. & Salmon, A. (2009) 'Medicine, morality and mothering: Public health discourses on foetal alcohol exposure, smoking around children and childhood overnutrition', *Critical Public Health*, 19(2), pp. 155–170.
Benoit, C., Westfall, R., Treloar, A., Phillips, R. & Jansson, S. M. (2007) 'Social factors linked with postpartum depression: A mixed-methods longitudinal study', *Journal of Mental Health*, 16(6), pp. 719–730.
Benoit, C., Wrede, S., Bourgeault, I., Sandall, J., Vries, R. D. & Teijlingen, E. R. V. (2005) 'Understanding the social organisation of maternity care systems: Midwifery as a touchstone', *Sociology of Health & Illness*, 27(6), pp. 722–737.
Benoit, C., Zadoroznyj, M., Hallgrimsdottir, H., Treloar, A. & Taylor, K. (2010) 'Medical dominance and neoliberalisation in maternal care provision: The evidence from Canada and Australia', *Social Science & Medicine*, 71(3), pp. 475–481.
Berger, J. (1972) *Ways of Seeing*, London: British Broadcasting Corporation.
Berlin, I. (1969) *Four Essays on Liberty*, Oxford: Oxford University Press.
Bessett, D. (2010) 'Pregnancy after "problems": Women's experiences of stigmatized reproductive careers', Paper presented at Eastern Sociological Society Annual Meeting, Boston, MA, 18–21 March.
Betts, K. (2009) 'Attitudes to abortion: Australia and Queensland in the twenty-first century', *People and Place*, 17(3), pp. 25–39.
Bhogal, A. K. & Brunger, F. (2010) 'Prenatal genetic counseling in cross-cultural medicine', *Canadian Family Physician*, 56(10), pp. 993–999.
Billig, M. (1991) *Ideology and Opinions*, London: Sage.
Bird, C. E. & Rieker, P. P. (2008) *Gender and Health: The Effects of Constrained Choices and Social Policies*, Cambridge: Cambridge University Press.
Birke, L. (1999) *Feminism and the Biological Body*, Edinburgh: Edinburgh University Press.
Björnberg, U. & Kollind, A. K. (2005) *Individualism and Families: Equality, Autonomy and Togetherness*, London: Routledge.
Blackston, M. (1993) 'Beyond brand personality: Building brand relationships', in A. Aaker & A. Biel (eds), *Brand Equity & Advertising: Advertising's Role in Building Strong Brands*, Hillside, NJ: Lawrence Erlbaum Associates, Inc., pp. 113–124.

Blanding, M. (2011) 'The Yelp factor: Are consumer reviews good for business?', 24 October, *Working Knowledge: The Thinking That Leads (Harvard Business School)*, http://hbswk.hbs.edu/item/6836.html, date accessed 15 July 2012.
Blood, S. K. (2005) *Body Work*, New York: Routledge.
Blum, L. M. (1999) *At the Breast: Ideologies of Breastfeeding and Motherhood in the Contemporary United States*, Boston: Beacon Press.
Blyth, E. (2004) 'Patient experiences of an "egg sharing" programme', *Human Fertility*, 7(3), pp. 157–162.
Blyth, E. & Farrand, A. (2005) 'Reproductive tourism – a price worth paying for reproductive autonomy?', *Critical Social Policy*, 25(1), pp. 91–114.
Bobel, C. (2002) *The Paradox of Natural Mothering*, Philadelphia: Temple University Press.
Bordo, S. (1990) 'Feminism, postmodernism, and gender-scepticism', in L. Nicholson (ed.), *Feminism/Postmodernism*, London: Routledge, pp. 133–156.
Bordo, S. (1993a) *Unbearable Weight: Feminism, Western Culture, and the Body*, Berkeley: University of California Press.
Bordo, S. (1993b) 'Feminism, Foucault and the politics of the body', in C. Ramazanoglu (ed.), *Up Against Foucault: Explorations of Some Tensions Between Foucault and Feminism*, London: Routledge, pp. 179–202.
Boston Women's Health Book Collective (2011) *Our Bodies, Ourselves*, New York: Touchstone Books.
Boswell-Penc, M. (2006) *Tainted Milk: Breastmilk, Feminisms, and the Politics of Environmental Degradation*, New York: State University of New York Press.
Böttcher, B., Radenbach, K., Wildt, L. & Hinney, B. (2012) 'Hormonal contraception and depression: A survey of the present state of knowledge', *Archives of Gynecology and Obstetrics*, 286(1), pp. 231–236.
Bourdieu, P. (1990) *Photography: A Middle-Brow Art*, Cambridge: Polity Press.
Bourgeault, I., Declercq, E., Sandall, J., Wrede, S., Vanstone, M., Van Teijlingen, E., DeVries, R. & Benoit, C. (2008) 'Too posh to push? Comparative perspectives on maternal request caesarean sections in Canada, the U.S., the U.K. and Finland', *Advances in Medical Sociology*, 10, pp. 99–123.
Boxer, M. J. (1982) 'For and about women: The theory and practice of women's studies in the United States', *Signs*, 7(3), pp. 661–695.
Brandeis, I. V. (2003) 'Comprehensive look at reproductive health of women with physical disabilities', *Obstetrics & Gynecology*, 101(4), pp. 31S–32S.
Brankovich, J. (2001) 'Constructing a feminist morality in the Western Australian abortion debate', *Journal of Australian Studies*, 25(67), pp. 86–94.
Braun, V. & Clarke, V. (2006) 'Using thematic analysis in psychology', *Qualitative Research in Psychology*, 3(2), pp. 77–101.
Breines, W. (2006) *The Trouble Between Us: An Uneasy History of White and Black Women in the Feminist Movement*, Oxford: Oxford University Press.
Briggs, L. (2010) '"The Pill" in Puerto Rico and mainland United States: Negotiating discourses of risk and decolonization', in L. Reed & P. Sukko (eds), *Governing the Female Body: Gender, Health, and Networks of Power*, Albany: SUNY Press, pp. 159–184.
British Columbia Perinatal Health Program (2008) 'Caesarean Birth Task Force Report', http://www.powertopush.ca/wp-content/uploads/2010/05/CBTF_REPORT.pdf, date accessed 28 May 2013.
Broom, D. H. (1991) *Damned If We Do: Contradictions in Women's Health Care*, Sydney: Allen & Unwin.

Broomhill, R. & Sharp, R. (2007) 'The problem of social reproduction under neoliberalism: Reconfiguring the male breadwinner model in Australia', in M. Griffin-Cohen & J. Brodie (eds), *Remapping Gender in the New Global Order*, Abingdon: Routledge, pp. 85–108.

Brown, J., Sorrell, J. H., McClaren, J. & Creswell, J. W. (2006) 'Waiting for a liver transplant', *Qualitative Health Research*, 16(1), pp. 119–36.

Brown, S., Small, R., Faber, B., Krastey, A. & Davis, P. (2002) 'Early postnatal discharge from hospital for healthy mothers and term infants', *Cochrane Database of Systematic Review*, 3, pp. 1–38.

Brown, W. (2005) 'Neoliberalism and the end of liberal democracy', in W. Brown (ed.), *Edgework: Critical Essays on Knowledge and Politics*, Princeton: Princeton University Press, pp. 37–59.

Browner, C. H. & Sargent C. F. (2011) 'Introduction: Toward global anthropological studies of reproduction: Concepts, methods, theoretical approaches', in C. H. Browner & C. F. Sargent (eds), *Reproduction, Globalization, and the State: New Theoretical and Ethnographic Perspectives*, Durham: Duke University Press, pp. 1–18.

Buckley, K. M. (2009) 'A double-edged sword: Lactation consultants' perceptions of the impact of breast pumps on the practice of breastfeeding', *Journal of Perinatal Education*, 18(2), pp. 13–22.

Burr, J. (2000) 'Repellent to proper ideas about the procreation of children', *Psychology, Evolution & Gender*, 2(2), pp. 105–117.

Butler, J. (1990) *Gender Trouble: Feminism and the Subversion of Identity*, London: Routledge.

Butler, J. (1993) *Bodies that Matter: On the Discursive Limits of 'Sex'*, New York: Routledge.

Canadian Association of Midwives (CAM) (2010) 'Media Release: The world needs Midwives now more than ever!', 4 May, http://www.marketwire.com/press-release/Canadian-Association-of-Midwives-The-World-Needs-Midwives-Now-More-Than-Ever-1158632.htm, date accessed 28 May 2013.

Canadian Association of Midwives (CAM) (2011) *Annual Report*, Montreal: Canadian Association of Midwives.

Canadian Institute of Health Information (CIHI) (2004) *Giving Birth in Canada*, Ottawa: Canadian Institute of Health Information.

Canadian Institute of Health Information (CIHI) (2007) *Giving Birth in Canada: Regional Trends from 2001–2002 to 2005–2006*, Ottawa: Canadian Institute of Health Information.

Cartwright, A. (1979) *The Dignity of Labour? A Study of Childbearing and Induction*, London: Tavistock.

Casper, M. (1994a) 'At the margins of humanity: Fetal positions in science and medicine', *Science, Technology & Human Values*, 19(3), pp. 307–323.

Casper, M. (1994b) 'Reframing and grounding nonhuman agency: What makes a fetus an agent?', *American Behavioral Scientist*, 37(6), pp. 839–856.

Centers for Disease Control and Prevention (2010) 'Birth data', http://www.cdc.gov/nchs/births.htm, date accessed 23 August 2013.

Chan, A., Scheil, W., Scott, J., Nguyen, A. & Sage, L. (2011) 'Pregnancy Outcomes in South Australia 2009', http://www.sahealth.sa.gov.au/wps/wcm/connect/349eb60047edf25c9d6a9df22c7c1033/Pregnancy+Outcome+SA+2009-Operations-POU-20110815.pdf?MOD=AJPERES&CACHEID=349eb60047edf25c9d6a9df22c7c1033, date accessed 30 September 2012.

Chanfreau, J., Gowland, S., Lancaster, Z., Poole, E., Tippingand S. & Toomse M. (2011) *Maternity and Paternity Rights and Women Returners Survey 2009/10. Department for Work and Pensions Research Report 777*, Sheffield: Department for Work and Pensions, Government UK.

Chase, S. & Rogers, M. (eds) (2001) *Mothers and Children: Feminist Analyses and Personal Narratives*, New Brunswick: Rutgers University Press.

Childs, S. & Evans, E. (2012) 'The revived debate on abortion is not simply dog whistle politics, but a threat to women's rights', http://blogs.lse.ac.uk/politics andpolicy/2012/10/15/abortion-womens-rights-childs-evans/, date accessed 2 December 2012.

Chin, R., Daiches, A. & Hall, P. (2011) 'A qualitative exploration of first-time fathers' experiences of becoming a father', *Community Practitioner*, 84(7), pp. 19–23.

Chitayat, D., Langlois, S. & Wilson, R. (2011) 'Prenatal screening for foetal aneuploidy in singleton pregnancies', *Journal of Obstetrics and Gynecology Canada*, 33(7), pp. 736–750.

Chodorow, N. J. (1999) *The Reproduction of Mothering: Psychoanalysis and the Sociology of Gender*, second edition, Berkeley: University of California Press.

Christie, J. & Bunting, B. (2011) 'The effect of health visitors' postpartum home visit frequency on first-time mothers', *International Journal of Nursing Studies*, 48(6), pp. 689–702.

Clark, A., Skouteris, H., Wertheim, E., Paxton, S. & Milgrom, J. (2009) 'My baby body: A qualitative insight into women's body-related experiences and mood during pregnancy and the postpartum', *Journal of Reproductive and Infant Psychology*, 27(4), pp. 330–345.

Clarke, A. E., Mamo, L., Fosket, J. R., Fishman, J. R. & Shim J. K. (eds) (2010) *Biomedicalization: Technoscience, Health, and Illness in the U.S.*, Durham: Duke University Press.

Clarke, A. E, Shim, J. K., Mamo, L., Fosket, J. R & Fishman, J. R (2003) 'Biomedicalization: Technoscientific transformations of health, illness, and U.S. biomedicine', *American Sociological Review*, 68(2), pp. 161–194.

Clarke, V. (2001) 'What about the children? Arguments against lesbian and gay parenting', *Women's Studies International Forum*, 24(5), pp. 555–570.

Cohen, R., Lange, L. & Slusser, W. (2002) 'A description of a male-focused breastfeeding promotion corporate lactation program', *Journal of Human Lactation*, 18(1), pp. 61–65.

Coleman, W. L., Garfield, C. & Committee on Psychosocial Aspects of Child Family Health (2004) 'Fathers and pediatricians: Enhancing men's roles in the care and development of their children', *Pediatrics*, 113(5), pp. 1406–1411.

Colen, S. (1986) '"With respect and feelings": Voices of West Indian child care workers in New York City', in J. B Cole (ed.), *All American Women: Lines That Divide, Ties That Bind*, New York: Free Press, pp. 46–70.

Collins, L. R. & Crockin, S. L. (2012) 'Fighting "personhood" initiatives in the United States', *Reproductive Biomedicine Online*, 24(7), pp. 689–691.

Collins, P. H. (1990) *Black Feminist Thought: Knowledge, Consciousness, and the Politics of Empowerment*, Boston: Unwin Hyman.

Conklin, B. & Morgan, L. (1996) 'Babies, bodies, and the production of personhood in North America and a native Amazonian society', *Ethos*, 24(4), pp. 657–694.

Connell, R. W. (2005) *Masculinities*, Berkeley: University of California Press.

Conrad, P. (1992) 'Medicalization and social control', *Annual Review of Sociology*, 18, pp. 209–232.

Conrad, P. (2007) *The Medicalization of Society: On the Transformation of Human Conditions into Treatable Disorders*, Baltimore: Johns Hopkins University Press.

Cooper, D. (1994) 'Productive, relational and everywhere? Conceptualising power and resistance within Foucauldian feminism', *Sociology*, 28(2), pp. 435–454.

Corea, G. (1977) *The Hidden Malpractice: How American Medicine Mistreats Women*, New York: Jove.

Corea, G. (1985) *The Mother Machine: Reproductive Technologies from Artificial Insemination to Artificial Wombs*, New York: Harper and Row.

Creswell, J. (1998) *Qualitative Inquiry and Research Design: Choosing Among Five Traditions*, London: Sage.

Crompton, R., Lewis, S. & Lyonette, C. (eds) (2007) *Women, Men, Work and Family in Europe*, Basingstoke: Palgrave Macmillan.

Crossley, M. (2009) 'Breastfeeding as a moral imperative: An autoethnographic study', *Feminism & Psychology*, 19(1), pp. 71–87.

Culley, L., Hudson, N., Blyth, E., Norton, W., Pacey, A. & Rapport, F. (2011) 'Transnational reproduction: An exploratory study of UK residents who travel abroad for fertility treatment', http://www.rcn.org.uk/__data/assets/pdf_file/0005/420296/TRANSREP_summary_report_FINAL_JuJu_2011.pdf, date accessed 5 July 2013.

Currie, D. (1988) 'Re-thinking what we do and how we do it: A study of reproductive decisions', *Canadian Review of Sociology and Anthropology*, 25(2), pp. 231–253.

Curthoys, A. (2000) 'Gender studies in Australia: A History', *Australian Feminist Studies*, 15(31), pp. 19–38.

Dahl, K., Hvidman, L., Jørgensen, F. S. & Kesmodel, U. S. (2011) 'Knowledge of prenatal screening and psychological management of test decisions', *Ultrasound in Obstetrics & Gynecology*, 38(2), pp. 152–157.

Daniels, C. (1993) *At Women's Expense: State Power and the Politics of Fetal Rights*, Cambridge, MA: Harvard University Press.

Davis-Floyd, R. (1992) *Birth as an American Rite of Passage*, Berkeley: University of California Press.

Dawson, D.S. (2011) 'Legal commentary on the internet sale of human milk', *Public Health Reports*, 126(2), pp. 165–166.

de Beauvoir, S. (1949) *The Second Sex*, London: Vintage.

De Laat, P. B. (2008) 'Online diaries: Reflections on trust, privacy, and exhibitionism', *Ethics and Information Technology*, 10(1), pp. 57–69.

Declerq, E. & Simmes, D. (1997) 'The politics of "drive-through deliveries": Putting early postpartum discharge on the legislative agenda', *Milbank Quarterly*, 75(2), pp. 172–202.

deCosta, C. (2010) 'Cairns abortion decision effectively decriminalises RU486 in Queensland', *Crikey*, 18 October, http://www.crikey.com.au/2010/10/18/cairns-abortion-decision-effectively-decriminalises-ru486-in-queensland/, date accessed 30 September 2012.

Deegan, M. J. & Brooks, N. A. (1985) *Women and Disability: The Double Handicap*, New Brunswick, NJ: Transaction Books.

Deeney, K., Lohan, M., Spence, D. & Parkes, J. (2012) 'Experiences of fathering a baby admitted to neonatal intensive care: A critical gender analysis', *Social Science & Medicine*, 75(6), pp. 1106–1113.

DeForge, D., Blackmer, J., Garrity, C., Yazdi, F., Cronin, V., Barrowman, N., Fang, M., Mamaladze, V., Zhang, L., Sampson, M. & Moher, D. (2005) 'Fertility following spinal cord injury: A systematic review', *Spinal Cord*, 43(12), pp. 693–703.

Dempsey, D. (2013) 'Surrogacy, gay male couples and the significance of biogenetic paternity', *New Genetics and Society*, 32(1), pp. 37–53.

Dennis, C., Fung, K., Grigoriardis, S., Robinson, G. E., Romans, S. & Ross, L. (2007) 'Traditional post-birth practices and rituals: A qualitative systematic review', *Women's Health*, 3(4), pp. 487–502.

Denzin, N. & Lincoln, K. (1998) 'The discipline and practice of qualitative research', in N. Denzin & K. Lincoln (eds), *Collecting and Interpreting Qualitative Materials*, London: Sage Publications, pp. 1–44.

Denzin, N. (1987) 'Postmodern children', *Society*, 24(3), pp. 32–36.

Department of Health and Ageing (2009) 'Healthy Eating Guidelines for Breastfeeding Women', http://www.health.gov.au/internet/healthyactive/publishing.nsf/Content/breast-feeding-women, date accessed 25 September 2012.

Dermott, E. (2008) *Intimate Fatherhood*, London: Routledge.

Deutsch, F. M. (1999) *Halving It All: How Equally Shared Parenting Works*, Cambridge, MA: Harvard University Press.

Deutsch, F. M. (2007) 'Undoing gender', *Gender & Society*, 21(1), pp.106–126.

Deveaux, M. (1994) 'Feminism and empowerment: A critical reading of Foucault', *Feminist Studies*, 20, pp. 223–247.

Dijkers, M. (2005) 'Quality of life of individuals with spinal cord injury: A review of conceptualization, measurement, and research findings', *The Journal of Rehabilitation Research and Development*, 42(3), pp. 87–110.

Dillaway, H., Cross, K., Lysack, C. & Schwartz, J. (2013) 'Normal and natural, or burdensome and terrible? Women with spinal cord injuries discuss ambivalence about menstruation', *Sex Roles: A Journal of Research*, 68(1–2), pp. 107–120.

Dixon-Woods, M. (2001) 'Writing wrongs? An analysis of published discourses about the use of patient information leaflets', *Social Science & Medicine*, 52(9), pp. 1417–1432.

Dolan, A. & Coe, C. (2011) 'Men, masculine identities and childbirth', *Sociology of Health & Illness*, 33(7), pp. 1019–1034.

DONA International (2014) 'What is a doula?', http://www.dona.org/mothers/index.php, date accessed 11 April 2014.

Donnellan-Fernandez, R. (2011) 'Having a baby in Australia: Women's business, risky business or big business?', *Outskirts*, 24, www.outskirts.arts.uwa.edu.au/volumes/volume-24/donnellan-fernandez , date accessed 23 June 2013.

Doucet, A. (2006) *Do Men Mother?*, Toronto: University of Toronto Press.

Draper, J. (2003a) 'Blurring, moving and broken boundaries: Men's encounters with the pregnant body', *Sociology of Health & Illness*, 25(7), pp. 743–767.

Draper, J. (2003b) 'Men's passage to fatherhood: An analysis of the contemporary relevance of transition theory', *Nursing Inquiry*, 10(1), pp. 66–78.

Dreyfus, H. L. & Rabinow, P. (1982) *Michel Foucault: Beyond Structuralism and Hermeneutics*, Chicago: University of Chicago Press.

Dubow, S. (2011) *Ourselves Unborn: A History of the Fetus in Modern America*, New York: Oxford University Press.

Duden, B. (1993) *Disembodying Women: Perspectives on Pregnancy and the Unborn*, trans. L. Hoinacki, Cambridge, MA: Harvard University Press.

Dudgeon, M. R. & Inhorn, M. C. (2003) 'Gender, masculinity, and reproduction: Anthropological perspectives', *International Journal of Men's Health*, 2(1), pp. 31–56.
Dudgeon, M. R. & Inhorn, M. C. (2004) 'Men's influences on women's reproductive health: Medical anthropological perspectives', *Social Science & Medicine*, 59(7), pp. 1379–1395.
Duster, T. (2003) *Backdoor to Eugenics*, New York: Routledge.
Dworkin, S. L. & Wachs, F. L. (2004) *Body Panic: Gender, Health and the Selling of Fitness*, New York: New York University Press.
Dykes, F. (2002) 'Western medicine and marketing: Construction of an inadequate milk syndrome in lactating women', *Health Care for Women International*, 23(5), pp. 492–502.
Dykes, F. (2006) *Breastfeeding in Hospital: Mothers, Midwives and the Production Line*, London: Routledge.
Earle, S. & Letherby, G. (2007) 'Conceiving time? Women who do or do not conceive', *Sociology of Health & Illness*, 29(2), pp. 233–250.
Earle, S., Foley, P., Komaromy, C. & Lloyd, C. E. (2008) 'Conceptualizing reproductive loss: A social sciences perspective', *Human Fertility*, 11(4), pp. 259–262.
Earle, S., Komaromy, C. & Layne, L. (eds) (2013) *Understanding Reproductive Loss: Perspectives on Life, Death and Fertility*, Surrey: Ashgate.
Ebert, T. L. (1996) *Ludic Feminism and After: Postmodernism, Desire, and Labor in Late Capitalism*, Ann Arbor: University of Michigan Press.
Edwards, R., Duncan, S. & Alexander, C. (2010) 'Conclusion: Hazard warning', in S. Duncan, R. Edwards & C. Alexander (eds), *Teenage Parenthood: What's the Problem?*, London: The Tufnell Press, pp. 188–202.
Ehrenreich, B. & English, D. (1973) *Complaints and Disorders: The Sexual Politics of Sickness*, London: Compendium.
Ehrensaft, D. (1990) *Parenting Together: Men and Women Sharing the Care of Their Children*, Urbana: University of Illinois Press.
Einarsdottir, K., Kemp, A., Haggar, F., Moorin, R., Gunnell, A., Preen, D., Stanley, F. & Holman, C. (2012) 'Increase in caesarean deliveries after the Australian private health insurance incentive policy reforms', *PLoS One*, 7(7), p. e41436.
Eldridge, L. (2010) *In Our Control: The Complete Guide to Contraceptive Choices for Women*, New York: Seven Stories.
Engels, F. ((1972)[1884]) *The Origin of the Family, Private Property, and the State*, trans. Alec West, New York: EB Leacock.
Erikson, S. L. (2007) 'Fetal views: Histories and habits of looking at the fetus in Germany', *Journal of Medical Humanities*, 28(4), pp. 187–212.
Ertman, M. M. (2003) 'What's wrong with a parenthood market? A new and improved theory of commodification', *North Carolina Law Review*, 28, pp. 1–60.
Estores, I. M. & Sipski, M. L. (2004) 'Women's issues after SCI', *Topics in Spinal Cord Injury Rehabilitation*, 10(2), pp. 107–125.
Evans, J. (2001) 'Photography', in F. Carson & C. Pajaczkowska (eds), *Feminist Visual Culture*, New York: Routledge, pp. 105–22.
Eysenbach, G. & Till, J. E. (2001) 'Ethical issues in qualitative research on internet communities', *British Medical Journal*, 323(7321), p. 1103–1105.
Ezzy, D. (2002) *Qualitative Analysis*, London: Routledge.
Fairclough, N. (1995) *Critical Discourse Analysis: The Critical Study of Language*, London: Longman.

Fallon, A. (2011) 'Human milk ice cream goes on sale', *The Guardian*, 25 February, http://www.guardian.co.uk/lifeandstyle/2011/feb/25/human-milk-ice-cream-sale, date accessed 6 August 2012.

Family & Community Services (2011) 'Thinking about adoption', NSW Government, http://www.community.nsw.gov.au/docswr/_assets/main/documents/adoption/adoption_thinking_about.pdf, date accessed 15 February 2013.

Featherstone, B. (2009) *Contemporary Fathering: Theory, Policy and Practice*, Bristol: The Policy Press.

Featherstone, L. (2008) 'Becoming a baby? The foetus in late nineteenth-century Australia', *Australian Feminist Studies*, 23(58), pp. 451–465.

Feeley, J. & Cronin Fisk, M. (2012) 'Bayer Yasmin lawsuit settlements climb to $142 million, 26 April, http://www.bloomberg.com/news/2012-04-26/bayer-yasmin-lawsuit-settlements-climb-to-142-million.html, date accessed 3 July 2012.

Fentiman, L. C. (2009) 'Marketing mothers' milk: The commodification of breastfeeding and the new markets in human milk and infant formula', *Pace Law Faculty Publications, Paper 566*, http://digitalcommons.pace.edu/cgi/viewcontent.cgi?article=1564&context=lawfaculty, date accessed 12 September 2012.

Fenwick, J., Butt, J., Dhaliwal, S., Hauck, Y. & Schmied, V. (2010) 'Western Australian women's perceptions of the style and quality of midwifery postnatal care in hospital and at home', *Women and Birth*, 23(1), pp. 10–21.

Feree, M. M. & Hess, B. (2000) *Controversy and Coalition: The New Feminist Movement*, Boston: G. K. Hall.

Field, F. (2010) *The Foundation Years: Preventing Poor Children Becoming Poor Adults*, London: UK Government.

Fildes, V. A. (1988) *Wet Nursing: A History from Antiquity to the Present*, Oxford: Basil Blackwell.

Finch, L. (1993) *The Classing Gaze*, Sydney: Allen & Unwin.

Finer, L. B., Frohwirth, L. F., Dauphinee, L. A., Singh, S. & Moore. A. M. (2005) 'Reasons U.S. women have abortions: Quantitative and qualitative perspectives', *Perspectives on Sexual and Reproductive Health*, 37(3), pp. 110–118.

Finkelstein, V. (1996) 'Outside, "inside out"', *Coalition*, April, pp. 30–36.

Firestone, S. (1970) *The Dialectic of Sex: The Case for Feminist Revolution*, New York: William Morrow.

Fisher, S. (1986) *In the Patient's Best Interest: Women and the Politics of Medical Decisions*, New Brunswick, NJ: Rutgers University Press.

Flacking, R., Nyqvist, K. H. & Ewald, U. (2007) 'Effects of socioeconomic status on breastfeeding duration in mothers of preterm and term infants', *The European Journal of Public Health*, 17(6), pp. 579–584.

Fletcher, R. (2006) 'Reproductive consumption', *Feminist Theory*, 7(1), pp. 27–47.

Flick, U. (2006) *An Introduction to Qualitative Research*, London: Sage.

Flood, A. (2010) 'Understanding phenomenology', *Nurse Researcher*, 17(2), pp. 7–15.

Florell, D. & Wilson, S. (2010) 'How should I parent', in L. S. Nease & M. W. Austin (eds), *Fatherhood Philosophy for Everyone: The Dao of Daddy*, Malden: Wiley-Blackwell, pp. 77–85.

Fogarty, K. & Augoustinos, M. (2008) 'Feckless fathers and monopolizing mothers: Motive, identity, and fundamental truths in the Australian public inquiry into child custody', *British Journal of Social Psychology*, 47(3), pp. 535–556.

Folbre, N. & Nelson, J. A. (2000) 'For love or money – or both?', *The Journal of Economic Perspectives*, 14(4), pp.123–140.

Forster, D., Mc Lachlan, H., Rayner, J., Yelland, J., Gold, L. & Rayner, S. (2008) 'The early postnatal period: Exploring women's views, expectations and experiences of care using focus groups in Victoria, Australia', *BMC Pregnancy and Childbirth*, 8(27), http://www.biomedcentral.com/1471-2393/8/27, date accessed 11 April 2014.

Foucault, M. (1972) *The Archaeology of Knowledge*, London: Tavistock Publications.

Foucault, M. (1977) *Discipline and Punish: The Birth of the Prison*, New York: Pantheon.

Foucault, M. (1990) *The History of Sexuality, vol. I*, New York: Vintage Books.

Foucault, M. (2003a) *Abnormal: Lectures at the College de France 1974–1975*, New York: Picador.

Foucault, M. (2003b) *The Birth of the Clinic: An Archaeology of Medical Perception*, trans. A.M. Sheridan, London: Routledge.

Fox Keller, E. (1985) *Reflections on Gender and Science*, New Haven, CT: Yale University Press.

Frankfort, E. (1972) *Vaginal Politics*, New York: Quadrangle Books.

Franklin, S. (1997) *Embodied Progress: A Cultural Account of Assisted Conception*, London: Routledge.

Franklin, S. (2006a) 'Embryonic economies: The double reproductive value of stem cells', *BioSocieties*, 1(1), pp. 71–90.

Franklin, S. (2006b) 'The cyborg embryo: Our path to transbiology', *Theory, Culture & Society*, 23(7–8), pp. 167–187.

Franklin, S. (2007) *Dolly Mixtures: The Remaking of Genealogy*, Durham: Duke University Press.

Fraser, N. & Nicholson, L. (1990) 'Social criticism without philosophy: An encounter between feminism and postmodernism', in L. Nicholson (ed.), *Feminism/Postmodernism*, London: Routledge, pp. 19–38.

Freeman, E. W., Grisso, J. A., Berlin, J., Sammel, M., Garcia-Espana, B. & Hollander, L. (2001) 'Symptom reports from a cohort of African American and white women in the late reproductive years', *Menopause*, 8(1), pp. 33–42.

Friese, C., Becker, G. & Nachtigall, R. D. (2006) 'Rethinking the biological clock: Eleventh-hour moms, miracle moms and meanings of age-related infertility', *Social Science & Medicine*, 63(6), pp. 1550–1560.

Fronek, P. & Cuthbert, D. (2012) 'The future of inter-country adoption: A paradigm shift for this century', *International Journal of Social Welfare*, 21(2), pp. 215–224.

Frow, J. (1997) *Time and Commodity Culture: Essays in Cultural Theory and Postmodernity*, Oxford: Oxford University Press.

Gagnon, M.A. & Lexchin, J. (2008) 'The cost of pushing pills: A new estimate of pharmaceutical promotion expenditures in the United States', *PLoS Medicine*, 5(1), p. e1.

Galtry, J. (1997) 'Suckling and silence in the USA: The costs and benefits of breastfeeding', *Feminist Economics*, 3(3), pp. 1–24.

Gannon, L. & Stevens, J. (1998) 'Portraits of menopause in the mass media', *Women & Health*, 27(3), pp. 1–15.

García, E., Timmermans, D. R. M. & Van Leeuwen, E. (2008) 'Rethinking autonomy in the context of prenatal screening decision-making', *Prenatal Diagnosis*, 28(2), pp. 115–120.

Gaskin, I. M. (1975) *Spiritual Midwifery*, Summertown, TN: Book Publishing Co.

Gerrits, T. (2008) 'Clinical encounters: Dynamics of patient-centred practices in a Dutch fertility clinic', Unpublished PhD Thesis, University of Amsterdam, Netherlands.

Ghazit, C. (1999) *The Pill*, PBS American Experience, Steward/Gazit Productions, Inc.

Giddens, A. (1991) *Modernity and Self-Identity: Self and Society in the Late Modern Age*, Stanford: Stanford University Press.

Giles, F. (2004) '"Relational and strange": A preliminary foray into a project to queer breastfeeding', *Australian Feminist Studies*, 19(45), pp. 301–314.

Giles, F. (2005) 'The tears of Lacteros: Integrating the meanings of the human breast', in C. E. Forth & I. Crozier (eds), *Body Parts: Critical Explorations in Corporeality*, London: Lexington Books, pp. 123–141.

Gill, R. (2007) 'Postfeminist media culture: Elements of a sensibility', *European Journal of Cultural Studies*, 10(2), pp. 147–166.

Gill, R. (2008) 'Culture and subjectivity in neoliberal and postfeminist times', *Subjectivity*, 25(1), pp. 432–445.

Gillespie, R. (2003) 'Childfree and feminine: Understanding the gender identity of voluntary childless women', *Gender & Society*, 17(1), pp. 122–136.

Gillies, V. (2009) 'Understandings and experiences of involved fathering in the United Kingdom: Exploring classed dimensions', *The ANNALS of the American Academy of Political and Social Science*, 624(1), pp. 49–60.

Gilligan, C. (1982) *In a Different Voice: Psychological Theory and Women's Development*, Cambridge, MA: Harvard University Press.

Ginsburg, F. (1990) 'The "word-made" flesh: The disembodiment of gender in the abortion debate', in F. Ginsburg & A. L. Tsing (eds), *Uncertain Terms: Negotiating Gender in American Culture*, Boston Beacon Press, pp. 59–73.

Ginsburg, F. & Rapp, R. (1991) 'The politics of reproduction', *Annual Review of Anthropology*, 20, pp. 311–343.

Ginsburg, F. & Rapp, R. (1995) 'Introduction', in F. Ginsburg and R. Rapp (eds), *Conceiving the New World Order: The Global Politics of Reproduction*, Berkeley: University of California Press, pp. 1–18.

Gjerdingen, D., Fontaine, P., Crow, S., McGovern, P., Center, B. & Miner, M. (2009) 'Predictors of mothers' postpartum body dissatisfaction', *Women & Health*, 49(6–7), pp. 491–604.

Glaser, B. G. (1978) *Theoretical Sensitivity: Advances in the Methodology of Grounded Theory*, Mill Valley: Sociology Press.

Gleeson, K. (2010) 'If it's legal, why is RU486 so hard to get?', *New Matilda*, 22 October, http://newmatilda.com/2010/10/22/its-legal-so-why-ru486-so-hard-get, date accessed 30 May 2013.

Gleeson, K. (2011) 'The strange case of the invisible woman in abortion law reform', in J. Jones, A. Grear, K. Stevenson & R. Fenton (eds), *Gender, Sexualities and Law*, London: Routledge, pp. 215–226.

Gleeson, K. (2014) 'Abortion and choice in the neoliberal aftermath', *Politics & Culture*, http://www.politicsandculture.org/2014/03/09/abortion-and-choice-in-the-neoliberal-aftermath-by-kate-gleeson/, date accessed 7 April 2014.

Goffman, E. (1959) *The Presentation of Self in Everyday Life*, New York: Anchor Books/Doubleday.

Goldberg, M. (2010) *The Means of Reproduction: Sex, Power, and the Future of the World*, New York: Penguin.

Golombok, S. (2000) *Parenting: What Really Counts?*, Hove: Routledge.

Goodman, J. H. (2005) 'Becoming an involved father of an infant', *Journal of Obstetric, Gynecologic, & Neonatal Nursing*, 34(2), pp. 190–200.

Gordon, T., Holland, J., Lahelma, E. & Thomson, R. (2005) 'Imagining gendered adulthood: Anxiety, ambivalence, avoidance and anticipation', *European Journal of Women's Studies*, 12(1), pp. 83–103.

Gottleib, A. (2000) 'Where have all the babies gone? Toward an anthropology of infants (and their caretakers)', *Anthropological Quarterly*, 73(3), pp. 121–132.

Government UK (2011) 'Paternity leave', *Government UK*, https://www.gov.uk/paternityleave/overview, date accessed 25 February 2013.

Gow, R. W., Lydecker, J. A., Lamanna, J. D. & Mazzeo, S. E. (2012) 'Representations of celebrities' weight and shape during pregnancy and postpartum: A content analysis of three entertainment magazine websites', *Body Image*, 9(1), pp. 172–175.

Grace, V. M., Daniels, K. R. & Gillett, W. (2008) 'The donor, the father, and the imaginary constitution of the family: Parents' constructions in the case of donor insemination', *Social Science & Medicine*, 66(2), pp. 301–314.

Greene, M. E. & Biddlecom, A. E. (2000) 'Absent and problematic men: Demographic accounts of male reproductive roles', *Population and Development Review*, 26(1), pp. 81–115.

Greer, A. (2009) 'You can't be just a little bit pro-choice', *New Matilda*, 2 July, http://newmatilda.com/2009/07/02/you-cant-be-just-little-bit-pro-choice, date accessed 11 April 2014.

Greer, G. (1970) *The Female Eunuch*, London: McGibbon & Kee.

Gregory, R. (2007) 'Hardly her choice: A history of abortion law reform in Victoria', *Women Against Violence*, 19, pp. 62–71.

Greil, A. L. (1991) *Not Yet Pregnant: Infertile Couples in Contemporary America*, New Brunswick, NJ: Rutgers University Press.

Greil, A. L. (2002) 'Infertile bodies: Medicalization, metaphor, and agency', in M. Inhorn & F. van Balen (eds), *Infertility Around the Globe: New Thinking on Childlessness, Gender, and Reproductive Technologies*, Berkeley: University of California Press, pp. 101–118.

Greil, A. L. & McQuillan, J. (2010) '"Trying" times: Medicalization, intent, and ambiguity in the definition of infertility', *Medical Anthropology Quarterly*, 24(3), pp. 137–152.

Greil, A. L., McQuillan, J., Johnson, K., Slauson-Blevins, K. & Shreffler, K. M. (2009) 'The hidden infertile: Infertile women without pregnancy intent in the United States', *Fertility & Sterility*, 93(6), pp. 2080–2083.

Greil, A. L., Slauson-Blevins, K. S. & McQuillan, J. (2011) 'The social construction of infertility', *Sociological Compass*, 5(8), pp. 736–746.

Griffiths, L. J., Tate, A. R., Dezateux, C. & the Millennium Cohort Study Child Health Group (2007) 'Do early infant feeding practices vary by maternal ethnic group?', *Public Health Nutrition*, 10(9), pp. 957–964.

Grigg-Spall, H. (2012a) 'Just how safe is Yaz? Women need to know!' *Ms. Magazine Blog*, 9 February, http://msmagazine.com/blog/blog/2012/02/09/just-how-safe-is-yaz-women-need-to-know/, date accessed June 28, 2012.

Grigg-Spall, H. (2012b) '"Lives will be saved" – The FDA decision not to ban Bayer's birth control pill', 18 April, *re:Cycling*, http://menstruationresearch.org/2012/04/18/%E2%80%9Clives-will-be-saved%E2%80%9D-%E2%80%93-the-fda-decision-not-to-ban-bayer%E2%80%99s-birth-control-pill/, date accessed July 8, 2012.

Grigg-Spall, H. (2013) *Sweetening the Pill Or How We Got Hooked on Hormonal Birth Control*, Hants: Zero Books.
Grimes, D. A. (1993) 'Over-the-counter oral contraceptives – An immodest proposal?', *American Journal of Public Health*, 83(8), pp. 1092–1094.
Grogan, S. (2008) *Body Image: Understanding Body Dissatisfaction in Men, Women, and Children*, London: Taylor and Francis.
Grøvslien, A. H. & Grønn, M. (2009) 'Donor milk banking and breastfeeding in Norway', *Journal of Human Lactation*, 25(2), pp. 206–210.
Guy, A. & Banim, M. (2000) 'Personal collections: Women's clothing use and identity', *Journal of Gender Studies*, 9(3), pp. 313–327.
Hagewen, K. J. & Morgan, S. P. (2005) 'Intended and ideal family size in the United States, 1970–2002', *Population and Development Review*, 31(3), pp. 507–527.
Haimes, E. (1993) 'Issues of gender in gamete donation', *Social Science & Medicine*, 36(1), pp. 85–93.
Halberstam, J. (1991) 'Automating gender: Postmodern feminism in the age of the intelligent machine', *Feminist Studies*, 17(3), pp. 439–446.
Hall, S. (1967) 'Cultural analysis', *Cambridge Review*, 89, pp. 154–157.
Hallgren, O., Aits, S., Brest, P., Gustafsson, L., Mossberg, A. K., Wullt, B. & Svanborg, C. (2008) 'Apoptosis and tumor cell death in response to Hamlet (Human A-Lactalbumin Made Lethal to Tumor Cells)', in Z. Bösze (ed.), *Bioactive Components of Milk*, New York: Springer, pp. 217–240.
Hamilton, B. E., Martin, J. A. & Ventura, S. J. (2007) 'Births: Preliminary data for 2007', *National Statistics Reports*, 57(12), pp. 1–23.
Harding, S. (1991) *Whose Science? Whose Knowledge?*, Ithaca, NY: Cornell University Press.
Harper, D. (2002) 'Talking about pictures: A case for photo-elicitation', *Visual Studies*, 17(1), pp. 13–26.
Harrison, B. (2002) 'Photographic visions and narrative inquiry', *Narrative Inquiry*, 12(1), pp. 87–111.
Hartmann, B. T., Pang, W. W., Keil, A. D., Hartmann, P. E. & Simmer, K. (2007) 'Best practice guidelines for the operation of a donor human milk bank in an Australian NICU', *Early Human Development*, 83(10), pp. 667–673.
Hartmann, M. (2011) 'The breast milk black market', *Jezebel*, http://jezebel.com/5803416/the-breast-milk-black-market, date accessed 6 August 2012.
Hartouni, V. (1991) 'Containing women: Reproductive discourse in the 1980s', in C. Penley & A. Ross (eds), *Technoculture*, Minneapolis: University of Minnesota Press, pp. 27–56.
Hartouni, V. (1992) 'Fetal exposures: Abortion politics and the optics of allusion', *Camera Obscura*, 10(2), pp. 130–149.
Haughton, J., Gregorio, D. & Pérez-Escamilla, R. (2010) 'Factors associated with breastfeeding duration among Connecticut Special Supplemental Nutrition Program for women, infants, and children (Wic) Participants', *Journal of Human Lactation*, 26(3), pp. 266–273.
Hausman, B. L. (2003) *Mother's Milk: Breastfeeding Controversies in American Culture*, London: Routledge.
Hausman, B. L. (2008) 'Women's liberation and the rhetoric of "choice" in infant feeding debates', *International Breastfeeding Journal*, 3(10), http://www.biomedcentral.com/content/pdf/1746-4358-3-10.pdf, date accessed 11 April 2014.

Hawkesworth, M. (2004) 'The semiotics of premature burial: Feminism in a postfeminist age', *Signs*, 29(4), pp. 961–985.

Hayden, S. & O'Brien Hallstein, D. L. (eds) (2010) *Contemplating Maternity in an Era of Choice*, Lanham: Lexington Books.

Haylett, J. (2012) 'One woman helping another: Egg donation as a case of relational work', *Politics and Society*, 40(2), pp. 223–247.

Health Canada (HC) (2003) *Canadian Perinatal Health Report 2003*, Ottawa: Ministry of Public Works and Government Services Canada.

Health Council of the Netherlands (2006) 'Launching a nationwide program for prenatal screening, down syndrome and neural tube defects', [*wet bevolkingsonderzoek: Aanzet tot een landelijk programma voor prenatale screening; downsyndroom en neuralebuisdefecten*], Report No.: 2006/03WBO, The Hague: Health Council of the Netherlands.

Hearn, J. & Pringle, K. (2006) *European Perspectives on Men and Masculinities: National and Transnational Approaches*, Basingstoke: Palgrave Macmillan.

Heaton, J. (2004) *Reworking Qualitative Data*, London: Sage.

Heinz, W. & Krüger, H. (2001) 'Life course: Innovations and challenges for social research', *Current Sociology*, 49(2), pp. 29–45.

Hertz, R. (2002) 'The father as an idea: A challenge to kinship boundaries by single mothers', *Symbolic Interaction*, 25(1), pp. 1–31.

Hertz, R. & Mattes, J. (2011) 'Donor-shared siblings or genetic strangers: New families, clans, and the internet', *Journal of Family Issues*, 32(9), pp. 1129–1155.

Ho, A. (2008) 'The individualist model of autonomy and the challenge of disability', *Journal of Bioethical Inquiry*, 5(2–3), pp. 193–207.

Hobson, B. (ed.) (2002) *Making Men into Fathers: Men, Masculinities and the Social Politics of Fatherhood*, Cambridge: Cambridge University Press.

Hobson, B. & Fahlén, S. (2009) 'Competing scenarios for European fathers: Applying Sen's capabilities and agency framework to work-family balance', *The ANNALS of the American Academy of Political and Social Science*, 624(1), pp. 214–233.

Hochschild, A. (1995) 'The culture of politics: Traditional, postmodern, cold-modern and warm-modern ideals of care', *Social Politics*, 2(3), pp. 331–346.

Hodgson, J., Hughes, E. & Lambert, C. (2005) '"SLANG": Sensitive language and the new genetics: An exploratory study', *Journal of Genetic Counseling*, 14(6), pp. 415–421.

Hoeyer, K. (2007) 'Person, patent and property: A critique of the commodification hypothesis', *BioSocieties*, 2, pp. 327–348.

Hofferth, S. L. & Curtin, S. C. (2006) 'Parental leave statutes and maternal return to work after childbirth in the United States', *Work and Occupations*, 33(1), pp. 73–105.

Hogle, L. (2010) 'Characterizing human embryonic stem cells: Biological and social markers of identity', *Medical Anthropology Quarterly*, 24(4), pp. 433–450.

Hookway, N. (2008) '"Entering the blogosphere": Some strategies for using blogs in social research', *Qualitative Research*, 8(1), pp. 91–113.

Hopkins, N., Zeedyk, S. & Raitt, F. (2005) 'Visualising abortion: Emotion discourse and fetal imagery in a contemporary abortion debate', *Social Science & Medicine*, 61(2), pp. 393–403.

Horton, E. S. (2007) 'Neoliberalism and the Australian healthcare system (factory)' *Proceedings of the 2007 Conference of the Philosophy of Education Society of Australasia*, 6–9 December, Wellington, New Zealand.

Hughes, B. & Patterson, K. (1997) 'The social model of disability and the disappearing body: Towards a sociology of impairment', *Disability & Society*, 12(3), pp. 325–340.

Human Fertilisation and Embryology Authority (HFEA) (2009) 'IVF in the UK: Statistics 1985, 1995 and 2005', London: HFEA.

Hunt, L. & De Voogd, K. (2003) 'Autonomy, danger, and choice: The moral imperative of an "at risk" pregnancy for a group of low-income Latinas in Texas', in B. Harthorn & L. Oaks (eds), *Risk, Culture, and Health Inequality: Shifting Perceptions of Danger and Blame*, Connecticut: Greenwood, pp. 37–56.

Hunt, L., De Voogd, K. & Castañeda, H. (2005) 'The routine and the traumatic in prenatal genetic diagnosis: Does clinical information inform patient decision-making?', *Patient Education and Counseling*, 56(3), pp. 302–312.

Hunt, S. (2005) *The Life Course: A Sociological Introduction*, Basingstoke: Palgrave Macmillan.

Hunter, M. & O'Dea, I. (1997) 'Menopause: Bodily changes and multiple meanings' in J. M. Ussher (ed.), *Body Talk: The Material and Discursive Regulation of Sexuality, Madness, and Reproduction*, London: Routledge, pp. 199–222.

Husbands, L. (2008) 'Blogging the maternal: Self-representations of the pregnant and postpartum body', *Atlantis*, 32(2), pp. 68–79.

Ikemoto, L. C. (1996) 'The in/fertile, the too fertile, and the dysfertile', *Hastings Law Journal*, 47, pp. 1007–1050.

Ikemoto, L. C. (2009) 'Eggs as capital: Human egg procurement in the fertility and stem cell industry enterprise', *Signs: Journal of Women in Culture and Society*, 34(4), pp. 763–781.

Inhorn, M. C. (1996) *Infertility and Patriarchy: The Cultural Politics of Gender and Family Life in Egypt*, Philadelphia: University of Pennsylvania Press.

Inhorn, M. C. (2003) 'Global infertility and the globalization of new reproductive technologies: Illustrations from Egypt', *Social Science & Medicine*, 56(9), pp. 1837–1851.

Inhorn, M. C. (2012) *The New Arab Man: Emergent Masculinities, Technologies, and Islam in the Middle East*, Princeton: Princeton University Press.

Inhorn, M. C., Ceballo, R. & Nachtigall, R. (2009) 'Marginalized, invisible, and unwanted: American minority struggles with infertility and assisted conception', in L. Culley, N. Hudson & F. van Rooij (eds), *Marginalized Reproduction: Ethnicity, Infertility, and Reproductive Technologies*, London: Earthscan, pp. 187–191.

Inhorn, M. C., Tjørnhøj-Thomsen, T., Goldberg, H. & la Cour Mosegaard, M. (eds) (2009) *Reconceiving the Second Sex: Men, Masculinity, and Reproduction*, New York: Bergaghn Books.

Inhorn, M. C. & van Balen, F. (eds) (2002) *Infertility Around the Globe: New Thinking on Childlessness, Gender, and Reproductive Technologies*, Berkeley: University of California Press.

Jackson, A. & Wadley, V. (1999) 'A multicenter study of women's self-reported reproductive health after spinal cord injury', *Archives of Physical Medicine & Rehabilitation*, 80(11), pp. 1420–1428.

Jackson, E. (2006) 'What is a parent?', in A. Diduck & K. O'Donovan (eds), *Feminist Perspectives on Family Law*, London: Routledge-Cavendish, pp. 59–74.

Jacob, M. C., McQuillan, J. & Greil, A. L. (2007) 'Psychological distress by type of fertility barrier', *Human Reproduction*, 22(3), pp. 885–894.

James, W. R. (2000) 'Placing the unborn: On the social recognition of new life', *Anthropology & Medicine*, 7(2), pp. 169–189.

Jamieson, L., Backett-Milburn, K., Simpson, R. & Wasoff, F. (2010) 'Fertility and social change: The neglected contribution of men's approaches to becoming partners and parents', *The Sociological Review*, 58(3), pp. 463–485.
Jenkin, W. & Tiggemann, M. (1997) 'Psychological effects of weight retained after pregnancy', *Women and Health*, 25(1), pp. 89–98.
Jenkins, H. (1998) 'Childhood innocence and other modern myths', in H. Jenkins (ed.), *The Children's Culture Reader*, New York: New York University Press, pp. 1–37.
Jensen, M., Kuehn, D., Amtmann, D. & Cardenas, D. (2007) 'Symptom burden in persons with spinal cord injury', *Archives of Physical Medicine and Rehabilitation*, 88(5), pp. 638–645.
Johansson, T. & Klinth, R. (2007) 'Caring fathers: The ideology of gender and equality and masculine positions', *Men and Masculinities*, 11(1), pp. 42–62.
Johns, N. & Gyimóthy, S. (2003) 'Postmodern family tourism at Legoland', *Scandinavian Journal of Hospitality and Tourism*, 3(1), pp. 3–23.
Johnson, K. J. (2012) 'Excluding lesbian and single women? An analysis of U.S. fertility clinic websites', *Women's Studies International Forum*, 35(6), pp. 394–402.
Johnson, S., Leeming, D., Williamson, I. & Lyttle, S. (2013) 'Maintaining the "good maternal body": Expressing milk as a way of negotiating the demands and dilemmas of early infant feeding', *Journal of Advanced Nursing*, 69(3), pp. 590–599.
Johnson, S., Williamson, I., Lyttle, S. & Leeming, D. (2009) 'Expressing yourself: A feminist analysis of talk around expressing breast milk', *Social Science & Medicine*, 69(6), pp. 900–907.
Jones, K. (1990) 'Citizenship in a woman-friendly polity', *Signs*, 15(4), pp. 781–812.
Jones, R. (2011) 'Beyond birth control: The overlooked benefits of oral contraceptive pills', *Guttmacher Institute*, http://www.guttmacher.org/pubs/Beyond-Birth-Control.pdf, date accessed 27 May 2013.
Jordan, K., Capdevila, R. & Johnson, S. (2005) 'Baby or beauty: A Q-study into postpregnancy body image', *Journal of Reproductive and Infant Psychology*, 23(1), pp. 19–31.
Jordanova, L. (1989) *Sexual Visions: Images of Gender in Science and Medicine Between the Eighteenth and Twentieth Centuries*, Madison: University of Wisconsin Press.
Kahn, R. P. (1989) 'Mother's milk: The "moment of nurture" revisited', *Resources for Feminist Research*, 18(3), pp. 29–36.
Kalfoflou, A.F. & Gittelsohn, J. (2000) 'A qualitative follow-up study of women's experiences with oocyte donation', *Medicine & Health*, 15(4), pp. 798–805.
Kaplan, C. (2006) 'Special issues in contraception: Caring for women with disabilities', *Journal of Midwifery & Women's Health*, 51(6), pp. 450–456.
Kapsalis, T. (1997) *Public Privates: Performing Gynecology from Both Ends of the Speculum*, Durham: Duke University Press.
Karpin, I. (1992) 'Legislating the female body: Reproductive technology and the reconstructed woman', *Columbia Journal of Gender and Law*, 3(1), pp. 325–348.
Karpin, I. (2010) 'Taking care of the "health" of preconceived human embryos or constructing legal harms', in J. Nisker, F. Baylis, I. Karpin, C. McLeod & R. Mykitiuk (eds), *The 'Healthy' Embryo: Social, Biomedical, Legal and Philosophical Perspectives*, Cambridge: Cambridge University Press, pp. 136–149.
Kaufman, S. & Morgan, L. (2005) 'The anthropology of the beginnings and ends of life', *Annual Review of Anthropology*, 34, pp. 317–341.

Kimmel, M. S., Hearn, J. & Connell, R. W. (eds) (2004) *Handbook of Studies on Men and Masculinities*, London: Sage.
King, L. & Meyer, M. H. (1997) 'The politics of reproductive benefits: U.S. insurance coverage of contraceptive and infertility treatments', *Gender & Society*, 11(1), pp. 8–30.
Kirkman, M. (2003) 'Egg and embryo donation and the meaning of motherhood', *Women & Health*, 38(2), pp.1–18.
Kirkman, M., Rosenthal, D., Mallett, S., Rowe, H. & Hardiman, A. (2010) 'Reasons women give for contemplating or undergoing abortion: A qualitative investigation in Victoria, Australia', *Sexual and Reproductive Healthcare*, 1(4), pp. 149–155.
Kissling, E. A. (2010) 'How the pill gave birth to the women's health movement', *re:Cycling: Society for Menstrual Cycle Research blog*, 25 May, http://menstruation research.org/2010/05/25/how-the-pill-gave-birth-to-the-women%E2%80%99s-health-movement/, date accessed 28 May 2013.
Kissling, E. A. (2013) 'Pills, periods, and postfeminism: The new politics of marketing birth control', *Feminist Media Studies*, 13(3), pp. 490–504.
Kitzinger, C. (1996) 'The token lesbian chapter', in S. Wilkinson (ed), *Feminist Social Psychologies: International Perspectives*, Buckingham: Open University Press, pp. 119–144.
Kitzinger, S. (1978) *Women as Mothers*, Oxford: Martin Robertson.
Kline, W. (2010) *Bodies of Knowledge: Sexuality, Reproduction, and Women's Health in the Second Wave*, Chicago: University of Chicago Press.
Kneale, D., Coast, E. & Stillwell, J. (eds) (2009) *Fertility, Living Arrangements, Care and Mobility. Understanding Population Trends and Processes – Volume 1*, Dordrecht: Springer.
Koeske, R. D. (1983) 'Lifting the curse of menstruation: Toward a feminist perspective on the menstrual cycle', *Women & Health*, 8(2–3), pp. 1–16.
Kornelsen, J. (2003) 'Midwifery: Building our contribution to maternity care', *Proceedings from the Working Symposium*, http://www.bccewh.bc.ca/publications-resources/documents/midwiferycontributions.pdf, date accessed 28 May 2013.
Koropeckyj-Cox, T., Romano, V. R. & Moras, A. (2007) 'Through the lenses of gender, race, and class: Students' perceptions of childless/childfree individuals and couples', *Sex Roles*, 56(7–8), pp. 415–428.
Krawiec, K. D. (2010) 'A woman's worth', *North Carolina Law Review*, 88, pp. 1739–1770.
Kroløkke, C. (2010) 'On a trip to the womb: Biotourist metaphors in fetal ultrasound imaging', *Women's Studies in Communication*, 33(2), pp. 138–153.
Kroløkke, C. (2011) 'Biotourist performances: Doing parenting during the ultrasound', *Text and Performance Quarterly*, 31(1), pp. 15–36.
Kroløkke, C., Foss, K. A., & Pant, S. (2012) 'Fertility travel: The commodification of human reproduction', *Cultural Politics*, 8(2), pp. 273–282.
Kukla, R. (2005) 'Pregnant bodies as public spaces', in S. Hardy & C. Wiedmer (eds), *Motherhood and Space: Configurations of the Maternal Through Politics, Home, and the Body*, New York: Palgrave Macmillan, pp. 283–305.
Kukla, R. (2010) 'The ethics and cultural politics of reproductive risk warnings: A case study of California's Proposition 65', *Health, Risk & Society*, 12(4), pp. 323–334.
Kulkarni, J. (2007) 'Depression as a side effect of the contraceptive pill', *Expert Opinion on Drug Safety*, 6(4), pp. 371–374.

Kuntsman, A. (2004) 'Cyberethnography as home-work', *Anthropology Matters Journal*, 6(2), pp. 1–10.

Kwon, Y. H. & Parham, E. S. (1994) 'Effects of state of fatness perception on weight conscious women's clothing practices', *Clothing and Textiles Research Journal*, 12(4), pp. 16–21.

Labiner-Wolfe, J., Fein, S. B., Shealy, K. R. & Wang, C. (2008) 'Prevalence of breast milk expression and associated factors', *Pediatrics*, 122(suppl. 2), pp. S63–S68.

LaDue, S. (2012) 'It happened to me: My birth control gave me a pulmonary embolism', *xoJane*, 12 January, http://www.xojane.com/it-happened-me/birth-control-gave-me-pulmonary-embolism-blood-clot, date accessed 2 July 2012.

Lauer, J. A., Betran, A. P., Merialdi, M. & Wojdyla, D. (2010) 'Determinants of caesarean section rates in developed countries: Supply, demand and opportunities for control', http://www.who.int/healthsystems/topics/financing/healthreport/29DeterminantsC-section.pdf, date accessed 14 June 2013.

Law, J. D. (2010) *The Social Life of Fluids: Blood, Milk, and Water in the Victorian Novel*, Ithaca: Cornell University Press.

Laws, S. (1990) *Issues of Blood: The Politics of Menstruation*, Basingstoke: Palgrave Macmillan.

Lee, D. H. (2005) 'Women's creation of camera phone culture', *The Fibreculture Journal*, 6, http://six.fibreculturejournal.org/fcj-038-womens-creation-of-camera-phone-culture/, date accessed 27 September 2012.

Lee, E. J. (2008) 'Living with risk in the age of "intensive motherhood": Maternal identity and infant feeding', *Health Risk & Society*, 10(5), pp. 467–477.

Lee, E. J. (2011) 'Infant feeding and the problems of policy', in P. Liamputtong (ed), *Infant Feeding Practices: A Cross-Cultural Perspective*, New York: Springer, pp. 77–91.

Leitner, S. (2003) 'Varieties of familialism: the caring function of the family in comparative perspective', *European Societies*, 5(4), pp. 353–375.

Letherby, G. (1999) 'Other than mother and mothers as others: The experience of motherhood and non-motherhood in relation to "infertility" and "involuntary childlessness"', *Women's Studies International Forum*, 22(3), pp. 359–372.

Letherby, G. (2002a) 'Challenging dominant discourses: Identity and change and the experience of "infertility" and "involuntary childlessness"', *Journal of Gender Studies*, 11(3), pp. 277–288.

Letherby, G. (2002b) 'Childless and bereft? Stereotypes and realities in relation to "voluntary" and "involuntary" childlessness and womanhood', *Sociological Inquiry*, 72(1), pp. 7–20.

Levesque-Lopman, L. (2000) 'Listen and you will hear: Reflections on interviewing from a feminist phenomenological perspective', in L. Fischer & L. Embree (eds), *Feminist Phenomenology*, Netherlands: Kluwer Academic Publishers, pp. 103–132.

Lewin, E. (2009) *Gay Fatherhood: Narratives of Family and Citizenship in America*, Chicago: University of Chicago Press.

Lewis, J. (2002) 'The problem of fathers: Policy and behaviour in Britain', in B. Hobson (ed.), *Making Men into Fathers*, Cambridge: Cambridge University Press, pp. 125–149.

Lewis, J. (2006) 'Perceptions of risk in intimate relationships', *Journal of Social Policy*, 35(1), pp. 39–58.

Li, R., Darling, N., Maurice, E., Barker, L. & Grummer-Strawn, L. M. (2005) 'Breastfeeding rates in the United States by characteristics of the child, mother, or family: The 2002 National Immunization Survey', *Pediatrics*, 115(1), pp. e31–e37.

Lippman, A. (1991) 'Prenatal genetic testing and screening: Constructing needs and reinforcing inequities', *American Journal of Law & Medicine*, 17(1–2), pp. 15–50.

Lippmann, A. (1999) 'Choice as a risk to women's health', *Health, Risk & Society*, 1(3), pp. 281–291.

Lippmann, A. & Wilfond, B. S. (1992) 'Twice-told tales: Stories about genetic disorders', *American Journal of Human Genetics*, 51(4), pp. 936–937.

Lloyd, M. (2001) 'The politics of disability and feminism: Discord or synthesis?', *Sociology*, 35(3), pp. 715–728.

Locock, L. & Alexander, J. (2006) '"Just a bystander"? Men's place in the process of fetal screening and diagnosis', *Social Science & Medicine*, 62(6), pp.1349–1359.

Loeben, G. L., Marteau, T. M. & Wilfond, B. S. (1998) 'Mixed messages: Presentation of information in cystic fibrosis-screening pamphlets', *The American Journal of Human Genetics*, 63(4), pp. 1181–1189.

Lohan, M., Cruise, S., O'Halloran, P., Alderdice, F. & Hyde, A. (2011) 'Adolescent men's attitudes and decision-making in relation to an unplanned pregnancy. Responses to an interactive video drama', *Social Science & Medicine*, 72(9), pp. 1507–1514.

Lopez, L. K. (2009) 'The radical act of "mommy blogging": Redefining motherhood through the blogosphere', *New Media and Society*, 11(5), pp. 729–747.

Lorber, J. (1989) 'Choice, gift, or patriarchal bargain? Women's consent to *in vitro* fertilization in male infertility', *Hypatia*, 4(3), pp. 23–36.

Lorentzen, J. (2008) '"I know my own body": Power and resistance in women's experience of medical interactions', *Body & Society*, 14(3), pp. 49–79.

Lublin, N. (1998) *Pandora's Box: Feminism Confronts Reproductive Technology*, Lanham: Rowman and Littlefield.

Lunau, K. (2009) 'Ditching the pill for good: New health concerns have women looking for different choices', *Macleans*, 23 November, http://www2.macleans.ca/2009/11/23/ditching-the-pill-for-good/, date accessed 14 July 2012.

Lupton, D. (1999) 'Risk and the ontology of pregnant embodiment', in D. Lupton (ed), *Risk and Sociocultural Theory: New Directions and Perspectives*, Cambridge: Cambridge University Press, pp. 59–85.

Lupton, D. (2003) *Medicine as Culture: Illness, Disease and the Body in Western Societies*, London: Sage.

Lupton, D. (2011) '"The best thing for the baby": Mothers' concepts and experiences related to promoting their infants' health and development', *Health, Risk & Society*, 13 (7–8), pp. 637–651.

Lupton, D. (2012) '"Precious cargo": Risk and reproductive citizenship', *Critical Public Health*, 22(3), pp. 329–340.

Lupton, D. & Barclay, L. (1997) *Constructing Fatherhood: Discourses and Experiences*, London: Sage.

Lyerly, A. D., Mitchell, L., Armstrong, E., Harris, L., Kukla, R., Kupperman, M. & Little, M. (2009) 'Risk and the pregnant body', *Hastings Center Report*, 39(6), pp. 34–42.

Macaluso, M., Wright-Schnapp, T. J., Chandra, A., Johnson, R., Satterwhite, C. L., Pulver, A., Berman, S. M., Wang, R. Y., Farr, S. L. & Pollack, L. A. (2010) 'A public health focus on infertility prevention, detection, and management', *Fertility and Sterility*, 93(1), pp. 1–10.

Macintyre, S. (1977) 'The management of childbirth: A review of sociological research issues', *Social Science & Medicine*, 11(8–9), pp. 477–484.

MacMillan, A. (2012) 'Study: IUDs, implants vastly more effective than the pill', 23 May, *CNN*, http://www.cnn.com/2012/05/23/health/iuds-implants-versus-pill-birth-control/index.html, date accessed 10 July 2012.

Macmillan, R. & Copher, R. (2005) 'Families in the life course: Interdependency of roles, role configurations, and pathways', *Journal of Marriage and Family*, 67(4), pp. 858–879.

Macvarish, J. & Billings, J. (2010) 'Challenging the irrational, amoral and anti-social construction of the "teenage mother"', in S. Duncan, R. Edwards & C. Alexander (eds), *Teenage Parenthood: What's the Problem?*, London: The Tufnell Press, pp. 47–69.

Maher, J. (2002) 'Visibly pregnant: Toward a placental body', *Feminist Review*, 72, pp. 95–107.

Maiden, S. (2006) 'The doctrine of choice', in N. Cater (ed), *The Howard Factor: A Decade that Changed the Nation*, Melbourne: Melbourne University Press, pp. 112–120.

Malson, H. & Swann, C. (2003) 'Re-producing "woman's" body: Reflections on the (dis)place(ments) of "reproduction" for (post)modern women', *Journal of Gender Studies*, 12(3), pp. 191–201.

Mamo, L. (2005) 'Biomedicalizing kinship: Sperm banks and the creation of affinity-ties', *Science as Culture*, 14(3), pp. 237–264.

Mamo, L. (2010) 'Fertility, Inc.: Consumption and subjectification in U.S. lesbian reproductive practices', in A. E. Clarke, L. Mamo, J. R. Fosket, J. R. Fishman & J. K. Shim (eds), *Biomedicalization: Technoscience, Health, and Illness in the U.S.*, Durham: Duke University Press, pp. 173–196.

Marcus, G. (2006) 'Assemblage', *Theory, Culture & Society*, 23 (2–3), pp. 101–106.

Markens, S. (2007) *Surrogate Motherhood and the Politics of Reproduction*, Berkeley: University of California Press.

Markens, S., Browner, C. H. & Preloran, H. M. (2003) '"I'm not the one they're sticking the needle into": Latino couples, fetal diagnosis, and the discourse of reproductive rights', *Gender & Society*, 17(3), pp. 462–481.

Marmot, M. G., Allen, J., Goldblatt, P., Boyce, T., McNeish, D., Grady, M., & Geddes, I. (2010) *Fair Society, Healthy Lives: A Strategic Review of Health Inequalities in England Post-2010*, London: Department of Health.

Marsiglio, W., Amato, P., Day, R., & Lamb, M. E. (2000) 'Scholarship on fatherhood in the 1990s and beyond', *Journal of Marriage and the Family*, 62(4), pp. 1173–1191.

Marsiglio, W., Lohan, M. & Culley, L. (2013) 'Framing men's experience in the procreative realm', *Journal of Family Issues*, 34(8), pp. 1011–1036.

Marsiglio, W. & Roy, K. (2012) *Nurturing Dads: Social Initiatives for Contemporary Fatherhood*, New York: Russell Sage Foundation.

Marteau, T. M., Dormandy, E. & Michie, S. (2001) 'A measure of informed choice', *Health Expectations*, 4(2), pp. 99–108.

Martin, E. (1992) *The Woman in the Body: A Cultural Analysis of Reproduction*, Boston: Beacon Press.

Martin, L. J. (2009) 'Reproductive tourism in the age of globalization', *Globalizations*, 6(2), pp. 249–263.

Martin, L. J. (2010) '"Anticipating infertility": Egg freezing, genetic preservation, and risk', *Gender & Society*, 24(4), pp. 526–545.

Marx, K. (1993[1867]) *Capital: Volume 1: A Critique of Political Economy*, London: Penguin.
Matthews, S. & Wexler, L. (2000) *Pregnant Pictures*, New York: Routledge.
Mauss, M. (2001 [1954]) *The Gift: The Form and Reason for Exchange in Archaic Societies*, London: Routledge.
May, E. T. (2010) *America and the Pill: A History of Promise, Peril, and Liberation*, New York: Basic Books.
McAndrew, F., Thompson, J., Fellows, L., Large, A., Speed M. & Renfrew, M. J. (2012) *Infant Feeding Survey 2010*, Leeds: The NHS Health and Social Care Information Centre.
McBride, B. A., Brown, G. L., Bost, K. K., Shin, N., Vaughn, B. & Korth, B. (2005) 'Paternal identity, maternal gate keeping, and father involvement', *Family Relations*, 54(3), pp. 360–372.
McCarthy, J. R., Edwards, R. & Gillies, V. (2000) 'Moral tales of the child and the adult: Narratives of contemporary family lives under changing circumstances', *Sociology*, 34(4), pp. 785–803.
McColl, M. (2002) 'A house of cards: Women, aging and spinal cord injury', *Spinal Cord*, 40(8), pp. 371–373.
McCoy, D., Storeng, K., Filippi, V., Ronsmans, C., Osrin, D., Matthias, B. & Manandhar, D. S. (2010) 'Maternal, neonatal and child health interventions and services: Moving from knowledge of what works to systems that deliver', *International Health*, 2(2), pp. 87–98.
McDonough, K. (2013) 'Popular birth control suspected in 23 deaths', *Salon*, 12 June, http://www.salon.com/2013/06/11/popular_birth_control_suspected_in_23_deaths/, date accessed 14 June 2013.
McKie, R. (2002) 'Men redundant? Now we don't need women either', *The Guardian*, 10 February, http://www.guardian.co.uk/world/2002/feb/10/medicalscience.research, date accessed 14 June 2013.
McLeod, C. (2002) *Self-Trust and Reproductive Autonomy*, Cambridge, MA: MIT Press.
McQuillan, J., Greil, A. L. & Schreffler, K. M. (2011) 'Pregnancy intentions among women who do not try: Focusing on women who are okay either way', *Maternal and Child Health Journal*, 15(2), pp. 178–187.
McRobbie, A. (2004) 'Post-feminism and popular culture', *Feminist Media Studies*, 4(2), pp. 255–264.
McRobbie, A. (2006) 'Yummy mummies leave a bad taste for young women', *The Guardian*, 2 March, http://www.guardian.co.uk/world/2006/mar/02/gender.comment, date accessed 13 September 2012.
McRobbie, A. (2009) *The Aftermath of Feminism: Gender, Culture and Social Change*, London: Sage.
Merleau-Ponty, M. (1962) *Phenomenology of Perception*, London: Routledge.
Millbank, J. (2011) 'The new surrogacy parentage laws in Australia: Cautious regulation or "25 Brick Walls"?', *Melbourne University Law Review*, 35, pp. 1–44.
Miller, T. (2005) *Making Sense of Motherhood: A Narrative Approach*, Cambridge: Cambridge University Press.
Miller, T. (2007) '"Is this what motherhood is all about?" Weaving experiences and discourse through transition to first-time motherhood', *Gender & Society*, 21(3), pp. 337–358.

Miller, T. (2010) *Making Sense of Fatherhood: Gender, Caring and Work*, Cambridge: Cambridge University Press.
Miller, T. (2011) 'Falling back into gender? Men's narratives and practices around first-time fatherhood', *Sociology*, 45(6), pp. 1094–1109.
Miller, T. (2013, in press) 'Maternal spheres: What are we doing with men?' *The Practising Midwife*, 16(1), p. 5.
Mills, C. (2008) 'Images and emotion in abortion debates', *The American Journal of Bioethics*, 8(12), pp. 61–62.
Mills, M. (2000) 'Providing space for time: The impact of temporality on life course research', *Time and Society*, 9(1), pp. 91–127.
Mitchell, L. (2001) *Baby's First Picture: Ultrasound and the Politics of Fetal Subjects*, Toronto: University of Toronto Press.
Mitchell, L. & Georges, E. (1997) 'Cross-cultural cyborgs: Greek and Canadian women's discourses on fetal ultrasound', *Feminist Studies*, 23(2), pp. 373–401.
Moore, L. J. (2007) *Sperm Counts: Overcome by Man's Most Precious Fluid*, New York: New York University Press.
Moos, M., Petersen, R., Meadows, K., Melvin, C. L. & Spitz, A. M. (1997) 'Pregnant women's perspectives on intendedness of pregnancy', *Women's Health Issues*, 7(6), pp. 385–392.
Morgan, K. P. (1998) 'Contested bodies, contested knowledges: Women, health, and the politics of medicalisation', in S. Sherwin (ed), *The Politics Of Women's Health: Exploring Agency and Autonomy*, Philadelphia: Temple University Press, pp. 83–121.
Morgan, L. M. (1996) 'Fetal relationality in feminist philosophy: An anthropological critique', *Hypatia*, 11(3), pp. 47–70.
Morgan, L. M. (2009) *Icons of Life: A Cultural History of Human Embryos*, Berkeley: University of California Press.
Morgan, L. M. & Michaels, M. W. (eds) (1999) *Fetal Subjects, Feminist Positions*, Philadelphia: University of Pennsylvania Press.
Morgan, W. (1997) 'A queer kind of law: The Senate inquires into sexuality', *International Journal of Discrimination and the Law*, 2(4), pp. 317–348.
Morris, J. (2001) 'Impairment and disability: Constructing an ethics of care that promotes human rights', *Hypatia*, 16(4), pp. 1–16.
Mosher, W. D. & Jones, J. (2010) 'Use of contraception in the United States', 1982–2008', *Vital Health Statistics*, 23(29), pp. 1–44.
Moustakas, C. (1994) *Phenomenological Research Methods*, Thousand Oaks: Sage.
Mulvey, L. (1975) 'Visual pleasure and narrative cinema', *Screen*, 16(3), pp. 6–18.
Murphy, M. (2012) *Seizing the Means of Reproduction: Entanglements of Feminism, Health, and Technoscience*, Durham: Duke University Press.
Murray, C. & Golombok, S. (2000) 'Oöcyte and semen donation: A survey of UK licensed centres', *Human Reproduction*, 15(10), pp. 2133–2139.
Murray, S. (2005) 'Doing politics or selling out: Living the fat body', *Women's Studies: An Interdisciplinary Journal*, 34(3–4), pp. 265–277.
Nahman, M. (2008) 'Nodes of desire: Romanian egg sellers, "dignity" and feminist alliances in transnational ova exchanges', *European Journal of Women's Studies*, 15(2), pp. 65–82.
Nash, M. (2007) 'From "bump" to "baby": Gazing at the foetus in 4D', *Philament*, 10, http://sydney.edu.au/arts/publications/philament/issue10_contents.htm, date accessed 3 December 2012.

Nash, M. (2008) 'Expanding seams: The geography of Australian maternity shopping', in T. Tunc & A. Babic (eds), *The Globetrotting Shopaholic: Consumer Spaces, Products, and their Cultural Places*, Newcastle: Cambridge Scholars Publishing, pp. 205–220.

Nash, M. (2011) '"You don't train for a marathon sitting on the couch": Performances of pregnancy "fitness" and "good" motherhood in Melbourne, Australia', *Women's Studies International Forum*, 34(1), pp. 50–65.

Nash, M. (2012a) *Making 'Postmodern' Mothers: Pregnant Embodiment, Baby Bumps and Body Image*, Basingstoke: Palgrave Macmillan.

Nash, M. (2012b) 'Weighty matters: Negotiating "fatness" and "in-betweenness" in early pregnancy', *Feminism & Psychology*, 22(3), pp. 307–323.

Nash, M. (2012c) '"Working out for two": Performances of "fitness" and femininity in Australian prenatal aerobics classes', *Gender, Place and Culture: A Journal of Feminist Geography*, 19(4), pp. 449–471.

Nash, M. (2013) 'Shapes of motherhood: Exploring postnatal body image through photographs', *Journal of Gender Studies*, DOI: 10.1080/09589236.2013.797340

Nash, M (2014) 'Breasted experiences in pregnancy: An examination through photographs', *Visual Studies*, 29(1), pp. 40–53.

National Health Service (NHS) (2013) 'How to make a birth plan', http://www.nhs.uk/Conditions/pregnancy-and-baby/pages/how-to-make-birth-plan.aspx%5C, date accessed 30 May 2013.

National Spinal Cord Injury Statistics Center (NSCISC) (2013) 'Facts and figures at a glance', https://www.nscisc.uab.edu/PublicDocuments/fact_figures_docs/Facts%202013.pdf, accessed 3 June 2013.

Newman, K. (1996) *Fetal Positions: Individualism, Science, Individuality*, Stanford: Stanford University Press.

News Limited (2006) 'Shock teenage abortion rate', *Adelaide Advertiser*, 3 January, http://www.adelaidenow.com.au/news/south-australia/shock-teen-abortion-rate/story-e6frea83-1111115251719, date accessed 30 September 2012.

Nicholson, L. (1990) *Feminism/Postmodernism*, London: Routledge.

Nosek, M. A. (2000) 'Overcoming the odds: The health of women with physical disabilities in the United States', *Archives of Physical Medicine and Rehabilitation*, 81(2), pp. 135–138.

Nosek, M. A., Howland, C., Rintala, D. H., Young, M. E. & Chanpong, G. F. (2001) 'National study of women with physical disabilities: Final report', *Sexuality and Disability*, 19(1), pp. 5–39.

O'Brien Hallstein, D. L. (2011) 'She gives birth, she's wearing a bikini: Mobilizing the postpregnant celebrity mom body to manage the post-second wave crisis in femininity', *Women's Studies in Communication*, 34(2), pp. 111–38.

Oakley, A. (1972) *Sex, Gender and Society*, London: Temple Smith.

Oakley, A. (1974a) *The Sociology of Housework*, London: Martin Robertson.

Oakley, A. (1974b) *Housewife*, London: Allen Lane.

Oakley, A. (1979) *Becoming a Mother*, Oxford: Martin Robertson.

Oakley, A. (1980) *Women Confined: Towards a Sociology of Childbirth*, Oxford: Martin: Roberston.

Office for National Statistics (ONS) (2008) 'Birth statistics: Births and patterns of family building England and Wales', http://www.official-documents.gov.uk/document/other/9787777148210/9787777148210.pdf, date accessed 20 August 2013.

Office for National Statistics (ONS) (2010) 'Fertility Summary, 2010', http://www.ons.gov.uk/ons/rel/fertility-analysis/fertility-summary/2010/uk-fertility-summary.html, date accessed 23 June 2013.

Ogbuanu, C., Glover, S., Probst, J., Liu, J. & Hussey, J. (2011) 'The effect of maternity leave length and time of return to work on breastfeeding', *Pediatrics*, 127(6), e1414–e1427.

Ogle, J. P., Tyner, K. E. & Schofield-Tomschin, S. (2011) 'Jointly navigating the reclamation of the "woman I used to be": Negotiating concerns about the postpartum body within the marital dyad', *Clothing & Textiles Research Journal*, 29(1), pp. 35–51.

Oinas, E. (1998) 'Medicalisation by whom? Accounts of menstruation conveyed by young women and medical experts in medical advisory columns', *Sociology of Health & Illness*, 20(1), pp. 52–70.

Okin, S. M. (1991) *Justice, Gender, and the Family*, New York: Basic Books.

Oliva, A., Spira, A. & Multigner, L. (2001) 'Contribution of environmental factors to the risk of male infertility', *Human Reproduction*, 16(8), pp. 1768–1776.

Olsson, A., Lundqvist, M., Faxelid, E. & Nissen, E. (2005) 'Women's thoughts about sexual life after childbirth: Focus group discussions with women after childbirth', *Scandinavian Journal of Caring Science*, 19(4), pp. 381–387.

Ontario Association of Midwives (2013) 'News room: FAQ. Toronto: Association of Ontario Midwives', http://www.ontariomidwives.ca/news-room/kit/faq, date accessed 17 February 2013.

Orloff, A. (1993) 'Gender and the social rights of citizenship: The comparative analysis of gender relations and welfare states', *American Sociological Review*, 58(3), pp. 303–328.

Orobitg, G. & Salazar, C. (2005) 'The gift of motherhood: Egg donation in a Barcelona infertility clinic', *Ethnos*, 70(1), pp. 31–52.

Page, S. & Harland, A. (2011) 'Tiptoe through the minefield: A state by state comparison of surrogacy laws in Australia', *Family Law Review*, 1(4), pp. 198–230.

Palmer, G. (2009) *The Politics of Breastfeeding: When Breasts Are Bad for Business*, London: Pinter and Martin.

Palmer, J. (2009a) 'Seeing and knowing: Ultrasound images in the contemporary abortion debate', *Feminist Theory*, 10(2), pp. 173–189.

Palmer, J. (2009b) 'The placental body in 3D: Everyday practices of non-diagnostic sonography', *Feminist Review*, 93, pp. 64–80.

Pande, A. (2010) 'Commercial surrogacy in India: Manufacturing a perfect mother-worker', *Signs*, 35(4), pp. 969–992.

Parry, D. C. (2005) 'Work, leisure, and support groups: An examination of the ways women with infertility respond to pronatalist ideology', *Sex Roles*, 53(5–6), pp. 337–346.

Pauls, R. N., Occhino, J. A. & Dryfhout, V. L. (2008) 'Effects of pregnancy on female sexual function and body image: A prospective study', *Journal of Sexual Medicine*, 5(8), pp. 1915–1922.

Pearson, H. (2006) 'Health effects of egg donation may take decades to emerge', *Nature*, 10 August, pp. 607–608.

Peatling, S. (2012) 'Health group gets green light to import abortion drug into Australia', *Sydney Morning Herald*, 30 August, http://www.smh.com.au/opinion/political-news/health-group-gets-green-light-to-import-abortion-drug-into-australia-20120830-2525s.html, date accessed 30 September 2012.

Peck, J., Theodore, N. & Brenner, N. (2009) 'Postneoliberalism and its malcontents', *Antipode*, 41(suppl. 1), pp. 94–116.
Peet, R. (2002) 'Ideology, discourse, and the geography of hegemony: From socialist to neoliberal development in postapartheid South Africa', *Antipode*, 34(1), pp. 54–84.
People (2009) 'Heidi Klum walks runway – six weeks after baby!', *People*, 20 November, http://celebritybabies.people.com/2009/11/20/heidi-klum-walks-runway-%E2%80%93-six-weeks-after-baby/, date accessed 25 Sep 2012.
Perz, J. & Ussher, J. M. (2008) '"The horror of this living decay": Women's negotiation and resistance of medical discourses around menopause and midlife', *Women's Studies International Forum*, 31(4), pp. 293–99.
Petchesky, R. (1987) 'Fetal images: The power of visual culture in the politics of reproduction', *Feminist Studies*, 13(2), pp. 263–292.
Petersen, A. (2003) 'Research on men and masculinities: Some implications of recent theory for future work', *Men and Masculinities*, 6(1), pp. 54–69.
Petersen, K. (2000) 'Abortion: Medicalisation and legal gatekeeping', *Journal of Law and Medicine*, 7(3), pp. 267–272.
Pfeffer, N. (2008) 'What British women say matters to them about donating an aborted fetus to stem cell research: A focus group study', *Social Science & Medicine*, 66(12), pp. 2544–2554.
Philipson, S., Wakefield, C. E. & Kasparian, N. A. (2011) 'Women's knowledge, beliefs, and information needs in relation to the risks and benefits associated with use of the oral contraceptive pill', *Journal of Women's Health*, 20(4), pp. 635–642.
Pilnick, A. (2004) '"It's just one of the best tests that we've got at the moment": The presentation of nuchal translucency screening for fetal abnormality in pregnancy', *Discourse & Society*, 15(4), pp. 451–465.
Pilnick, A. (2008) '"It's something for you both to think about": Choice and decision making in nuchal translucency screening for Down's syndrome', *Sociology of Health & Illness*, 30(4), pp. 511–530.
Plummer, K. (2003) *Intimate Citizenships: Private Decisions and Public Dialogues*, Seattle: University of Washington Press.
Pocock, B. (2006) *The Labour Market Ate My Babies: Work, Children and a Sustainable Future*, Sydney: Federation Press.
Pollack, A. (2003) 'Complicating power in high-tech reproduction: Narrative of anonymous paid egg donors', *Journal of Medical Humanities*, 24(3–4), pp. 241–263.
Pope, C., Ziebland, S. & Mays, N. (2000) 'Analysing qualitative data', *BMJ*, 320(7227), pp. 114–116.
Pope, D. (2012) 'Birth choice', *Monday Magazine*, 4 May, http://www.mondaymag.com/news/150188235.html, date accessed 1 September 2012.
Pratt, A., Biggs, A. & Buckmaster, L. (2005) 'How many abortions are there in Australia?', A discussion of abortion statistics, their limitations, and options for improved statistical collection', *Parliamentary Library Research Brief*, 9, http://parlinfo.aph.gov.au/parlInfo/search/display/display.w3p;query=Id%3A%22library%2Fprspub%2FCF7F6%22, date accessed 3 July 2013.
Price, K. (2010) 'What is reproductive justice? How women of color activists are redefining the pro-choice paradigm', *Meridians: Feminism, Race, Transnationalism*, 10(2), pp. 42–65.

Public Health Agency of Canada (PHAC) (2008) *Canadian Perinatal Health Report*, Ottawa: Public Health Agency of Canada.
Public Health Agency of Canada (PHAC) (2009) *Mothers' Voices*, Ottawa: Public Health Agency of Canada.
R v Bayliss and Cullen (1986) 9 Qld Lawyer Reps 8.
Ragoné, H. (1994) *Surrogate Motherhood: Conception in the Heart*, Boulder: Westview Press.
Rallis, S., Skouteris, H., Wertheim, E. H. & Paxton, S. J. (2007) 'Predictors of body image during the first year postpartum: A prospective study', *Women & Health*, 45(1), pp. 87–104.
Ramazanoglu, C. (ed) (1993) *Up Against Foucault: Explorations of Some Tensions Between Foucault and Feminism*, London: Routledge.
Raphael, D. (1955) *The Tender Gift: Breastfeeding*, New York: Schocken Books.
Rapp, R. (2000) *Testing Women, Testing the Fetus: The Social Impact of Amniocentesis in America*, New York: Routledge.
Rapp, R. (2001) 'Gender, body, biomedicine: How some feminist concerns dragged reproduction to the center of social theory', *Medical Anthropology Quarterly*, 15(4), pp. 466–477.
Rawls, A. W. (2005) 'Garfinkel's conception of time', *Time and Society*, 14(2–3), pp. 163–190.
Ray, R. (2006) 'Is the revolution missing or are we looking in the wrong places?', *Social Problems*, 53(4), pp. 459–465.
Raymond, E. G., Burke, A. E. & Espey, E. (2012) 'Combined hormonal contraceptives and venous thromboembolism: Putting the risks into perspective', *Obstetrics & Gynecology*, 119(5), pp. 1039–1044.
Raymond, J. (1994) *Women as Wombs: Reproductive Technologies and the Battle Over Women's Freedom*, North Melbourne: Spinifex Press.
Reed, R. K. (2005) *Birthing Fathers: The Transformation of Men in American Rites of Birth*, New Brunswick, NJ: Rutgers University Press.
Reiger, K. (2006) 'A neoliberal quickstep: Contradictions in Australian maternity policy', *Health Sociology Review*, 15(4), pp. 330–340.
Reinharz, S. (1992) *Feminist Methods in Social Research*, New York: Oxford University Press.
Reitz, A., Tobe, V., Knapp, P. A. & Schurch, B. (2004) 'Impact of spinal cord injury on sexual health and quality of life', *International Journal of Impotence Research*, 16(2), pp. 167–174.
Remennick, L. (2000) 'Childless in the land of imperative motherhood: Stigma and coping among infertile Israeli women', *Sex Roles*, 43(11–12), pp. 821–841.
Retsinas, J. (1987) 'Nature versus technology: The breast-feeding decision', *Sociological Spectrum*, 7(2), pp. 121–139.
Rich, A. (1977) *Of Woman Born: Motherhood as Experience and Institution*, New York: Bantam Books.
Riessman, C. K. (2000) 'Stigma and everyday resistance practices: Childless women in South India', *Gender & Society*, 14(1), pp. 111–135.
Riggs, D. W. (2010) 'Becoming parent: Lesbians, gay men, and family', *Journal of GLBT Family Studies*, 6, pp. 341–348.
Riggs, D. W., Delfabbro, P. H. & Augoustinos, M. (2010) 'Foster fathers and carework: Engaging alternate models of parenting', *Fathering: A Journal of Theory, Research, and Practice about Men as Fathers*, 8(1), pp. 24–36.

Riggs, D. W. & Due, C. (2012) 'Representations of surrogacy in submissions to a parliamentary inquiry in New South Wales', *Techne*, 16(1), pp. 71–84.
Rindfuss, R. Morgan, S. & Swicegood, C. (1984) 'The transition to motherhood: The intersection of structural and temporal dimensions', *American Sociological Review*, 49(3), pp. 359–72.
Ripper, M. (2001) 'Abortion: The shift in stigmatisation from those seeking abortion to those providing it', *Health Sociology Review*, 10(2), pp. 65–77.
Roberts, D. (1997) *Killing the Black Body: Race, Reproduction, and the Meaning of Liberty*, New York: Pantheon.
Roberts, H. (1985) *Patient Patients: Women and their Doctors*, London: Pandora.
Roberts, J. (2012) '"Wakey wakey baby": Narrating four-dimensional (4-D) bonding scans', *Sociology of Health & Illness*, 34(2), pp. 219–314.
Robertson, C. (2009) *Legislation on Altruistic Surrogacy in NSW*, Sydney: New South Wales Parliamentary Library.
Rody, C. (1995) 'Toni Morrison's beloved: History, "rememory", and a "clamor for a kiss"', *American Literary History*, 7(1), pp. 92–119.
Roe, B., Whittington, L. A., Fein, S. B. & Teisl, M. F. (1999) 'Is there competition between breast-feeding and maternal employment?', *Demography*, 36(2), pp. 157–171.
Rojjanasrirat, W. (2004) 'Working women's breastfeeding experiences', *MCN: The American Journal of Maternal/Child Nursing*, 29(4), pp. 222–227.
Rose, G. (2010) *Doing Family Photography: The Domestic, the Public and the Politics of Sentiment*, Farnham: Ashgate.
Rosenthal, D., Rowe, H., Mallett, S., Hardiman, A. & Kirkman, M. (2009) 'Understanding women's experiences of unplanned pregnancy and abortion', *Key Centre for Women's Health in Society*, Melbourne School of Population Health, The University of Melbourne, http://www.cwhgs.unimelb.edu.au/__data/assets/pdf_file/0006/135834/UPAP_Final_Report.pdf, date accessed 30 September 2012.
Rothman, B. K. (1982) *In Labor: Woman and Power in the Birthplace*, New York: W.W. Norton.
Rothman, B. K. (1989) *Recreating Motherhood: Ideology and Technology in a Patriarchal Society*, New York: W.W. Norton.
Rothman, B. K. (1993) *The Tentative Pregnancy: How Amniocentesis Changes the Experience of Motherhood*, New York: W.W. Norton.
Rothman, B. K. (2006) 'Marketing maternity: Consumer ideologies and the making of mothers', in W. Ernst (ed) *Naturbilder und Lebensgrundlagen: Konstruktionen von Geschlecht*, Hamburg: Lit Verlag, pp. 107–119.
Rothman, B. K. (2008) 'New breast milk in old bottles', *International Breastfeeding Journal*, 3(9), pp. e1–e4.
Rott, H. (2012) 'Thrombotic risks of oral contraceptives', *Current Opinions in Obstetrics & Gynecology*, 24(4), pp. 235–240.
Rouse, R. (2010) 'Thousands of women now choosing medical abortion', *Medical Observer*, 8 October, http://www.medicalobserver.com.au/news/thousands-of-women-now-choosing-medical-abortion, date accessed 30 September 2012.
Rowland, R. (1985) 'Motherhood, patriarchal power, alienation and the issue of "choice" in ex preselection', in R. D. Klein, J. Hanmer, H. B. Holmes, B. Hoskins, M. Kishwar, J. Raymond & R. Rowland (eds), *Man-Made Women: How New Reproductive Technologies Affect Women*, London: Hutchinson, pp. 74–87.

Royal Australian and New Zealand College of Obstetricians and Gynaecologists (RANZCOG) (2010) 'Prenatal screening for foetal abnormalities', *The Royal Australian and New Zealand College of Obstetricians and Gynaecologists*, Report No.: C-Obs 35, http://www.ranzcog.edu.au/documents/doc_view/968-c-obs-35-prenatal-screening-for-fetal-abnormalities.html, date accessed 23 June 2013.

Rubin, H. & Rubin, I. (1995) *Qualitative Interviewing: The Art of Hearing Data*, London: Sage.

Ruddick, S. (2007) 'At the horizons of the subject: Neo-liberalism, neo-conservatism and the rights of the child Part One: From "knowing" fetus to "confused" child', *Gender, Place & Culture*, 14(5), pp. 513–527.

Rutberg, L., Friden, B. & Karlsson, A. K. (2008) 'Amenorrhea in newly spinal cord injured women: An effect of hyperprolactinaemia?', *Spinal Cord*, 46(3), pp. 189–191.

Ryan, K., Team, V. & Alexander, J. (2013) 'Expressionists of the 21st Century: The commodification and commercialization of expressed breast milk', *Medical Anthropology: Cross-Cultural Studies of Health and Illness*, 32(5), pp. 467–486.

Sandelowski, M. (1993) *With Child in Mind: Studies of the Personal Encounter with Infertility*, Philadelphia: University of Pennsylvania Press.

Sandelowski, M. & de Lacey, S. (2002) 'The uses of a "disease": Infertility as rhetorical vehicle', in M. C. Inhorn & F. van Balen (eds), *Infertility Around the Globe: New Thinking on Childlessness, Gender, and Reproductive Technologies*, Berkeley: University of California Press, pp. 33–51.

Saraiya, M., Berg, C. J., Shulman, H., Green, C. A. & Atrash, H. K. (1999) 'Estimates of the annual number of clinically recognized pregnancies in the United States, 1981–1990', *American Journal of Epidemiology*, 149(11), pp. 1025–1029.

Sarvas, R. & Froelich, D. M. (2011) *From Snapshots to Social Media: The Changing Picture of Domestic Photography*, London: Springer.

Scherger, S. (2009) 'Social change and the timing of family transitions in West Germany: Evidence from cohort comparisons', *Time and Society*, 18(1), pp. 106–129.

Schmidt, J. (2008) 'Gendering in infant feeding discourses: The good mother and the absent father', *New Zealand Sociology*, 23(2), pp. 61–74.

Schmidt, M. & Moore, L. J. (1998) 'Constructing a "good catch," picking a winner: The development of technosemen and the deconstruction of the monolithic male', in R. Davis-Floyd & J. Dumit (eds), *Cyborg Babies: From Techno-Sex to Techno-Tots*, New York: Routledge, pp. 21–39.

Schmied, V., Mills, A., Kruske, S., Kemp, L., Fowler, C. & Homer, C. (2010) 'The nature and impact of collaboration and integrated service delivery for pregnant women, children and families', *Journal of Clinical Nursing*, 19(23–4), pp. 3516–3526.

Schoen, J. (2005) *Choice & Coercion: Birth Control, Sterilization, and Abortion in Public Health and Welfare*, Chapel Hill: University of North Carolina Press.

Schwarten, E. & Agius, K. (2010) 'Pressure for new laws as couple free over abortion', *The Age*, 15 October, http://www.theage.com.au/national/pressure-for-new-laws-as-couple-free-over-abortion-20101014-16lsr.html, date accessed 30 September 2012.

Scully, D. (1980) *Men Who Control Women's Health: The Miseducation of Obstetrician-Gynecologists*, Boston: Houghton Mifflin.

Scully, D. & Bart, P. (1973) 'A funny thing happened on the way to the orifice: Women in gynaecology textbooks', *American Journal of Sociology*, 78(4), pp. 1045–1051.
Seavilleklein, V. (2009) 'Challenging the rhetoric of choice in prenatal screening', *Bioethics*, 23(1), pp. 68–77.
Segal, L. (2007) *Slow Motion: Changing Masculinities, Changing Men*, Basingstoke: Palgrave Macmillan.
Seidler, V. J. (2006) *Transforming Masculinities: Men, Cultures, Bodies, Power, Sex and Love*, London: Routledge.
Serfaty, V. (2003) 'Me, myself and I: Online embodied identity in America', *Recherches Anglaises et Nord-Americaines* (RANAM), 36(3), pp. 35–47.
Shakespeare, T. (1999) '"Losing the plot"? Medical and activist discourses of contemporary genetics and disability', *Sociology of Health & Illness*, 21(5), pp. 669–688.
Shakespeare, T. (2005) 'The social context of individual choice', in D. T. Wasserman, R. S. Wachbroit & J. E. Bickenbach (eds), *Quality of Life and Human Difference: Genetic Testing, Health Care, and Disability*, Cambridge: Cambridge University Press, pp. 217–236.
Shakespeare, T. (2006) *Disability Rights and Wrongs*, London: Routledge.
Shaw, E., Levitt, C., Wong, S. & Kaczorowski, J. (2006) 'Systematic review of the literature on post-birth care', *Birth*, 33(3), pp. 210–220.
Shaw, R. (2007) 'The gift-exchange and reciprocity of women in donor-assisted conception', *The Sociological Review*, 55(2), pp. 293–310.
Shaw, R. (2008a) 'Rethinking reproductive gifts as body projects', *Sociology*, 42(1), pp. 11–28.
Shaw, R. (2008b) 'The notion of the gift in the donation of body tissues', *Sociological Research Online*, 13(6), http://www.socresonline.org.uk/13/6/4.html, date accessed 13 June 2013.
Shaw, R. (2012) 'Thinking and reciprocating under the New Zealand organ donation system', *Health: An Interdisciplinary Journal for the Social Study of Health, Illness & Medicine*, 16(3), pp. 295-310.
Sherriff, N., Hall, V. & Pickin, M. (2009) '"Fathers" perspectives on breastfeeding: Ideas for intervention', *British Journal of Midwifery*, 17(4), pp. 223–227.
Sherwin, S. (1998) 'Relational approach to autonomy in health care', in S. Sherwin (ed), *The Politics of Women's Health: Exploring Agency and Autonomy*, Philadelphia: Temple University Press, pp. 19–47.
Shildrick, M. (2002) *Embodying the Monster: Encounters with the Vulnerable Self*, London: Sage Publications.
Shildrick, M. & Price, J. (1998) *Vital Signs: Feminist Reconfigurations of the Biological Body*, Edinburgh: Edinburgh University Press.
Shirani, F. (2010) 'Researcher change and continuity in a qualitative longitudinal study: The impact of personal characteristics', in F. Shirani & S. Weller (eds), *Conducting Qualitative Longitudinal Research: Fieldwork Experiences. Timescapes Working Paper Series No. 2.*, http://www.timescapes.leeds.ac.uk/assets/files/WP2-final-Jan-2010.pdf, date accessed 27 May 2013.
Shirani, F. (2011) 'The "right time" for fatherhood? A temporal study of men's transition to parenthood', Unpublished PhD Thesis, Cardiff University, Wales, UK.
Shirani, F. (2013) 'The spectre of the wheezy dad: Masculinity, fatherhood and ageing', *Sociology*, 47(6), pp. 1104–1119.

Shirani, F. & Henwood, K. (2011) 'Taking one day at a time: Temporal experiences in the context of unexpected life course transitions', *Time and Society*, 20(1), pp. 49–68.

Shirani, F., Henwood, H. & Coltart, C. (2012) 'Meeting the challenges of intensive parenting culture: Gender, risk management and the moral parent', *Sociology*, 46(1) pp. 25–40.

News Limited (2006) 'Shock teenage abortion rate', *Adelaide Advertiser*, 3 January, http://www.adelaidenow.com.au/news/south-australia/shock-teen-abortion-rate/story-e6frea83-1111115251719, date accessed 30 September 2012.

Showalter, E. (1990) *Sexual Anarchy: Gender and Culture at the Fin de Siècle*, New York: Viking.

Shreffler, K. M., McQuillan, J. & Greil, A. L. (2011) 'Pregnancy loss and distress among U.S. women', *Family Relations*, 60(3), pp. 342–355.

Silverstein, L. (1996) 'Fathering is a feminist issue', *Psychology of Women Quarterly*, 20(1), pp. 3–37.

Simic, Z. (2010) 'Fallen girls? Plumpton High and the "problem" of teenage pregnancy', *Journal of Australian Studies*, 34(4), pp. 429–445.

Simpson, R. (2010) 'Single subjects: Representations of autonomy and agency in discourses on singleness and childlessness', Paper presented at Centre for Research on Families and Relationships International Conference: Changing Families in a Changing World, 16–18 June, Edinburgh, Scotland.

Skeggs, B. (2008) 'The dirty history of feminism and sociology: Or the war of conceptual attrition', *The Sociological Review*, 56(4), pp. 670–690.

Skoog Svanberg, A., Lampic, C., Berg, T. & Lundkvist, Ö. (2003) 'Public opinion regarding oöcyte donation in Sweden', *Human Reproduction*, 18(5), pp. 1107–1114.

Slater, D. (1995) 'Domestic photography and digital culture', in M. Lister (ed), *The Photographic Image in Digital Culture*, London: Routledge, pp. 49–59.

Small, M. L. (2011) 'How to conduct a mixed methods study: Recent trends in a rapidly growing literature', *Annual Review of Sociology*, 37, pp. 57–86.

Smith, J. (2003) '"Suitable mothers": Lesbian and single women and the "unborn" in Australian parliamentary discourse', *Critical Social Policy*, 23(1), pp. 63–88.

Smith, J. (2004) 'Mothers' milk and markets', *Australian Feminist Studies*, 19(45), pp. 369–379.

Solinger, R. (1998) *Abortion Wars: A Half Century of Struggle 1950–2000*, Berkeley: University of California Press.

Sontag, S. (1977) *On Photography*, London: Penguin Books.

Sorin, R. & Galloway, G. (2006) 'Constructs of childhood: Constructs of self', *Children Australia*, 31(2), pp. 12–21.

Spar, D. L. (2006) *The Baby Business: How Money, Science, and Politics Drive the Commerce of Conception*, Boston: Harvard Business School Press.

Stabile, C. A. (1992) 'Shooting the mother: Fetal photography and the politics of disappearance', *Camera Obscura*, 10(1), pp. 178–205.

Stacey, J. & Thorne, B. (1985) 'The missing feminist revolution in sociology', *Social Problems*, 32(4), pp. 301–316.

Statistics Canada (2006) 'Community profiles: Victoria', http://www12.statcan.gc.ca/census-recensement/2006/dp-pd/prof/92-591/details/page.cfm?Lang=E&Geo1=CMA&Code1=935&Geo2=PR&Code2=59&Data=Count&SearchText=Victoria&SearchType=Begins&SearchPR=01&B1=All&GeoLevel=PR&GeoCode=935, date accessed 14 June 2013.

Steinberg, D. L. (1997) 'A most selective practice: The eugenics logic of IVF', *Women's Studies International Forum*, 20(1), pp. 33–48.
Stevens, E. E., Patrick, T. E. & Pickler, R. (2009) 'A history of infant feeding', *Journal of Perinatal Education*, 18(2), pp. 32–39.
Stewart, D. (2012) '109 Headlines about Jessica Simpson's baby weight', *Jezebel*, http://jezebel.com/5940780/109-headlines-about-jessica-simpsons-baby-weight, date accessed 13 Sep 2012.
Stewart, D., Robertson, E., Dennis, C. & Grace, S. (2004) 'An evidence-based approach to post-partum depression', *World Psychiatry*, 3(2), pp. 97–98.
Stilwell, F. (2006) *Political Economy: The Contest of Economic Ideas*, South Melbourne: Oxford University Press.
Stormer, N. (2008) 'Looking in wonder: Prenatal sublimity and the commonplace "life"', *Signs*, 33(3), pp. 647–673.
Strauss, A. & Corbin, J. (1994) 'Grounded theory methodology: An overview', in N. K. Denzin & Y. S. Lincoln (eds), *Handbook of Qualitative Research*, Thousand Oaks: Sage Publications, pp. 273–285.
Stuhmcke, A. (2011) 'The criminal act of commercial surrogacy in Australia: A call for review', *Journal of Law and Medicine*, 18(3), pp. 601–613.
Summers, A., Langlois, S., Wyatt, P. & Wilson, R. (2007) 'Prenatal screening for fetal aneuploidy', *Journal of Obstetrics and Gynecology of Canada*, 29(2), pp. 146–179.
Summers, M. (2007) 'Rhetorically self-sufficient arguments in Western Australian parliamentary debates on lesbian and gay law reform', *British Journal of Social Psychology*, 46(4), pp. 839–858.
Sundby, J. (2002) 'Infertility and health care in countries with less resources', in M. C. Inhorn & F. van Balen (eds), *Infertility Around the Globe: New Thinking on Childlessness, Gender and Reproductive Technologies*, Berkeley: University of California Press, pp. 247–259.
Swann, C. (1997) 'Reading the bleeding body: Discourses of premenstrual syndrome', in J. M. Ussher (ed), *Body Talk: The Material and Discursive Regulation of Sexuality, Madness, and Reproduction*, London: Routledge, pp. 176–198.
Tabet, P. (1996) 'Natural fertility, forced reproduction', in D. Leonard & L. Adkins (eds), *Sex in Question: French Materialist Feminism*, London: Taylor & Francis, pp. 109–77.
Tasker, Y. & Negra, D. (eds) (2007) *Interrogating Postfeminism: Gender and the Politics of Popular Culture*, Durham: Duke University Press.
Taylor, J. (2000) 'Of sonograms and baby prams: Prenatal diagnosis, pregnancy, and consumption', *Feminist Studies*, 26(2), pp. 391–418.
Taylor, J. (2008) *The Public Life of the Fetal Sonogram: Technology, Consumption and the Politics of Reproduction*, New Brunswick, NJ: Rutgers University Press.
Taylor, J., Layne, L. & Wozniak, D. (2004) *Consuming Motherhood*, New Brunswick, NJ: Rutgers University Press.
Teman, E. (2010) *Birthing a Mother: The Surrogate Body and the Pregnant Self*, Berkeley: University of California Press.
The Royal College of Paediatrics & Child Health (RCPCH) (2011) 'Position statement: Breastfeeding, June 2011', http://www.rcpch.ac.uk/system/files/protected/page/RCPCH%20breastfeeding%20PS%20FINAL.pdf, date accessed 3 June 2013.
Thomas, C. (1999) *Female Forms: Experiencing and Understanding Disability*, Buckingham: Open University Press.

Thompson, C. (2002) 'Fertile ground: Feminists theorize infertility', in M. C. Inhorn & F. van Balen (eds), *Infertility Around the Globe: New Thinking on Childlessness, Gender and Reproductive Technologies*, Berkeley: University of California Press, pp. 52–78.

Thorne, B. (2006) 'How can feminist sociology sustain its critical edge?', *Social Problems*, 53(4), pp. 473–478.

Throsby, K. & Gill, R. (2004) '"It's different for men": Masculinity and IVF', *Men and Masculinities*, 6(4), pp. 330–348.

Tickell, A. & Peck, J. (2003) 'Making global rules: Globalization or neoliberalization?' in J. Peck & H. Wai-chung Yeung (eds), *Remaking the Global Economy*, London: Sage, pp. 163–181.

Titmuss, R. (1997 [1970]) *The Gift Relationship: From Human Blood to Social Policy*, New York: The New Press.

Todorova, I. L. G. and Kotzeva, T. (2003) 'Social discourses, women's resistive voices: Facing involuntary childlessness in Bulgaria', *Women's Studies International Forum*, 26(2), pp. 139–151.

Tomori, C. (2009) 'Breastfeeding as men's "kin work" in the United States', *Phoebe (Oneonta NY)*, 21(2), pp. 31–44.

Tone, A. (2001) *Devices and Desires: A History of Contraceptives in America*, New York: Hill and Wang.

Tong, R. (1996) 'Toward a feminist perspective on gamete donation and reception policies', in C. B. Cohen (ed), *New Ways of Making Babies: The Case of Egg Donation*, Bloomington: Indiana University Press, pp. 138–155.

Toobin, J. (2012) 'The Republicans' lost privacy', *The New Yorker*, 10 January, http://www.newyorker.com/online/blogs/comment/2012/01/the-republicans-lost-privacy.html, date accessed 10 July 2012.

Townsend, N. (2002) *The Package Deal: Marriage, Work and Fatherhood in Men's Lives*, Philadelphia: Temple University Press.

Treanor, P. (2005) 'Neoliberalism: Origins, theory, definition, *Paul Treanor*, 2 December, http://web.inter.nl.net/users/Paul.Treanor/neoliberalism.html, date accessed 2 July 2012.

Tremain, S. (2005) *Foucault and the Government of Disability*, Ann Arbor: University of Michigan Press.

Tseelon, E. (1997) *The Masque of Femininity*, London: Sage.

Tully, M. R. (2000) 'Cost of establishing and operating a donor human milk bank', *Journal of Human Lactation*, 16(1), pp. 57–59.

Turkmendag, I. (2012) 'The donor-conceived child's "right to personal identity": The public debate on donor anonymity in the United Kingdom', *Journal of Law and Society*, 39(1), pp. 58–75.

Turner, B. S. (2001) 'Disability and the sociology of the body', in G. L. Albrecht, K. D. Seelman, & M. Bury (eds), *Handbook of Disability Studies*, Thousand Oaks: Sage, pp. 252–264.

Tyler, I. (2008) 'Chav mum chav scum', *Feminist Media Studies*, 8(1), pp. 17–34.

Tyler, I. (2011) 'Pregnant beauty: Maternal femininities under neoliberalism', in R. Gill & C. Scharff (eds), *New Femininities: Postfeminism, Neoliberalism and Subjectivity*, Basingstoke: Palgrave Macmillan, pp. 21–36.

Ulrich, M. & Weatherall, A. (2000) 'Motherhood and infertility: Viewing motherhood through the lens of infertility', *Feminism & Psychology*, 10(3), pp. 323–336.

United Kingdom National Screening Committee (2010) 'NHS foetal anomaly screening programme, UK NSC policy recommendation 2007–2010: Model

of best practice', http://fetalanomaly.screening.nhs.uk/mobpreview, date accessed 24 May 2013.
United Nations Millennium Project (2005) *Who's Got the Power? Transforming Health Systems for Women and Children*, London: Earthscan.
United States Census Bureau (2011) 'Race', http://www.census.gov/population/race/, date accessed 21 August 2013.
United States Conference of Catholic Bishops (2012) 'U.S. bishops vow to fight HHS edict', http://usccb.org/news/2012/12-012.cfm, date accessed 3 July 2012.
United States Food and Drug Administration (FDA) (2011) 'Birth control pills containing drospirenone: Possible increased risk of blood clots', 27 October, http://www.fda.gov/Safety/MedWatch/SafetyInformation/SafetyAlertsforHuman MedicalProducts/ucm257337.htm, date accessed 9 July 2012.
United States Food and Drug Administration (FDA) (2012a) 'Updated external questions and answers – Ongoing safety review of birth control pills containing drospirenone and a possible increased risk of blood clots', 20 April, http://www.fda.gov/Drugs/DrugSafety/ucm299348.htm, date accessed 9 July 2012.
United States Food and Drug Administration (FDA) (2012b) 'FDA drug safety communication: Updated information about the risk of blood clots in women taking birth control pills containing drospirenone', 10 April, http://www.fda.gov/Drugs/DrugSafety/ucm299305.htm, date accessed 4 July 2012.
Upton, R. L. & Han, S. S. (2003) 'Maternity and its discontents: "Getting the body back" after pregnancy', *Journal of Contemporary Ethnography*, 32(6), pp. 670–692.
Us (2012) 'Hollywood's hottest post-baby bods!', *Us*, http://www.usmagazine.com/celebrity-body/pictures/hollywoods-hottest-post-baby-bods-201231/19688, date accessed 26 Sep 2012.
Ussher, J. M. (1989) *The Psychology of the Female Body*, London: Routledge.
Ussher, J. M. (1992) 'Reproductive rhetoric and the blaming of the body', in P. Nicholson & J. M. Ussher (eds), *The Psychology of Women's Health and Health Care*, London: Macmillan, pp. 31–61.
Ussher, J. M. (2006) *Managing the Monstrous Feminine: Regulating the Reproductive Body*, London: Routledge.
van Balen, F. & Inhorn, M. C. (2002) 'Interpreting infertility: A view from the social sciences,' in M. C. Inhorn & F. van Balen (eds), *Infertility Around the Globe: New Thinking on Childlessness, Gender, and Reproductive Technologies*, Berkeley: University of California Press, pp. 79–118.
Van Esterik, P. (1996) 'Expressing ourselves: Breast pumps', *Journal of Human Lactation*, 12(4), pp. 273–274.
Van Esterik, P. (2009) 'Vintage breast milk: Exploring the discursive limits of feminine fluids', *Canadian Theatre Review Performance Art*, 137(Winter), pp. 20–23.
Vanstone, M. (2012) 'The process of informed decision-making about prenatal screening: Policy, patient education and pregnant women's perspectives', Unpublished PhD Thesis, University of Western Ontario, London, Ontario.
Vanstone, M. & Kinsella, E. A. (2010) 'Critical reflection and prenatal screening public education materials: A metaphoric textual analysis', *Reflective Practice: International and Multidisciplinary Perspectives*, 11(4), pp. 451–465.
Vanstone, M., Kinsella, E. A. & Nisker, J. (2012) 'Information sharing to promote informed choice about prenatal screening in the spirit of the SOGC Clinical Practice Guideline: A proposal for an alternative model', *Journal of Obstetrics and Gynecology Canada*, 34(3), pp. 269–275.

van Teijlingen, E., Wrede, S., Benoit, C., Sandall, J., & DeVries, R. (2009) 'Born in the USA: Exceptionalism in maternity care organisation among high-income countries', *Sociological Research Online*, 14(1), http://www.socresonline.org.uk/14/1/5.html/, date accessed 7 April 2014.

Vincent, C. & Ball, S. J. (2006) 'Inside the "black box" of the family: Gender relations and childcare", in C. B. Vincent & J. Stephen (eds), *Childcare, Choice, and Class Practices. Middle-Class Parents and Their Children*, London: Routledge, pp. 69–110.

Vogel, L. (2011) 'Milk sharing: Boon or biohazard?', *Canadian Medical Association Journal*, 183(3), pp. e155–e156.

Waldby, C. & Cooper, M. (2008) 'The biopolitics of reproduction: Post-fordist biotechnology and women's clinical labour', *Australian Feminist Studies*, 23(55), pp. 57–73.

Waldby, C. & Mitchell, R. (2006) *Tissue Economies: Blood, Organs and Cell Lines in Late Capitalism*, Durham: Duke University Press.

Walker, J. (2006) 'Abortion pill "within law"', *Courier Mail*, 8 June, p. 4.

Walker, J. (2009) 'Exposed: Black-market abortion drugs being used in terminations', *The Australian*, 4 August, p. 3.

Walker, J. & Hyde, V. (2009) '"Scared kids" in abortion case first', *The Australian*, 1 August, p. 1.

Walker, S. (2011) 'Attitudes to a male contraceptive pill in a group of contraceptive users in the UK', *Journal of Men's Health*, 8(4), pp. 267–273.

Wall, G. & Arnold, S. (2007) 'How involved is involved fathering? An exploration of the contemporary culture of fatherhood', *Gender & Society*, 21(4), pp. 508–527.

Wang, C. (1999) 'Photovoice: A participatory action strategy applied to women's health', *Journal of Women's Health*, 8(2), pp. 185–192.

Wang, C. & Burris, M. (1997) 'Photovoice: Concept, methodology and use for participatory needs assessment', *Health Education & Behavior*, 24(3), pp. 369–387.

Wang, C. & Pies, C. (2004) 'Family, maternal, and child health through photovoice', *Maternal and Child Health Journal*, 8(2), pp. 95–102.

Wang, Y. A., Macaldowie, A., Hayward, I., Chambers, G. M. & Sullivan, E. A. (2011) 'Assisted reproductive technology in Australia and New Zealand 2009', http://www.aihw.gov.au/publication-detail/?id=10737420465, date accessed 28 May 2013.

Warnock, M. (1984) *Report of the Committee of Inquiry into Human Fertilisation and Embryology*, London: Her Majesty's Stationery Office.

Warren, N. & Blood, J. (2003) 'Who donates? Why donate? An exploration of the characteristics and motivations of known egg donors: The Victoria, Australia experience', *Journal of Fertility Counselling*, 10(3), pp. 20–24.

Weil, J. (2003) 'Psychosocial genetic counseling in the post-nondirective era: A point of view', *Journal of Genetic Counseling*, 12(3), pp. 199–211.

Weir, L. (1998) 'Cultural intertexts and scientific rationality: The case of pregnancy ultrasound', *Economy and Society*, 27(2–3), pp. 249–253.

Weiten, W., Dunnand, D. S. & Hammer E. Y. (2010) *Psychology Applied to Modern Life: Adjustment in the 21st Century*, Belmont: Cengage Learning.

Welner, S. (1999) 'Contraceptive choices for women with disabilities', *Sexuality and Disability*, 17(3), pp. 209–214.

Wendell, S. (1996) *The Rejected Body: Feminist Philosophical Reflections on Disability*, New York: Routledge.
Wetherell, M. & Potter, J. (1992) *Mapping the Language of Racism: Discourse and the Legitimation of Exploitation*, London: Harvester Wheatsheaf.
Whittaker, S. (2007) 'Demand sparking growth. Private clinics expanding', *The Montreal Gazette*, 13 January, http://www2.canada.com/montrealgazette/news/story.html?id=8c1a98a1-c70e-4a65-85c0-e3ec5ec2d04a, date accessed 12 July 2012.
Wilkin, A. & Liamputtong, P. (2010) 'The photovoice method: Researching the experiences of Aboriginal health workers through photographs', *Australian Journal of Primary Health*, 16(3), pp. 231–239.
Williams, C. (2005) 'Framing the fetus in medical work: Rituals and practices', *Social Science & Medicine*, 60(9), pp. 2085–2095.
Williams, C., Alderson, P. & Farsides, B. (2002) 'What constitutes balanced information in the practitioners portrayals of Down's syndrome?', *Midwifery*, 18(3), pp. 230–237.
Williams, C., Sandall, J., Lewando-Hundt, G., Heyman, B., Spencer, K. & Grellier, R. (2005) 'Women as moral pioneers? Experiences of first trimester antenatal screening', *Social Science & Medicine*, 61(9), pp. 1983–1992.
Williams, S. (2008) 'What is fatherhood? Searching for the reflexive father', *Sociology*, 42(3), pp. 487–502.
Winner, B., Peipert, J. F., Zhao, Q., Buckel, C., Madden, T., Allsworth, J. E. & Secura, G. M. (2012, May 24) 'Effectiveness of long-acting reversible contraception', *The New England Journal of Medicine*, 366(21), pp. 1998–2007.
Woo, K. & Spatz, D. (2007) 'Human milk donation: What do you know about it?', *MCN: The American Journal of Maternal/Child Nursing*, 32(3), pp. 150–155.
World Health Organization (WHO) (1985) 'Appropriate technology for birth', *Lancet*, 2(8452), pp. 436–437.
World Health Organization (WHO) & United Nations Children's Fund (UNICEF) (2003) 'Global strategy for infant and young child feeding. World Health Organization', http://whqlibdoc.who.int/publications/2003/9241562218.pdf, date accessed 3 June 2013.
Wright, C. M., Parkinson, K. & Scott, J. (2006) 'Breast-feeding in a UK urban context: Who breast-feeds, for how long and does it matter?', *Public Health Nutrition*, 9(6), pp. 686–691.
(2012) 'Yaz lawsuit irony', *Seedol.com*, 25 June, http://www.seedol.com/blog/2012/06/25/yaz-lawsuit-irony-281/, date accessed 10 July 2012.
Yeatman, A. (1990) 'A feminist theory of social differentiation', in L. Nicholson (ed), *Feminism/Postmodernism*, London: Routledge, pp. 281–299.
Yeatman, A. (1991) 'The epistemological politics of postmodern feminist theorizing', *Social Semiotics*, 1(1), pp. 30–48.
Yee, S., Hitkari, J. & Greenblatt, E. M. (2007) 'A follow-up study of women who donated oöcytes to known recipient couples for altruistic reasons', *Human Reproduction*, 22(7), pp. 2040–2050.
Young, I. M. (1984) 'Pregnant embodiment: Subjectivity and alienation', *Journal of Medicine and Philosophy*, 9(1), pp. 45–62.
Young, I. M. (1990) *Throwing Like a Girl and Other Essays in Feminist Philosophy and Social Theory*, Bloomington: Indiana University Press.

Zadoroznyj, M. (2006) 'Surveillance, support and risk in the postnatal period', *Health Sociology Review*, 15(4), pp. 353–363.

Zadoroznyj, M., Benoit, C. & Berry, S. (2012) 'Motherhood, medicine and markets: The changing cultural politics of postnatal care provision,' *Sociological Research Online*, 17(3), http://www.socresonline.org.uk/17/3/24.html, date accessed 10 July 2012.

Zimmerman, M., Litt, J. & Bose, C. (2006) *Global Dimensions of Care Work and Gender*, Stanford: Stanford University Press.

Zinn, J. O. (2004) 'Health, risk and uncertainty in the life course: A typology of biographical certainty constructions', *Social Theory and Health*, 2(3), pp. 199–221.

Zlotkowski, M. (2009) 'Accused man thought abortion pills were legal', *Cairns Post*, 14 September, http://www.cairns.com.au/article/2009/09/14/63631_local-news.html, date accessed 30 September 2012.

Index

A
Abortion (*see also* Termination; Chapter 4)
　anti-abortion discourses, 78, 103, 109, 113, 239
　'a woman's right to choose', 6, 73 (*and see* Choice)
　feminist perspective, 73
　fertility barriers, 34–5
　human/non-human binary, 103, 112–3, 188
　informed consent , 74
　Marie Stopes International, 72
　medicalisation of abortion governance, 71
　Melbourne School of Population Health study, 77–8
　neoliberal responses, 78, 103, 109, 113, 239
　'permission', 80
　Personhood Pledge, 236
　politics, 10, 71–4, 80
　pre-natal screening and 'termination', 61, 66
　privatisation, 74
　pro-abortion legal protection of doctors, 72
　self-abortion, 70 (*and see* RU-486)
　social indications for, 69–72, 77, 80
　statistics, 34–5, 72, 76
　teenage abortion, 76
Adoption, 11, 42, 45
　alternative to abortion, 78
　hetero- or homosexual parents, 46
　social solution to infertility, 37
　social need for children, 48
　surrogacy and the *Adoption Act 2000*, 40, 45–8
Altruism (*see also* Donor)
　altruistic surrogacy, 40–1
　altruism and the universal stranger, 204, 205–6, 216, 219
　breast milk as gift, 221
　feminist constructions, 207–8
　power and reciprocity, 206
　tissue donation, 203–4
Antenatal (*see also* Postpartum/Postnatal)
　fathers attending antenatal classes, 167, 170, 228, 232
　fathers' preparation, 165, 171–5, 178–9

B
Baby (*see also* Breastfeeding; Childbirth)
　embodying fatherhood, 171
　implied personhood, 60–1, 102, 103–7, 110–2
　meeting a social need, 48
　perfect/imperfect, 63–4, 110–1
　product in oöcyte donation/market, 185, 189, 193, 196
Beck, Ulrich, 14
Beck-Gernsheim, Elisabeth, 14
Binary
　altruism/instrumentality, 214
　fatherhood/paternity, 199
　gender/sex, 12–3
　human/non-human (and abortion), 103
　medical/social need for children, 48,
　mind/body, bodily change, *133*
　normal/abnormal (able-bodied/disabled), 136–7, 144, 148
　perfect/imperfect baby, 63–4, 110–1
　public/private decision-making, 75
　public/private experience of pregnancy, 107
　self/other, 105, 108
　surrogacy debates, 52
Birth
　and labour, 175, 178
　caesarean section experience, 1, 6, 118, 126, 175

Birth – *continued*
  caesarean section rates, 11, 84–6, 137
  fathers' presence, 173, 175–9
  interventions, 6, 56, 63, 85, 137, 146,
  'lying in', 85
  maternity services, 84–5, 103, 110, 175
  men as mediators, 177, 180
  'natural' , 6, 175–6
  point of separation, 105, 108, 216
  post-birth care, 87–8, 96–7
  premature, 106, 223
  role of professionals, 87, 89, 177
Bodies
  agency, 124
  asexual, disabled, 136–7, 139, 146–7
  asexual, maternal, 129
  reproductive self-discipline, 75, 117, 123, 129
  bodily ambivalence, 118, 122, 132–3
  body shaping, 112, 123
  body projects, 115, 117, 249
  'bouncing back', 112, 115, 123, 126, 131, 133
  corporeal boundaries, 138
  fat, 129
  fit, 117–8
  hysteria, hysterisation, 4
  maternal/foetal, 108, 112, 253
    (*and see* Foetus; Embryo)
  monstrous, 138–9
  'normal', 15, 57, 60
  postpartum, 115, 118, 122–6, 129
  pregnant, 111, 113
  regulating, 4
  self/selfhood, 119–20
  societal definitions of impairment, 138
  unreliable, uncertain, 124, 132–3, 135
  weight, 112, 118, 123, 126, 129, 213, 240, 247
Body image
  ambivalent, 16, 118, 133
  body work, a 'third shift', 115, 118, 133
  corporeal boundaries, 138
  dissatisfaction, 116, 122–6
  exercise, 112, 116, 121
  fat, 129 (*and see* Bodies: weight)
  fitness, 118
  identity, 122, 132
  pre-pregnancy, 115,
  postpartum dissatisfaction, 116, 122
  socially and culturally constructed, 4, 138, 249 (*and see* Chapter 3)
Body projects (*see also* Giddens, Anthony)
  yummy-mummies, 117
  postfeminism, 249
Bordo, Susan, 4
Boston Women's Health Collective
  Our Bodies, Our Selves, 6, 239
Bourdieu, Pierre
  theorising photography, 119–20
Breastfeeding, 18 (*see also* Chapter 13)
  advice or support, 84, 87, 94–6
  as labour or production, 225
  breastfeeding, process or product, 221
  breastfeeding and workforce participation, 18, 226–7, 231, 233
  EBM values, 224–6, 231
  mastitis, 230
  neoliberal body project, 117, 122–3
  paternal involvement, 221, 227–33 (*and see* Chapter 13)
  social status and breastfeeding rates, 227
  wet-nurses, 221
Breast milk
  breast milk banks, 221–3
  breast milk measurability, 225
  breast pump, 225–6, 228, 232–3
  exclusive breastfeeding, 221
  expressed breast milk (EBM), *see* Chapter 13
  breast or formula, 95, 221–2, 228, 232–3
  'involved fatherhood', *see* Chapter 13
  measurability, 225
  Prolacta Bioscience breast milk, 221
  social value, 225
  storage and maternal work, 225–7, 233

C

Cartesian dualism
  bodily change, 133
Celebrity (*see* Pregnancy: celebrity pregnancy)
Choice, 6, 13, 36–8, 252
  (*see also* Chapter 4)
  bioethical principle of autonomy, 57
  consumer choice in reproductive marketplace, 10–11, 50, 252
  contexts of choice, 67
  'do-it-yourself' biographies, 151
  eugenics, 67, 200
  expressed breast milk, 233
  individual choice in egg donation, 218
  informed choice, 56, 65
  parenthood and 'lifestyle' choice, 15, 50, 49, 73–4
  morality, 3, 12, 25, 151
  personal responsibility rather than collective action, 240, 243, 244–9
  postfeminism, 239–40
  prenatal screening, *see* Chapter 3; Screening
  reproductive choice and neoliberalism, 9, 73–4, 186, 200, 239, 242–8
  semen, 189–90, 195, 196–9
  social and financial resources, 57, 156–7
  sterilisation, 28, 144
Children
  'best interests', *see* Chapter 2
  childbearing, 124, 152, 154
  social need for children, 47–9, 159
Childlessness (*see also* Fertility; Infertility, or Fertility barriers)
  involuntarily childless, 24, 28–9, 34–6
  defining childlessness, 26–7
  normative fertility, 11, 13, 26, 155
Class
  blood donation, 205
  body image, 116, 129
  breastfeeding rates, 227
  fertility control, 238
  gift of milk, 222
  maternity clothing and social status, 129

  in sperm marketplace, 200
  stratified reproduction, 10–5, 23, 32–7, 167, 251–2 (*and see* Chapter 1)
  timing of parenthood, 150–1, 154–6
  working-class women as mothers, 71
Clothes/clothing
  dressing the foetus, 110
  expressions of motherhood, 128
  identity, 126–33
  social status, 129
  track pants, 128–9, 134
Commodification, 11–2 (*see also* Chapter 11; Neoliberalism)
  breast milk, *see* Chapter 13
  cycle of consumerism, 243
  ethics, 12
  genetic material (sperm/semen/ova), *see* Chapter 11
  health, 79
  human milk markets, 222–3, 232–3
  impression management and sperm banking, 196–7
  'lacto-porn', 223
  post-birth care, 86–8
  objectification and loss of dignity, 204
  of men, 188–91, 199
  reproductive marketplace, 40, 189, 192–3, 200, 234, 252
  sperm and social media, 185, 191
  sperm banks and technosemen, 189, 197
  sperm branding, 192–6
  sperm donor profiles, 189–91, 195–6
  surrogacy as act of consumption, 50
  'taboo trades', 221
  the unborn, 102, 109–10
Conception, 139, 236 (*see also* Chapter 1, Fertility; Infertility, or Fertility barriers)
  assisted, 203, 213 (and *see* Chapters 2 and 11)
  delaying, 37, 154, 161
  human embryonic stem cell (hESC), 103, 113
  in vitro fertilisation, 103
  'normal', 135
  personhood, 36

Conception – *continued*
  surrogacy, 44, 46, 50
  women with spinal cord injury, 144–5
Contraception, 8, 10 (*see also* Chapter 14; Reproductive technologies)
  abortion and personhood, 113
  Depo-Provera, 145
  for 'economically active female citizens', 75
  Nelson Pill Hearings (United States of America), 238–9
  Personhood Pledge, 236
  the Pill, 6, 18, 144 (*and see* Chapter 14)
  the Pill and non-contraceptive purposes, 239, 243
  RU-486, *see* Chapter 4
  statistics, 236
  sterilisation, 28, 144, 145
  women with spinal cord injury, 137, 144–5
Control (*see also* Medicalisation; Policy and legislation; Regulation)
  postnatal bodies, 116, 124, 126
  reproductive bodies and breastfeeding, 234
Culture
  reproductive (bio)tourism, 11–2, 110
  values and pre-natal screening, 57

D
Moore, Demi
  celebrity pregnancy, 117
Disability (*see also* Screening; Eugenics; Chapter 3; Chapter 8)
  'birth defect', 56, 62–3, 65
  disabled women and reproductive health, 136–7, 142–3
  Down syndrome, 56, 61–2
  medically constructed, 62–3, 65
  objectification, 64
  perfect/imperfect babies, 63–4
  pregnant women's discourses, 61–2, 63–4
  reproductive legitimacy, 145–6, 147
  social barriers, 136, 147
  social model of disability, 136
  spina bifida, 56

  spinal cord injury (SCI), *see* Chapter 8
  statistics, spinal cord injury, 135–6
Discourse analysis (*see* Research methodologies)
Donor (*and see* Chapter 11; Chapter 12)
  altruism, 40–1, 203–8, 214, 218–219
  anonymity, 189, 204, 205–6, 208, 212–3, 216, 219
  Bauman, Mauss and the 'gift', 206
  breastmilk banks, 221–3
  choosing recipients, 205–6, 208, 210–13, 217–9
  'compensation', instrumental donation, 12, 203, 207–8, 214, 224
  donors, disembodied and objectified, 190, 196–9
  donor-recipient relationship, 208, 213, 214–6 (*and see* Kinship)
  egg, *see* Chapter 12 ethics, 205
  egg extraction, 212
  gift, 18, 204, 206–8, 214, 216–7, 219, 221
  gratitude, 197, 206
  historical, 203, 222
  not-for-profit human milk banks, 221
  oöcytes, *see* Chapters 6 and 12
  reciprocity, 214, 217–9
  sperm, 9, 12, 17–8, 103, 207, 217, 219 (*and see* Chapter 11)
Drugs (*see also* Contraception; Reproduction technologies)
  Australian Therapeutic Goods Administration, 72
  Commonwealth Pharmaceutical Benefits Scheme, 72
  Depo-Provera, 145
  Drospirenone, 238
  Fosamax, 144
  gonadotrophins, 212–3
  Mifolian, 69
  RU486, *see* Chapter 4
  Seasonique, 239
  United States Food and Drug Administration, 238–9, 244, 248
  Yaz/Yasmin, 238, 248
Durkheim, Emil, 8

E

Eggs (*see also* Genetic material)
  donation process, 204–5, 212–3
  Somatic Cell Nuclear Transfer (SNCT), 208
  surplus ova, 105
Embodiment, 253 (*see also* Chapters 6, 7, 8, 13)
  alternative—Margrit Shildrick, 138
  (bio)medical/medicalised, 253
  control and excess, 115, 133
  disabled women, 138–9, 143
  embryo and foetus, 106
  'fatness', 129
  motherhood, 132
  'normality', 122
  personhood, 103–4
  pre- and post-pregnancy, 116–7, 118, 122–6
  recipients as 'good parents', 219
  self and 'other', 4, 65
  structuring embodied experience, 138
  the 'yummy mummy', 117
  unborn, 16, 103–4, 106–7, 109–10
Embryo
  histories of the unborn, 105–7
  human embryonic stem cell (hESC), 103, 112–3
  in vitro fertilisation, 10, 11, 25, 103, 150, 212
  medicalisation, 106
Empowerment, 252
  bodily control, 115, 119
  disability and reproductive lives, 148
  making reproductive decisions, 25, 45, 124, 242–4, 252–3
  men, childbirth and disempowerment, 176
  photovoice, 120
Eugenics (*see also* Disability; 'Normal'; Screening; Chapter 3)
  choice discourses, 67
  screening, 55, 64
  sperm marketplaces, 200
  sterilisation, 28
Exercise
  as maternal responsibility, 112
  bouncing back, 116
  fit, 117–8

F

Facebook (*see* Social Media)
Fairclough, Norman, 55 (*see also* Research methodologies)
Family (*see also* Kinship)
  'affinity-ties' and sperm donation, 190, 195
  changing norms, 13, 167, 253–4
  donor-created families, 191, 217
  'lifestyle choice', 47
  moral identity, 151
  non-traditional family forms, 46, 51–3
  normative, 42, 44–50, 215, 217, 218–9
  postmodern, 13
Fathers and fatherhood, 8–9 (*see also* Chapter 9)
  age, physical ability and masculinity, 160–1
  ambivalent, 172, 179
  'bonding' with baby, 229–30
  'breadwinner', 8, 13, 156–7, 163, 166
  culturally acceptable age, 154
  donor profiles, 189–91, 195–6
  expressed breast milk, *see* Chapter 13
  fatherhood/paternity binary, 199–201
  'ideals' in sperm advertising, 185–6, 189
  imagined, in sperm donation, 17, 188, 195
  infant feeding, 221, 228–33
  'involved' father, 166–8, 172, 180, 227–34
  mediator role of women, 167–8
  partners in reproduction, 168–9
  power and powerless in the maternal arena, 175–6
  preparation for, 168, 170–1
  reflexive fathers, 162
  responsible or irresponsible, 158
  role in labour and birth, 175–9
  social need for children, 159
  statistics, 170
  timing, 150–1, 154–5, 160–1
Femininities
  biomedical power and subordination, 4

Femininity
  in egg donation and receipt, 207
  normative, 115, 124, 133–4, 207, 249
Feminism, 252 (*see also* Postfeminism)
  abortion, 73
  altruism, 207–8
  consciousness-raising, 6–7,
  Equality feminism, 231
  Liberation feminism, 231
  poststructuralist, 1–2, 8, 12–3
  'personal is political', 251, 253
  Second wave, 73
Fertility (*see also* Infertility, or Fertility barriers)
  'biological clock', 13, 17, 153–5, 160, 163
  control and decision making, 237, 249
  fertility and masculinity, 155
  fertility brokers, 205
  services, sperm banks, 188
  spinal cord injury, 137
  statistics, 32–6
Firestone, Shulamith
  The Dialectic of Sex, 5
Fitness (*see also* Exercise)
  postpartum bodies, 118
Foetus (*see also* Chapters 3 and 6)
  anomalous, 56, 60, 63, 65, 81
  becoming 'human', 104
  (bio)medicalisation, 60, 65–6, 106
  biotourism, 11–2, 110
  commodified, 109–10
  enigmatic and hidden
  foetal citizens, 111
  histories of the unborn, 105–7
  mother as host, 106, 109–10
  mother's perceptions, 105, 107
  personhood, 102–4, 108
  public foetus, 11, 107
  rights of the unborn, 151
  risk, 111–2
  the 'royal foetus', 101
  unborn assemblages, 102, 104, 110
  visualising the unborn, 102, 106–9 (*and see* Nilsson, Lennart; Tsiaras, Alexander; Screening: ultrasound)

Foucault, Michel
  bio-power, 5, 65
  discourse, 55, 58
  health systems, 79
  medical gaze, 80
  self-regulating 'techniques of the self', 67, 74
  The Birth of the Clinic, 70, 79
Frankfort, Ellen
  Vaginal Politics, 5

G
Gaskin, Ina May, 6
Gaze (*see also* Surveillance)
  medical, 6, 79–80
  photography and control, 119
  public, 102, 107–8, 109–10, 147
  women as object of, 124–5 (and see 'Normal')
Gender
  destabilising gender/sex binary, 12
  maternal and paternal actors, 167–9
Genetic material (*see also* Eggs; Semen; Sperm; Chapter 11; Chapter 12)
  as commodity, 185
  donor profiles, sperm, 189–91, 195–6
  disconnected from kinship, 214–6
  genetics and motherhood, 214–6
  marketing strategies, 193
  maternity and paternity in genetic material donation, 216–7
  silenced men/sperm donors, 196, 200
  'taboo trades', 221
Giddens, Anthony, 14
  body projects,
Gill, Rosalind
  postfeminism, 239–40
  woman as neoliberal subject, 249
Gilligan, Carol
  In a Different Voice, 77
  decision making, 80
Globalisation
  reproductive choice, 3, 251–2
Goffman, Irving
  impression management (sperm donors), 196

Greer, Germaine
  The Female Eunuch, 5
Grigg-Spall, Holly
  Sweetening the Pill, 249
Grounded theory (*see* Research Methodologies)
Guilt (*see also* Body image; 'Normal')
  body control, 123
  breastfeeding mothers, 229–30
  personal time, 228–9

H
Health
  disabled women and reproductive health, 136, 143, 147
  hierarchies in health care, 79–80
  monitoring mothers and children, 84
  public or private health systems, 73, 79, 83, 242–3 (*and see* Chapter 5)
  women's health movements, 5, 238–9

I
Identity
  biology, 4
  embodiment, 122, 138, 253
  essentialist, gendered, 2, 5–6, 12–3, 124, 253 (*and see* Chapter 10)
  fathers, *see* Chapter 10
  moral, 151
  mother as host, 106, 109
  pre-pregnancy, 115
  social need for children, 159
  sperm donors', 195
  technoscientific identities, 188
Infertility, or Fertility barriers (*see also* Chapter 1)
  abortion, 34–5
  defining infertility, 26, 37
  'hidden infertile', 30
  infertility and blame, 24
  medicalised phenomenon, 65
  structural barriers, 29
  surrogacy, 42
Information
  ante-natal classes, 169, 171, 175, 179, 180, 228
  in prenatal screening, 56–8, 60–1, 63
  post-birth information provision, 91–2
Interviewing (*see* Research methodologies)

K
Kinship
  biomedicalised, 188–91, 195–9
  genetics and motherhood, 214–6
  negotiating relatedness, 203
  virtual kinship, 186, 191, 195, 198–201
Klum, Heidi
  celebrity pregnancy, 115, 117

L
Legal proceedings
  foetus and mother, 111–2, 113
  lawsuits regarding the Pill, 237–8
  Project on Government Oversight, 238
  R v Bayliss and Cullen, see Chapter 4
  R v Leach and Brennan, see Chapter 4

M
McRobbie, Angela
  post-feminism and neoliberalism, 70– 75, 239
Marriage (*see also* Family)
  'best interests' of the child, 44, 212
Masculinity and masculinities
  age, physical ability and fatherhood, 160
  'breadwinner' discourse, 8, 13, 156–7, 163, 166
  challenging and reinforcing hegemonic, 175, 180
  disrupted by fatherhood, 151, 172, 177
  fertility and masculinity, 155
  impact of technosemen, 189
  'involved' father, 166–8, 172, 180, 227–34 (and *see* Chapter 13)
  marginalised in sociology of reproduction, 253
  masculinities in sperm advertising, 185–6, 188–91, 195–6
  masculinisation of the home sphere, 167

Masculinity and masculinities – *continued*
  men as procreative beings, 9
  multiple and fluid masculinity, 168, 180
  paternal success at birth, 174
  'second sex', 8
Mauss, Marcel
  anonymous donation, 216
  The Gift, 206
  gift economy, 18
  debts of reciprocity, 219
Maternity
  clothes, 126
  maternity and paternity in genetic material donation, 216
  maternity leave, 224, 231, 234,
  midwifery services, 87, 91–2
Media
  Demi Moore, 117
  Jessica Simpson, 118
  Kate Middleton, 101
  Queensland RU486 case, see Chapter 4
Medicalisation (*see also* Screening)
  biomedical discourses, 37, 58–9, 112, 188, 253
  biomedicalised kinship, 186, 188–91, 195, 198
  biomedical power, 4, 57
  bio/social need for children, 48
  creating distance, 62
  'disability;, 64–5
  'jurisprudential medicalisation' of abortion, 70–1
  models of infertility, 37
  pathologising reproductive bodies, 4, 73
  postpartum bodies, 116–7
  pregnancy and birth, 71, 91, 105
  sperm, 199–200
Menopause, 4, 139, 140
Menstruation, 4,
  spinal cord injury, 139, 140, 143
Middleton, Kate (Duchess of Cambridge)
  the 'royal foetus', 101
Miscarriage and Stillbirth, 24, 28, 31–8
Moore, Lisa Jean
  Sperm Counts: Overcome by Man's Most Precious Fluid, 186

Motherhood
  conflated with womanhood, 25
  disabled mothers, see Chapter 8
  disembodied, 227, 234
  embodied, 132
  genetics and kinship, 214–6
  gestational not genetic, 215–6
  'good', 111, 124, 227
  in surrogacy, 48
  mother as host, 106, 109–10, 112
  older, 150, 154
  'questionable' mothers, 145
  risk and the unborn, 111
  surveillance of mothers, 84
  unborn-maternal assemblage, 105, 253 (*and see* Chapter 6)

N
Narrative research (*see* Research methodologies)
Neoliberalism, 9–10
  abortion discourses, 78, 103, 109, 113, 239
  care deficit, 85, 96
  discourses of 'patient as client' and 'individual choice', 69–70, 74
  genetic material marketplace, 200, 204
  human milk markets, 232–3
  in Australian healthcare, 73, 79
  in Canadian post-birth care, 96
  outsourcing infant feeding, 221
  personal responsibility rather than collective action, 240, 243, 244–9
  postfeminism, 240–2, 248–9
  reframing parental reproductive roles, 234
  self-regulating 'techniques of the self', 67, 74
  women as entrepreneurial or selfish, 78
Nilsson, Lennart,
  visualising the unborn, 109
'Normal' (*see also* Chapter 3)
  able-bodied and disabled women, 136–7, 143–4, 148
  bio-power and normalisation, 65
  'ideal', 65

masculinity, disrupted by fatherhood, 151
medicalised definition, 65–6
normative motherhood, 25, 215 (*and see* Chapter 8)
normative timing of fatherhood, 17, 162 (*and see* Chapter 9)
norms of fatherhood, 158–60
resistance to, 66
versions of normal, 143, 149

O
Oakley, Ann,
Becoming a Mother, 8
Women Confined, 8

P
Parsons, Talcott, 8
Parenthood (*see also* Fathers and fatherhood; Motherhood; Family)
gay/lesbian, 190
'good' parent (in non-traditional conception), 52, 208, 210–15, 218–9, 228
impact on relationships, 161
intended and 'proper' surrogate parents, 41–2
prerequisites, 155–8, 163 (*and see* Chapter 2)
qualifying for parenthood, 28–9, 46, 145
surveillance of, 84
teenage, 150
timing and class, 154
timing, 150–1, 154, 156, 160
Partners (*see also* Fathers and fatherhood; Breastfeeding)
abortion decision, 74
as mediators in reproductive journeys, 167, 169, 173–4
Paternity
agency, 166–7, 169, 172–3, 175
biological contribution, 199, 216
maternity and paternity in genetic material donation, 216–7
paternity leave, 170, 231
social contribution, 234

Performance (*see also* Gender; Identity; 'Normal')
femininity and the postpartum body, 117–8
gender, 12
men as actors during child birth, 174
Photography (*see also* Chapter 7; Research methodologies)
ambivalent cultural practice, 120
digital, 119, 133
everyday, 119
family photographs, 120
headless, 118–9, 133
portraying 'self', 133
postpartum bodies, 118
resisting normative discourses, 126
theorising, 119
visualising the unborn, 102, 106–9 (*and see* Nilsson, Lennart; Tsiaras, Alexander; Screening: ultrasound
Policy and legislation (*see also* Regulation)
abortion, 80–2 (*and see* Chapter 4)
Adoption Act 2000, 40
Assisted Reproductive Technology Act 2007, 40
Australian maternity and reproductive policy, 73
breastmilk, 223
egg donation, 216
fertility and infertility, 26
genetic material, 216
influence of pharmaceutical companies, 248
lack of regulation in Indian surrogacy, 12
maternity and paternity in genetic material donation, 216–7
oöcyte donation, *see* Chapters 6 and 12
'personal is political', 251, 253
'Personhood Pledge', 236
post-birth care, 96
Sex Discrimination Act 1984
Status of Children Act 1996, 40, 216
surrogacy, and *Surrogacy Act (NSW) 2010*, see Chapter 2
West Australian Acts Amendment (Lesbian and Gay Reform) Bill 2001, 42

Postfeminism, 13–4 (*see also* Feminism)
  anti-feminism, 75
  body projects, 249
  contraception, 239, 243
  'entrepreneurial female self', 75–7
  personal responsibility rather than collective action, 240, 243, 244–9
  subjectivity and the Pill, 248–9
  woman as neoliberal subject, 70, 249
Postmodernism (*see also* Introductory chapter)
  postmodern reproduction, 36–7, 139, 149, 203–4, 208, 218–9, 251–2, 253
  male and female reproductive roles, 23–4, 133, 139, 151, 163, 221, 231, 234–5
Postpartum/Postnatal (*see also* Chapter 5)
  body work, 117
  'bouncing back', 115, 123, 126, 131, 133
  care, *see* Chapter 5
  care statistics, 86
  doulas, 87
  embodiment, 116–7, 253
  hospital services, 93
  informal supports, 94
  midwifery services, 87, 91–2
  norms of the postpartum body, 118
  postnatal depression, 93
  private post-natal care services, 87
  role of the state, 87
  stratification of post-birth care, 88
Poststructuralist
  gender theory, 253
Power, 252 (*see also* Control; 'Normal'; Policy and legislation; Regulation; Surveillance)
  hierarchies in health care, 79–80
Pregnancy
  bonding with foetus, 107, 110
  celebrity, 101, 115, 117, 118, 126
  class and the body project, 16, 129
  clothes, 126
  commodification, 11
  medicalisation, 60, 105
  miscarriage, 24, 28, 31–8
  mother as host, 106
  'natural', 11
  'normal' or 'healthy', 62
  planned, and the Pill, *see* Chapter 14
  'publishing' pregnancy, 108
  'quickening', 105
  risk, 12, 28, 58, 60–1, 111–2, 136–7, 145, 154
  teenage, stigmatised, 76
  teenage, 76, 78, 150
  unplanned, 77, 153–4, 157–8
  unwanted, 31–2, 34, 36–7, 104

Q
Qualitative research (*see* Research methodologies)
Queer
  queer families, 13, 42, 46, 190, 193, 212, 253

R
Reflexive research (*see* Research methodologies)
Regulation (*see also* Control; Neoliberalism, Policy and legislation; Surveillance)
  Foucault's 'techniques of the self', 74
  patriarchal medical practices, 4, 12, 14, 24, 167, 180, 254
  regulating the body, 2, 4, 12, 15, 111,
Reproductive health (*see also* Chapters 1 and 14)
  barriers and opportunities, 142–3
  'seizing' control of, 5–7, 242–8
Reproductive technologies, 6, 10–1 (*see also* Donors; Surrogacy; Screening)
  ambivalence to, 25
  assisted reproduction, 15, 24, 40, 50–1, 69, 199–200, 203, 213, 252
  in-vitro fertilisation, 212
  kinship and donation, 214 (*and see* Kinship)
  patriarchal medical practices, 4, 12, 14, 24, 167, 180, 254
  prenatal screening, *see* Screening
  reproductive tourism, 11–2, 110
  medicalising pregnancy and birth, 71, 91, 105
  social discourses, 37

Reproductive rights
  consumer or citizen, 10, 38, 42
  intended surrogate parents, 41
Research methodologies
  cultural analysis, 241
  cyberethnography, 187
  discourse analysis, 43, 55, 57–8
  feminist qualitative research, 241
  grounded theory, 187
  interviewing, 140, 152, 170, 224
  interviewing, feminist, 140–1
  interviewing, phenomenological, 121, 140–1
  mixed methods, 88–90
  narrative, 118–9, 125–6, 132–3, 141, 208, 216
  photovoice, 120–1
  qualitative research, 116, 152, 187, 208–10, 223
  reflexive, 224
  sampling, 30, 57–8, 88–9, 120–1, 139–40, 152, 170, 192, 209, 223
  thematic analysis, 121, 224
  theoretical analysis, 209
  transcripts, 42, 64, 121, 141, 170, 209, 224
  visual sociology, *see* Chapter 7
Resistance
  dominant images of infertile women, 25
  fathers' resistance and ambivalence, 172
  fertility control, 249
  medicalised normalisation, 66
  technologised bodies, 11
Rich, Adrienne, 5
Risk
  caesarean section, 85, 137
  contraceptives, 144, 236
  mothers and unborn, 111–2, 136–7, 253
  older fathers, 153
  Patient Package Inserts (PPIs), 239
  post-birth care, 85
  prenatal screening, 61
  screening for childhood risks, 84
  sterilisation and osteoporosis, 145
  the Pill, 237–40, 242

S
Sampling (*see* Research methodologies)
Screening (*see also* Disability; Donors; Eugenics; 'Normal'; Chapter 3)
  amniocentesis, 56
  breastmilk donors, 221
  prenatal, 56
  ultrasound , 107
  ultrasound art, 110
  ultrasound and bonding, 107
  magnetic resonance imaging (MRI), 109
  values in, 65–7
Seaman, Barbara
  The Doctors' Case Against the Pill, 238–9
Semen
  personifying semen, 192
  technosemen, 189–90, 198
Sexuality
  asexuality, 129, 136–7. 139, 146–7, 217
  hetero- or homosexual parents, 42, 46, 49, 51, 212, 253–4
Shakespeare, Tom, 54, 64
Shildrick, Margrit
  embodiment, 138
  normativity, 148
Simpson, Jessica
  celebrity pregnancy, 118
Skeggs, Beverley, 3, 252
Social media
  blogs, 118, 188, 192–4, 237, 240–1 (*and see* Chapter 14)
  Facebook for donor-family interaction, 191, 196
  homosocial spaces for women, 197, 200
  mediated sperm donation communities, 198
  online profiles of the unborn, 101
  sperm online, Facebook, 185, 191–2
  Twitter, 193, 196
  virtual kinship, 186, 191, 195, 198–201
  YouTube, 193

Sperm (*see also* Chapter 11)
  advertising, 186, 193
  banks, 189–91, 192
  (bio)medicalisation, 188
  California Cyrobank, 192
  Cryos International, 192
  donors, disembodied and objectified, 190, 196–9
  Fairfax Cryobank, 192
  The Cryobank of California, 192
  technosemen, 189–90, 198
  Xytex Cryo International Sperm Bank, 192
Sterilisation (*see also* Disability; Infertility, or Fertility barriers; Screening; Chapter 1)
  desire for, 144
  risk of osteoporosis, 145
  sterilisation regret, 28, 31, 37
Stillbirth (*see* Miscarriage)
Stratified reproduction, 10–5, 23, 37 (*see also* Chapter 1)
Studies, 7–9
  Community living after spinal cord injury: Models and outcomes, 139
  Melbourne School of Population Health, 77
  Men-as-Fathers, 152
  National Survey of Fertility Barriers, *see* Chapter 1
  Oöcytes for stem cell research: Donation and regulation in Australia 2008–2011, *see* Chapter 12
  UK Infant Feeding Survey, 227
Surrogacy, 12 (*see also* Chapter 2)
  'commonsense' attitudes, 51–2
  gestational carrier, 25, 207, 216–7
  'proper' intended parent, 42
  social justice, 12
  'taboo trades', 221
Surveillance (*see also* Gaze; 'Normal')
  of parents and parenting, 84
  postfeminist self-monitoring, 204
  pregnant women, 113
  pregnant and postpartum bodies, 124
  public observation of the unborn, 107
  self-regulating 'techniques of the self', 67, 74
  visualising the unborn, 102, 106–9 (*and see* Nilsson, Lennart; Tsiaras, Alexander; Screening: ultrasound)

T
Titmuss, Richard
  altruism and the 'right to give', 204–5, 219, 220
  altruism and the universal stranger, 204, 205–6, 216, 218
  debts of reciprocity, 214, 219
Transcripts (*see* Research methodologies)
Tsiaras, Alexander
  visualising the unborn, 109
Twitter (*see* Social media)

V
Visual sociology (*see* Research methodologies)

W
Web 2.0 (*see* Social media)
Websites and blogs
  Aphrodite Women's Health, 240
  Donor Sibling Registry, 191
  Dr. Sugar, 240, 246–7
  Healthtalkonline, 223
  No More Dirty Looks, 240, 242–3, 247
  Our Bodies Our Blog, 240
  re:Cycling, 240
  The Shape of a Mother (SOAM), 118
  The Visible Embryo, 110
  xoJane, 240, 243, 244, 247
Weight (*see also* Body image; Chapter 7)
  bouncing back, 115, 123, 126, 131, 133
  postpartum, 123–4, 129
Wolfson, Alice
  Nelson Pill Hearings, 238

Y
YouTube (*see* Social Media)
Yummy mummies
  postpartum norms, 117

GPSR Compliance
The European Union's (EU) General Product Safety Regulation (GPSR) is a set of rules that requires consumer products to be safe and our obligations to ensure this.

If you have any concerns about our products, you can contact us on

ProductSafety@springernature.com

In case Publisher is established outside the EU, the EU authorized representative is:

Springer Nature Customer Service Center GmbH
Europaplatz 3
69115 Heidelberg, Germany

www.ingramcontent.com/pod-product-compliance
Lightning Source LLC
Chambersburg PA
CBHW061806110426
42873CB00042B/55